Mathematical Modeling in Agriculture

Scrivener Publishing
100 Cummings Center, Suite 541J
Beverly, MA 01915-6106

Publishers at Scrivener
Martin Scrivener (martin@scrivenerpublishing.com)
Phillip Carmical (pcarmical@scrivenerpublishing.com)

Mathematical Modeling in Agriculture

Edited by

Sabyasachi Pramanik

Niranjanamurthy M.

Ankur Gupta

and

Ahmed J. Obaid

Scrivener Publishing

WILEY

This edition first published 2024 by John Wiley & Sons, Inc., 111 River Street, Hoboken, NJ 07030, USA and Scrivener Publishing LLC, 100 Cummings Center, Suite 541J, Beverly, MA 01915, USA

For more information about Scrivener publications please visit www.scrivenerpublishing.com.

Wiley Global Headquarters
111 River Street, Hoboken, NJ 07030, USA

For details of our global editorial offices, customer services, and more information about Wiley products visit us at www.wiley.com.

Library of Congress Cataloging-in-Publication Data

ISBN 978-1-394-23369-4

Front cover images supplied by Pixabay.com
Cover design by Russell Richardson

Set in size of 11pt and Minion Pro by Manila Typesetting Company, Makati, Philippines

Printed in the USA

10 9 8 7 6 5 4 3 2 1

Contents

Preface

The research in the book looks at the likelihood and level of use of implemented technological components with regard to the adoption of different precision agricultural technologies. To identify the variables affecting farmers' choices to embrace more precise technology, zero-inflated Poisson and negative binomial count data regression models were utilized. Outcomes from the count data analysis of a random sample of various farm operators show that various aspects, including farm dimension, farmer demographics, soil texture, urban impacts, farmer position of liabilities, and position of the farm in a state, were significantly associated with the approval severity and likelihood of precision farming technologies.

Farm management information systems (FMIS) have constantly advanced in complexity as they have incorporated new technology, the most recent of which is Internet connection. However, few FMIS have fully tapped into the Internet's possibilities, and the newly developing idea of precision agriculture receives little or no support in the FMIS that are now being sold. FMIS for precision agriculture must meet a few more criteria beyond those of regular FMIS, which increases the technological complexity of these systems' deployment in a number of ways. In order to construct an FMIS that meet these extra needs, the authors here evaluated various cutting-edge web-based methods. The goal was to determine the requirements that precision agriculture placed on FMIS.

1

Analyzing the Impact of Food Safety Regulations on Agricultural Supply Chains: A Mathematical Modeling Perspective

Nimit Kumar[1], Shwetha M.S.[2], Govind Shay Sharma[3], Nitin Ubale[4], Nuzhat Fatima Rizvi[5] and Dharmesh Dhabliya[6*]

[1]College of Agriculture Sciences, Teerthanker Mahaveer University, Moradabad, Uttar Pradesh, India
[2]Department of Food Technology, Faculty of Engineering and Technology, JAIN (Deemed-to-be University), Karnataka, India
[3]Department of Mathematics, Vivekananda Global University, Jaipur, Rajasthan, India
[4]Department of Horticulture, College of Agriculture, Parul University, Limda, Dist- Vadodara, Gujarat, India
[5]Symbiosis Law School, Nagpur Campus, Symbiosis International (Deemed University), Pune, India
[6]Department of Information Technology, Vishwakarma Institute of Information Technology, Pune, Maharashtra, India

Abstract
With several mathematical models developed for diverse food subjects, mathematical modeling plays a significant role in the field of food engineering. There is, however, little data available on the utilization of statistical models in the foodstuff industry. This essay intends to analyze the scope and circumstances of mathematical model use in connection to the North American foodstuff and beverage sector. It examines the understanding, characteristics, and present utilization of modeling

**Corresponding author*: dharmesh.dhabliya@viit.ac.in
Nimit Kumar: Orcid: https://orcid.org/0009-0005-4341-0565
Shwetha M.S.: Orcid: https://orcid.org/0000-0002-9651-8095
Govind Shay Sharma: Orcid: https://orcid.org/0000-0002-5884-4674
Nitin Ubale: Orcid: https://orcid.org/0000-0002-1270-481x

Sabyasachi Pramanik, Niranjanamurthy M., Ankur Gupta and Ahmed J. Obaid (eds.) Mathematical Modeling in Agriculture, (1–16) © 2024 Scrivener Publishing LLC

methodologies in connection to the key features of the industry. This study covered 203 food firms overall from 12 different North American nations. The size of the firm and the nation in which it operates are shown to be more relevant determinants of the utilization of statistical models than the kind of food industry. The most advanced nations are located with a better degree of knowledge and model use. Similar patterns were seen at the micro level, demonstrating that smaller and mid-category businesses had limited resources, expertise, and model adoption.

***Keywords*:** Food industry mathematical models, modeling expertise using models

1.1 Introduction

A valuable technique for determining the influence of various system and process features on the output of a process is mathematical modeling. Modeling diverse food items and/or processes is difficult, mostly because the phenomena are not well understood, modeling experiments are complex, and there are unknowns about trustworthy data and food attributes.

While statistical models on foodstuff nature through the foodstuff chain facilitate knowledge based on food features and various events which take place, every activity, grade and foodstuff safety properties is a huge worry of both clients and the foodstuff industry.

The study of competing options is improved by foodstuff process modeling and/or multiscale simulations from foodstuff components up to the complete foodstuff supply chain (SC). It is important to remember that foodstuff tissues are multiscale aggregation having unique properties at every spatial scale, necessitating the use of multiscale modeling. Because of this, the primary goals in technological food procedures are to comprehend a specific phenomena using current theoretical knowledge and accessible data, create methods, and regulate those processes. According to [1], there are two main purposes for using foodstuff technique modeling: (i) to well realize a method, and (ii) to test potential "what-if" situations. Additionally, modeling of food products and processes might be done using sophisticated model-based methodologies. These methods could involve model-based production control or mathematically based product/process optimization. Locating "good" data points or feature representations is the goal of knowledge transfer techniques in order to improve the target model's predictability and believability.

The use of models in the foodstuff sector often depends on static, simplistic models which don't provide a meaningful assessment of observable methods, conditions for quality or safety, or environmental effect.

Additionally, these models streamline the explanations of the food system's mechanics and rate equations for change. Models may be divided into three categories, analytical models, numerical models, and observational models. Models may also be divided into three types based on the point of view: product, process, and product-process interactions. A newer model called multiscale [2] modeling has emerged to address the difficulty of modeling at several geographical scales. According to [3] the complexity of modeling depends on the fact that a variety of skills are required, including knowledge in food technology, applied math and statistics, technology, information technology, etc. It's crucial to remember that every modeling technique has its drawbacks. Even though many models have been developed, [4] claims that there is very little evidence of their use in actual situations. The authors of [5] provide some scientific guidelines in modeling information extraction and formalization for addressing the function of foodstuff operators in little businesses.

Therefore, the goal of the current study was to assess the utilization of statistical models in the foodstuff sector related to the understanding of statistical approaches, the extent to which these tools are used by businesses, and obstacles to adopting mathematical modeling. Environmental aims and indicators as well as modeling of environmental consequences were also studied. This study determined the need for food modeling across a range of application sectors. According to the nation in which the firms operate, the functions of the organizations in the food chain, and the dimension of the organizations, the results were distributed.

Review of the literature

By examining published work utilizing the academic databases Web of Science, Scopus, a critical literature review was carried out. These databases located the best scientific publications about the quality and safety of food items as well as environmental models for food processes and products. No topographical limitations were used, and the hunt was only allowed to turn up research that had been published during the previous ten years. Modeling foodstuff products and dangers from a foodstuff safety belief was the main emphasis of the bulk of papers linked to modeling in the food industry.

The degree of understanding of a certain phenomenon is correlated with how difficult it is to analyze this issue. For instance, during the processing of food, a variety of heat transfer activities, like chilling, disinfecting, chilling, making, roasting, etc., take place in numerous unit operations. Mass transfer is a typical subject discussed in many publications in addition to heat transfer. Analysis of foodstuff processing processes such as baking, freezing, hydrating, filtering, dissipation, draining, osmosis, membrane separation, remoistening, aggregating, removal and repository requires modeling of mass movement. In food processing, simultaneous heat and mass transmission may be seen in roasting processes, baking processes, and food drying models. As the developing automations have distinct forms of action relying on the origin of energy transfer, modeling assumptions might vary depending on whether a food process occurs using traditional or non-thermal technologies. To ensure food safety while maintaining food quality, non-thermal processing has been developed. [6] gave one of the most recent updates on modeling heat transport in traditional and cutting-edge technology.

[7] developed quality modeling of many foodstuff grade characteristics, like taste, complexion, presence, and nutritive value. Some of the most recent efforts to model quality index were addressed by [8]. Foodstuff safety models range from risk assessment to food security to modeling in order to improve transportation and shelf life. Modeling of ecological effects and environmental indicators in the foodstuff industry is becoming more important in light of the significance of the Sustainable Development Goals established by the UN (UNESCO, 2022). Food firms, food processes, and food products are the three views that make up the scale of environmental models in the food chain.

The authors of this work recognized this as a research need since they found that examination of the deployment of statistical models [9] in foodstuff firms was not the center of the research. This study's working premise was that food corporations don't often employ mathematical models.

1.2 Resources and Techniques

Specifications of the survey

The investigation took place in the first half of 2018. An English-language questionnaire was created and then translated into the native tongues of the participating nations. There were 203 food firms overall, from 12 different European nations, and they were split into two groups: Asian Nations (AN) and Other North American (ONA) countries. The System Partnership Accord between the American Organization in Technology defines ITC as less research-intensive nations. Companies were picked from every region of the tested nations. The authors acknowledge that this approach only provides a "convenience sample" of food firms rather than a genuinely random sample of them. The specimen is compared to numerous produced surveys on the execution of specific tools in various countries having fewer than 80 food companies in a nation like quality management, hygiene practices, pest control, or food fraud. This is true despite the sample's small number of companies per country. Under certain conditions of care, the broader European food industry might benefit from our findings.

The authors wanted to disseminate a questionnaire on the use of statistical models in the foodstuff business, so they contacted firms in advance and indicated that the survey is anonymous.

Questionnaire

To assess the food industry's use of modeling and optimization techniques from computer science and mathematics, a questionnaire has been created. The collection of responses offered the chance to examine how tools are now used for a variety of purposes, including the creation of products and processes, process control, foodstuff safety, decision assistance, and ecological effects.

The firms' basic characteristics (nation of origin, size, industry of operation, and adopted enterprise systems) were covered in the first part. The understanding of different mathematical procedures, the extent of tool usage in businesses, and challenges associated with applying mathematical modeling were all examined in the second portion. A 5-point Likert scale [10] having the choice of one for "strongly disagree," two for "disagree," three for "no opinion," four for "agree," and five for "strongly agree" allowed the respondents to express their level of agreement. The analysis of modeling requirements in different application domains made up the third segment. The respondents were given the chance to rank their needs on a scale of 0 ("not applicable in our company"), 1 ("there is no need for mathematical models in my company"), 2 ("we would like to use some mathematical models"), 3 ("there is some (limited) use, we would like to expand our knowledge in this area"), and 4 ("we have an extensive use of models in this area"). The fourth section of the survey asked respondents to rate their awareness of environmental impacts on a scale of 0 (no analysis of this impact is done), 1 (organization analyses fundamental ecological data), 2 (organization computes particular ecological measures for this influence), 3 (organization changes basic data to compute ecological affect per technique/functional unit), and 4 (organization computes ecological footprints connected to this thrust). Using the similar 5-point Likert scale from 1 (firmly disagree) to 5 (strongly agree), organizations were asked to indicate if they had recorded environmental aims or indicators in the last section.

Statistical analysis

Since the data were not normally distributed, Likert scale results were treated as ordinal estimates and non-parametric statistical tests were utilized. We classified the noticed remarks using a cluster analysis. The food industry, firm size, and nation type were used as categorical variables in a two-step cluster analysis. For finding statistically major differences between the clusters, the Mann-Whitney U test was performed.

In order to better understand the general relationships in the two data sets, a PCA [11] was performed on 10 statements assessing knowledge of environmental implications and 16 statements indicating desired requirements for modeling in diverse application areas. Prior to analysis, the appropriateness of PCA was determined. With specific KMO measures all more than 0.8 and ratings ranging from "meritorious" to "marvelous," the entire Kaiser-Meyer-Olkin (KMO) [12] estimate relating to the demands for modeling was 0.925. Data were probably factorable since Bartlett's test [13] of sphericity was statistically significant (p .0005). The total KMO score [14] for awareness of environmental effects was 0.747, and all every KMO scores were over 0.75, earning them the designation "meritorious." Statistics were significant for Bartlett's test of sphericity (p .0004). The PCA retrieved two elements explaining 73.7% of the cumulative variance for the analysis of modeling requirements and two components showing 81.4% of the summed variance for recognition of ecology effects based on the criterion of eigen values [15] above one. MS Excel and SPSS [16] were used to process the statistical data. A 0.05 threshold for statistical significance was established.

1.3 Results and Analysis

1.3.1 Knowledge, Application, and Obstacles to Food Modeling

The food industry, firm size, and nation type were all used as categorical variables in a two-cluster analysis. Entire findings indicate that the greatest degree of agreement across businesses was connected to their understanding of transportation phenomena and mechanics, production planning, real-time process control [17], and their regular usage of production planning models. They don't agree that response surface modeling should be used often, and they don't think their product is too basic to benefit from any modeling. Food modeling calls a solid knowledge of applied mathematics, computer science, and food technology, stressing the requirement.

1.3.2 Obstacles to Our Company's Use of Mathematical Modeling

It is important to note that they did not express a position on environmental issues such as the need to save energy and water and to prevent ecosystem and greenhouse gas degradation. Because most models are not "user-friendly," this occurs as a consequence. In the past, models have been created for scientific objectives, and then customized to fit the demands of the user. As a result, many users still find it challenging to acquire model results or to otherwise utilize models. Two modeling clusters, referred to as "developing" and "developed," were identified via cluster analysis [18]. Cluster 1 (100 firms - "developing") is made up of ITC nations, small and midsized businesses, and businesses that primarily work in the food producing industry. Respondents acknowledged their lack of understanding (scores ranging from 1.8 to 3.4) and use of models (scores ranging from 1.8 to 3.1) for the majority of the models in this cluster. Insufficient resources (knowledge and infrastructure) were also verified by them.

When it comes to business success, knowledge transmission is crucial, and both scientists and industry professionals are always looking for the best ways to innovation, a result of quality and environmental management. According to a recent research on environmental models in the food chain, it is important to establish simpler models for evaluating environmental performance so that they may be widely and easily used in the food industry. To meet different corporate profiles, a broad variety of models created at the research level need better knowledge transfer.

Cluster 2 (125 respondents; "developed") included OEC nations, large corporations, and businesses engaged in the foodstuff service industry. In this cluster, respondents agreed that they knew about modeling (scores ranging from 3.4 to 4.1) and used these models (scores ranging from 2.9 to 4.2), but they disagreed that there weren't enough resources for modeling. When it comes to environmental concerns, the responses that were most often given in this cluster revealed that businesses seek to enhance ecological performance and prevent pollution (rating ranging from 3.5 to 4.2). It demonstrates that environmental factors have a favorable impact on company performance as a whole. Pollution [19] prevention was identified as the primary impetus for developing any environmental management system in a study on the impact of certified enterprises' environmental management. The fact that many of the businesses in this cluster have an accredited environmental management system is noteworthy. Outcomes from the cluster were better than those from Cluster 1, having 26 out of

28 assertions exhibiting statistically significant differences across clusters (p .05). The first set of findings demonstrates that larger organizations are, predictably, best equipped, having sophisticated modeling abilities and with knowledge of environmental impact concerns.

Level of modeling requirements

In order to assess the scale's dependability, the reliability of 18 elements was assessed by computing the Cronbach's alpha coefficient. Our scale has a Cronbach's alpha [20] of 0.941 that is a good degree of internal consistency. Our findings indicated that there is little modeling [21] of foodstuff quality and safety, having modeling of microbial development. The most typical response to these claims was that businesses employ models extensively. This is understandable as product quality control is a specific industrial application where the approach of knowledge transmission is used. The usage of quality tools in foodstuff industries has been examined in previous research, ranging from simple tools to more complicated tools that need specialized training. According to several rules on food safety, food businesses must use mathematical modeling that predicts the development or survival of the microorganisms that are of concern. Lowest results, however, indicate a lack of any environmental mathematical modeling. The most common response was that businesses would want to utilize certain mathematical models [22]. The usage of ecological models in the foodstuff business depends on many factors, including the model's applicability, user-friendliness, cost (if applicable), and need for specialized environmental knowledge.

Figure 1.1 displays the PCA result for the data matrix. Dimension reduction by PCA divided the observed factors into two specific directions which were accepted as two dimensions: a "product-based dimension" (PC1) that was driven toward modeling numerous product-dependent models and a "risk-based dimension" (PC2) that was a dimension focused on environmental or foodstuff safety risks. Regulators and policymakers continually propose standards in response to hazards with the aim of detecting and reducing risks. Food safety management standards (BRC, 2018; IFS, 2014; ISO, 2018) and environmental management standards (ISO, 2015b) are examples of risk-based standards. In our research, 41.4% of the companies examined had an environmental system in place, while 76.8% had a food safety system (Table 1). Confirmation of the two aspects, the product-dependent and the risk-dependent, can enhance the study of foodstuff modeling by expanding on the body of research already supporting modeling in the food sector.

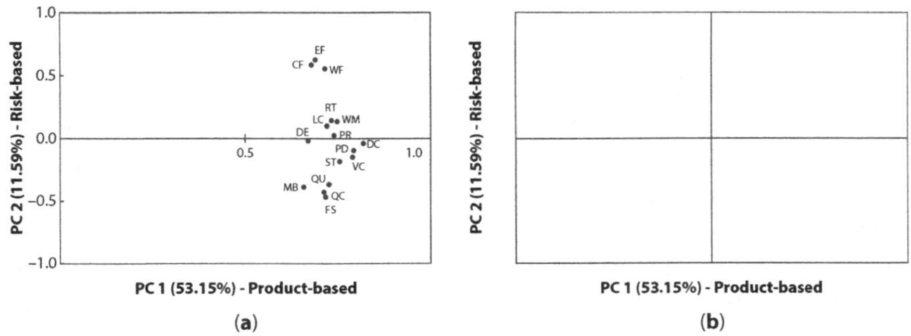

Figure 1.1 The 16 parameters that affect the requirement for modeling in different application areas are shown by PCA loadings (a) and scores (b) plots according to the type of nation, the size of the enterprises, and their activity in the food industry. 67.7% of the total variation is explained by the two retrieved components. Factors include: product development (DE), process development (PR), real-time process control (RT), food storage optimization and control (ST), food quality control (QC), microbial growth modeling (MB), food safety (FS), characterizing food quality (QU), value chain management (VC), decision control (DC), productivity analysis (PD), life cycle assessment (LC), carbon footprint (CF), water footprint (WF), energy footprint (EF), and waste management (WM). Country type: OEC - Other European countries (Denmark, France, Germany, Greece, United Kingdom); Company size - small, medium, and large.

A loading plot (Figure 1.1a) shows the outcomes in detail. In Figure 1.1a, it is clear that every outcome has positive loadings, indicating that it has a significant positive impact on the "product-based" components. All statements loaded the "product-based dimension" (PC1) strongly (> 0.65). In terms of the "risk-based" dimension (PC2), environmental modeling (carbon footprint, water footprint, and energy footprint) has the largest positive loading, whereas modeling of food safety, microbiological growth, and quality control has the highest negative loading.

Nine of the statements (item and process development, real-time process control, decision support, food storage, value-chain, productivity, life-cycle evaluation, and waste management) have loadings that may be changed. Productivity and value chain are examples of quality-oriented approaches where risks are mostly related to (not) meeting customer expectations.

Scoring guidelines: "0" indicates that it does not apply to our business; "1" indicates that we do not need to use mathematical models; "2" indicates that we would like to utilize a few statistical models; "3" indicates that we have few models (limited utilization) and would like to explore our expertise in the domain; and "4" indicates that we utilize models extensively in this business. In the first extracted components item loadings the raw data was used to get the Mean values [23], Standard deviations [24], and modes [25].

Customer satisfaction is achieved when a product satisfies customer wants and expectations (ISO). In order to identify significant sources, pollutants, receptors, and exposure routes across a product's life cycle, life cycle assessment from a risk viewpoint may be helpful. In addition, waste management puts human health and groundwater in danger.

The scores plot (Figure 1.1b) summarizes the connections between the nations and the businesses. Large and small businesses, each representing a company with a different modeling strategy, were in opposition to one another. Companies were clustered towards the center according to their activity and nation type, suggesting that they had comparable average modeling practice ratings. The management of safety-related hazards is a particularly sensitive area of the food sector in accordance to customer expectations, as seen by the second family of findings.

Our scale has a good degree of internal uniformity, as shown by the Cronbach's alpha value of 0.934 for the dependability of 10 items. A summary of the findings (Table 4) revealed that enterprises only had a rudimentary understanding of environmental information about trash, water use, and electric energy use. This is because rising energy and water costs are being attributed to economic and legal problems. Thus, in order to comply with regulatory obligations, businesses are primarily focused on reducing costs as well as tracking waste production. [26] also support the economic component of environmental performance.

Factors, Items, and Results

According to the polluter-pays concept defined in EU law, corporations are obligated to have some information about the trash they produce (EC, 2008). Fig. displays the PCA result for the data matrix. The detected components were reduced in size by using PCA to split them into two independent dimensions: a "level of awareness dimension" (PC1) and a "type of impact dimension" (PC2). Figure 1.2a's loading plot displays the outcomes in detail. All of the findings had positive loadings, indicating that they have a significant positive impact on the component measuring "level of awareness." Outcomes that provide related information are classified as below.

With all assertions, the "level of awareness dimension" (PC1) had a high load (> 0.69). According to their degree of environmental consciousness, [27] distinguished between two categories of businesses: those that develop the necessary competencies to just comply with environmental regulations and those that consider environmental performance while making business decisions. In relation to the kind of negative loading is representing

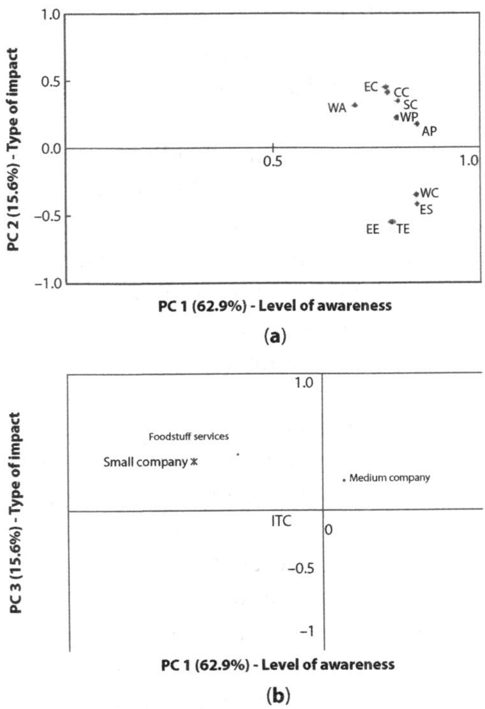

Figure 1.2 By nation type, company size, and their involvement in the food industry, PCA loadings (a) and scores (b) plots for the 10 variables impacting knowledge of environmental implications are shown. 76.8% of the total variation is explained by the two retrieved components. Factors include the following: EE stands for electrical energy, TE for thermal energy, ES for energy sources, WC for water use, AP for air pollution, WP for water pollution, SC for soil contamination, EC for ecosystem, CC for climate change, and WA for waste produced. Country type: OEC - Other European countries.

resource depletion consequences like energy and water consumption pacts (atmospheric pollution, water pollution, soil contamination, climate alteration and waste disposal). Environmental effects of the food chain affect how much natural resources (mostly water and energy) are used and how the environment is purged.

The scores plot (Figure 1.1b) summarizes the connections between the nations and the businesses. Big firms were opposed by small businesses and those in the food service industry, while those in the plant origin industry were in opposition to the former.

Scoring guidelines: "0" means that the environmental impact has not been analyzed; "1" means that the company has done so; "2" means that the company has calculated particular environmental measures for this consequence; "3" means that the organization has converted the basic data to compute environmental consequences per process or functional unit; and "4" means that the organization has calculated environmental footprints based on this impact.

The first extracted component's item loadings. The raw data was used to calculate the Mean values and Standard deviations.

There is a lot of untapped potential for using data and models in different "knowledge products" to improve their efficiency. This third family of data therefore seems to suggest that, in addition to the company's size (as was previously noted above), the company's sector, animal, or plant influences its environmental consciousness.

Implications for food chain stakeholders on a practical level

This method of examining the on-site implementation of mathematical models adds value in terms of the study of existing methods in the food chain and the degree of comprehension of food models in the foodstuff industry. These results encourage everyone involved in the food chain, particularly foodstuff businesses and academics, to step up knowledge transfer initiatives. Results show variations when company size, primary business, and nation of origin are taken into consideration. Managers may find opportunities to enhance their goods and procedures by taking into account the advantages of modeling in the food business. These findings could be useful for food consultants looking to increase the range of services they provide to the food industry. Last but not least, our results may be used as a guide to create a variety of user-friendly approaches designed for certain foodstuff industries and for smaller and midsized businesses to boost their competitive edge.

1.4 Conclusion

At the international level, nations use models in many ways. The more advanced nations utilize the available modeling and are positioned at a better degree of understanding. Similar patterns may be seen at the micro level, where smaller and midsized businesses show a restricted usage of models as a result of their shortage of resources and understanding. Most of the models reviewed here did not show significant differences across nations, and the usage of statistical models in the foodstuff business may be regarded as lower to medium. It is crucial because huge corporations may have operations in many nations, while small and medium-sized businesses function on a national basis. In light of the complexity of food matrices, the results show low to medium levels of knowledge in relation to the numerous models used by the food industry, with greater levels of knowledge being connected with certain food processes (and related foodstuff safety/environmental issues) than food items. Having restricted utilization of environmental models in foodstuff production, the most often employed models relate to comprehending and enhancing different elements of food safety and food quality. To achieve a greater level of proficiency and understanding of its commercial potential, the researchers feel that education attempts regarding modeling and simulation tools must be strengthened in both the industry (particularly in SMEs) and academics. The study's limitations include that no on-site evaluations were carried out to determine if the findings reported by the firms were accurate since the study was focused on the attitudes and beliefs of companies on the usage of mathematics. Additional restrictions are connected to the survey's sample size, corporate profiles, and the number of participating European nations. With a certain amount of care, the findings of this poll may be extrapolated to the whole European food industry.

References

1. D. Serrano *et al.*, "Why MagNet: Quantifying the Complexity of Modeling Power Magnetic Material Characteristics," in *IEEE Transactions on Power Electronics*, doi: 10.1109/TPEL.2023.3291084.
2. D. Mondal, A. Ratnaparkhi, A. Deshpande, V. Deshpande, A. P. Kshirsagar and S. Pramanik, Applications, Modern Trends and Challenges of Multiscale Modelling in Smart Cities, in Data-Driven Mathematical Modeling in Smart Cities, IGI Global, 2023, doi: 10.4018/978-1-6684-6408-3.ch001
3. N. Nayakanti, R. Al-Rfou, A. Zhou, K. Goel, K. S. Refaat and B. Sapp, "Wayformer: Motion Forecasting via Simple & Efficient Attention Networks," *2023 IEEE International Conference on Robotics and Automation (ICRA)*, London, United Kingdom, 2023, pp. 2980-2987, doi: 10.1109/ICRA48891.2023.10160609.
4. D. Sui, X. Zeng, Y. Chen, K. Liu and J. Zhao, "Joint Entity and Relation Extraction with Set Prediction Networks," in *IEEE Transactions on Neural Networks and Learning Systems*, doi: 10.1109/TNNLS.2023.3264735.
5. Wang, Y.; Wang, W.; Liu, J.; Chen, T.; Wang, S.; Yu, B.; Qin, X. Framework for Geometric Information Extraction and Digital Modeling from LiDAR Data of Road Scenarios. *Remote Sens.* 2023, *15*, 576. https://doi.org/10.3390/rs15030576

6. J. Li, S. Feng and B. Chiu, "Few-Shot Relation Extraction with Dual Graph Neural Network Interaction," in *IEEE Transactions on Neural Networks and Learning Systems*, doi: 10.1109/TNNLS.2023.3278938.
7. J. Zhou *et al.*, "HiNet: Novel Multi-Scenario & Multi-Task Learning with Hierarchical Information Extraction," 2023 *IEEE 39th International Conference on Data Engineering (ICDE)*, Anaheim, CA, USA, 2023, pp. 2969-2975, doi: 10.1109/ICDE55515.2023.00227.
8. Md Galal Uddin, Stephen Nash, Azizur Rahman, and Agnieszka I. Olbert, Performance analysis of the water quality index model for predicting water state using machine learning techniques, *Process Safety and Environmental Protection*, Vol. 169, 2023, pp. 808-828, ISSN 0957-5820, https://doi.org/10.1016/j.psep.2022.11.073.
9. S. Choudhary, V. Narayan, M. Faiz and S. Pramanik, "Fuzzy Approach-Based Stable Energy-Efficient AODV Routing Protocol in Mobile Ad hoc Networks", in *Software Defined Networking for Ad Hoc Networks*, M. M. Ghonge, S. Pramanik and A. D. Potgantwar, Eds., Springer, 2022, https://doi.org/10.1007/978-3-030-91149-2_6
10. F. Peng, S. Wang and X. Wang, "A Survey of Digital Literacy Requirements for Students in Vocational Education based on Likert Scale using the R Language," *2023 IEEE 6th Information Technology, Networking, Electronic and Automation Control Conference (ITNEC)*, Chongqing, China, 2023, pp. 1795-1798, doi: 10.1109/ITNEC56291.2023.10082499.
11. A. Mandal, S. Dutta, S. Pramanik, "Machine Intelligence of Pi from Geometrical Figures with Variable Parameters using SCILab", in *Methodologies and Applications of Computational Statistics for Machine Learning*, D. Samanta, R. R. Althar, S. Pramanik and S. Dutta, Eds., IGI Global, 2021, pp. 38-63, DOI: 10.4018/978-1-7998-7701-1.ch003
12. W. Erpurini, A. G. Putrada, N. Alamsyah, S. F. Pane and M. Nurkamal Fauzan, "Confirmatory Factor Analysis for the Impact of Students' Social Medial on University Digital Marketing," *2023 International Conference on Computer Science, Information Technology and Engineering (ICCoSITE)*, Jakarta, Indonesia, 2023, pp. 615-620, doi: 10.1109/ICCoSITE57641.2023.10127744.
13. V. Sharma, T. Manocha, S. Garg, A. Luthra, S. Dixit and M. Sharma, "Factors Affecting Awareness and Practices of Green Technology," *2023 3rd International Conference on Innovative Practices in Technology and Management (ICIPTM)*, Uttar Pradesh, India, 2023, pp. 1-5, doi: 10.1109/ICIPTM57143.2023.10118326.
14. A. Bhattacharya, A. Ghosal, A. J. Obaid, S. Krit, V. K. Shukla, K. Mandal and S. Pramanik, "Unsupervised Summarization Approach with Computational Statistics of Microblog Data", in *Methodologies and Applications of Computational Statistics for Machine Learning*, D. Samanta, R. R. Althar, S. Pramanik and S. Dutta, Eds, IGI Global, 2021, pp. 23-37, doi: 10.4018/978-1-7998-7701-1.ch002

15. B. K. Pandey, D. Pandey, V. K. Nassa, A. S. George, S. Pramanik and P. Dadheech, Applications for the Text Extraction Method of Complex Degraded Images, in *The Impact of Thrust Technologies on Image Processing*, Nova Publishers, 2023.
16. S. Pramanik, Intelligent Farming, Utilizing a Soil Tracking Device, in *Convergence of Cloud Computing, AI and Agricultural Science*, IGI Global, 2023, doi: 10.4018/979-8-3693-0200-2.ch009
17. Y. Meslie, W. Enbeyle, B. K. Pandey, S. Pramanik, D. Pandey, P. Dadeech, A. Belay, A. Saini, "Machine Intelligence-based Trend Analysis of COVID-19 for Total Daily Confirmed Cases in Asia and Africa", in *Methodologies and Applications of Computational Statistics for Machine Learning*, D. Samanta, R. R. Althar, S. Pramanik and S. Dutta, Eds., IGI Global, 2021, pp. 164-185, doi: 10.4018/978-1-7998-7701-1.ch009
18. S. Ahamad, V. Veeraiah, J.V.N. Ramesh, R. Rajadevi, Reeja S R, S. Pramanik and A. Gupta, Deep Learning-based Cancer Detection Technique, *Handbook of Research on Thrust Technologies' Effect on Image Processing*, IGI Global, 2023.
19. A. Roy and S. Pramanik, A Review of the Hydrogen Fuel Path to Emission Reduction in the Surface Transport Industry, *International Journal of Hydrogen Energy*, 2023.
20. Y. Ahmad Antouz, I. A. Akour, M. Turki Alshurideh, H. M. Alzoubi and E. K. Alquqa, "The impact of Internet of Things (IoT) and Logistics Activities on Digital Operations," *2023 International Conference on Business Analytics for Technology and Security (ICBATS)*, Dubai, United Arab Emirates, 2023, pp. 1-5, doi: 10.1109/ICBATS57792.2023.10111287.
21. A. Shafee *et al.*, "Techno-Economic Feasibility Analysis of Hybrid System in Bangladesh - A Case Study for Higher Learning Institution," *2023 3rd International Conference on Robotics, Electrical and Signal Processing Techniques (ICREST)*, Dhaka, Bangladesh, 2023, pp. 275-279, doi: 10.1109/ICREST57604.2023.10070053.
22. R. Jayasingh, J. Kumar R.J.S, D. B. Telagathoti, K. M. Sagayam and S. Pramanik, "Speckle noise removal by SORAMA segmentation in Digital Image Processing to facilitate precise robotic surgery", *International Journal of Reliable and Quality E-Healthcare*, vol. 11, issue 1, doi: 10.4018/IJRQEH.295083, 2022
23. K. Dushyant, G. Muskan, A. Gupta and S. Pramanik, "Utilizing Machine Learning and Deep Learning in Cyber security: An Innovative Approach", in *Cyber Security and Digital Forensics*, M. M. Ghonge, S. Pramanik, R. Mangrulkar, D. N. Le, Eds., Wiley, 2022, https://doi.org/10.1002/9781119795667.ch12
24. G. Taviti Naidu, KVB Ganesh, V. Vidya Chellam, S Praveenkumar, D. Dhabliya, S. Pramanik, A. Gupta, Technological Innovation Driven by Big Data, in Advanced Bioinspiration Methods for Healthcare Standards,

Policies, and Reform, Hadj Ahmed Bouarara IGI Global, 2023, doi: 10.4018/978-1-6684-5656-9

25. D. K.aushik, M. Garg, Annu, A. Gupta and S. Pramanik, Application of Machine Learning and Deep Learning in Cyber security: An Innovative Approach, in *Cybersecurity and Digital Forensics: Challenges and Future Trends*, M. Ghonge, S. Pramanik, R. Mangrulkar and D. N. Le, Eds., Wiley, 2022, doi: 10.1002/9781119795667.ch12
26. S. Ghosh, R. D. Raut, M. C. Mandal and A. Ray, "Revisiting the Nexus Between Sustainable Supply Chain Management and the Performance of Steel Manufacturing Organizations: A Case Study from an Emerging Economy," in *IEEE Engineering Management Review*, vol. 51, no. 1, pp. 189-213, 1 First quarter, March 2023, doi: 10.1109/EMR.2022.3233484.
27. M. C. Tavares, G. Azevedo, J. Vale and A. M. Bandeira, "The Accountant in the New Era: Renewing the Profession for Society 5.0," *2023 18th Iberian Conference on Information Systems and Technologies (CISTI)*, Aveiro, Portugal, 2023, pp. 1-6, doi: 10.23919/CISTI58278.2023.10211382.

2

Modeling the Effects of Land Degradation on Agricultural Productivity: Implications for Legal and Policy Interventions

Amit Verma[1], Istita Auddy[2], Murli Manohar Gour[3], Dhwani Bartwal[4], Sukhvinder Singh Dari[5] and Ankur Gupta[6]*

[1]College of Law, Teerthanker Mahaveer University, Moradabad, Uttar Pradesh, India
[2]Department of Food Technology, Faculty of Engineering and Technology, JAIN (Deemed-to-be University), Karnataka, India
[3]Department of Mathematics, Vivekananda Global University, Jaipur, Rajasthan, India
[4]Department of Agronomy, College of Agriculture, Parul University, Limda, Dist- Vadodara, Gujarat, India
[5]Symbiosis Law School, Nagpur Campus, Symbiosis International (Deemed University), Pune, India
[6]Department of Computer Science and Engineering, Vaish College of Engineering, Rohtak, Haryana, India

Abstract
Using seaweed extracts as a natural plant growing energizer might be a beneficial farming strategy in the development of sustainable and organic crops. Thus, during the winter growth seasons of 2020 and 2021, the present research was conducted in a private farm in Manchester, UK. The impact of BSE on the vegetable growth, production, and some nutritive uses of the leaves and roots of Celesta F1 hybrid RR plants was evaluated. BSE was utilized as a pre-sowing seed soaking for 10 hours at a rate of 5 ml/l, as well as foliar spraying once at rates of 2, 3 or 5 ml/l after 30 days from the sowing date. Three duplicates of the implementation were used

**Corresponding author*: ankurdujana@gmail.com
Amit Verma: Orcid: https://orcid.org/0009-0006-5521-1800
Istita Auddy: Orcid: https://orcid.org/0000-0001-9414-0028
Murli Manohar Gour: Orcid: https://orcid.org/0000-0003-2907-8226

Sabyasachi Pramanik, Niranjanamurthy M., Ankur Gupta and Ahmed J. Obaid (eds.) Mathematical Modeling in Agriculture, (17–32) © 2024 Scrivener Publishing LLC

in a split plot design. The findings clearly showed that seaweed-immersed seeds greatly outperformed water-immersed seeds in terms of all examined features. It was quite clear that, when compared to the other treatments, the foliage spraying therapy with the greatest concentration of SE (5 ml/l) produced considerably the finest results for all evaluated features. With the exception of leaves and roots dry weight, leaves Fe and roots Fe and Zn concentrations in all seasons, chlorophyll an in the first season and chlorophyll b and carotenoids in the second season, the interaction significantly affected all measured parameters. Seaweed-soaked seeds that were then sown and foliar spraying SE at a rate of 5 ml/l once 15 days after sprinkling were determined to be the most successful treatments.

Keywords: Mineral contents, phenols, flavonoids, phytochemicals

2.1 Introduction

The annual cool-season root vegetable crop is known as the RR. RR roots, leaves, and sprouting seeds may all be eaten raw or in salads right away. The radish root's skin may be white, red, pink, purple, or yellow, but the root's meat is a crisp white hue having a strong taste. The pigment anthocyanin is what gives root skin its red color. Vitamin C, glucosinolates, polyphenolic chemicals, S, Ca, K, and P are all present in sufficient amounts in root. According to [1], the Brassicaceae family of vegetables includes potent anti-oxidants that are physiologically active and have been linked to positive health impacts. According to [2], these compounds may help reduce or prevent the risk of a number of illnesses, including cardiovascular, some malignancies, hypertension, stroke and various chronic disorders.

SE is regarded as a natural and organic plant growth energizer according to [3] definition of a plant growth stimulant from 2015. Depending on their colour, marine macroalgae may be divided into three groups: green algae, brown algae, and red algae.

Ulva species (Chlorophyta); red algae, Lithophyllum spp.; brown algae, Sargassum spp., and Ascophyllum spp. and Cladophora spp. and Asparagopsis varieties. The extracts of brown algae, a kind of sea algae, have gained significant attention and are often employed in contemporary, sustainable, and farming as bio-fertilizers or plant growing stimulants. This may be explained by the fact that these algae have larger concentrations of natural phytohormones, micronutrients, and macronutrients than other varieties of algae.

According to several studies, SE contains a variety of functional organic active materials, including polyphenols, alginates, pigments, free amino acids, vitamins, micro and macronutrients, and natural phytohormones.

According to [4], the stimulant impact of SE can be connected to every compound existing which may alter the cellular metabolic techniques of treated vegetations. This might explain all familiar positive benefits of SE. Many researchers have shown that SE stimulates seed germination, early seedling development and vigor and enhancing seedling foundation in a range of crops.

Additionally, the utilization of SEs expanded the amount of cumulative chlorophyll in the leaves that had an effect on the extent and proficiency of the photosynthetic technique and the amount of organic C in the soil, nutrient availability, and nutritional absorption. Using SEs improved plant tolerance to various environmental stress conditions; it also made plants more resistant to diseases and nematodes; and it increased plant resilience to nematodes. Overall, SE application shows growth-energizing activation of the plant growing, production, and condition in various farming crops.

SE may be administered using one approach, such as seed priming before sowing, root dipping for seedlings before transplantation, root or soil soaking, foliar sprinkling, or a combination of two or additional ways. According to the plant type and development phases, the suggested application technique, timing, and rate varied substantially. To address this issue, [5] showed that applying SE early in a plant's development cycle resulted in more prominent stimulatory effects.

In contrast to control and other treatments, BSE used as a pre-sowing seed soaking for 14 hours at 0.6 and 3%, respectively, resulted in the highest seed fertilization percentage and growth, bio-chemical components of radish plants. Although there were negligible consequences on the production parameter of both crops, the application of SE increased the aggregation of phytochemicals compounds, improving the nutrient standard of cabbage plants and broccoli plants. In a similar vein, [6] demonstrated that leafage administration of BSE enhanced spinach leaves' phenolic levels and their anti-oxidant schemes. In accordance with [7], sprinkling cowpea plants having BSE at a rate of 4% substantially enhanced shoot size, quantity of leaves/plant, production, cumulative phenolic, total flavonoid contents, and antioxidant actions compared to the control handling. Additionally [8] described that leaf spraying "Alga 600" seaweed on celeriac plants had positive results.

In order to increase RR plant vegetation growth, yields of leaves and branches, mineral contents of leaves and branches, and few biologically active materials, this study looked at the effects of utilizing BSE of Sargassum vulgare as a pre-sowing seed soak inclusive to a foliar sprinkling once in 15 days after sowing at various rates.

2.2 Materials and Procedures

Testing ground

The present research was conducted throughout the winters of 2021 and 2022 at a private farm in the UK's Manchester area. On the vegetation growth, leaves and roots produces, and few nutritional estimates of the leaves and branches of RR plants, the consequence of BSE of Sargassum vulgare as a natural plant growing stimulant utilized as a pre-sowing seed soaking at rate of 5 ml/l for 14 h and foliar sprinkling once at rates of 2, 3 or 5 ml/l for 20 DAS was studied. The analytical procedures outlined by [9] were used to examine the experimental soil. Table 2.1 displays the experimental soil's physical characteristics and chemical analyses. Table 2.2 displays the chemical make-up of the commercial BSE utilized in the research.

Three duplicates of the implementation were used in a split plot [10] design. As opposed to the foliar sprinkling nursing of RR plants 15 DAS having SE at rates of 2, 3, or 5 ml/l besides water composed of nursing, which were arbitrarily dispersed within the sub-plots, the seed drenching nursing, seeds soaked in SE at a rate of 5 ml/l or in distilled water, for 14 hour prior to sowing, were displayed in the main plots. Every experimental sub-plot included three rows, each measuring 2 meters long and 0.7 meters wide, for a total net area of 4.2 meters. The greatest significant levels of the extract were typically found at a rate of 2000 ppm.

In comparison to the control nursing, the following values for plant height, fresh and dry weight, cumulative chlorophyll, and Phosphorus and Potassium constituents of leaves. In addition, yields of green leaves and roots were much higher than those of the other treatments preparation cultivation of experimental soil.

Table 2.1 The experimental soil's physical characteristics and chemical analyses.

Physical characteristics											
Clay (%)		Silt (%)		Sand (%)							
62.8		29.4		14.3							
Chemical Nature											
pH		Cations (meq/l)						Anions (meq/l)			
		Ca^+	Mg^+	Na^+	K^{++}	P^{++}		SO_4^-	CO_3^{--}	HCO_3^-	Cl^-
		2.5	0.7	1.6	0.42	0.82		3.6	NA	1.63	1.4

The soil for the experiment was prepared using a land plow, leveling, and row building. Following that, rows were separated into three implemented duplicates, any of which has two main plots and four sub-plots. Inorganic fertilizer, phosphorus in the form of calcium superphosphate (15.5% P2O5), was applied to all experimental sub-plots at a rate of 200 kg/ha and organic manure in the form of cow dung at a rate of 25 m3/ha. During the experimental soil preparation, full dosages of calcium superphosphate and manure from cattle were applied and fully merged and integrated into the soil. Additionally, $(NH_4)_2SO_4$ (22.5% N) at a rate of 300 kg/ha was applied as a nitrogen fertilizer. Two equal doses of nitrogen were administered; the first dosage was provided 20 days after sowing, and the second dose was administered two weeks later. K_2SO_4 (50% K_2O), a potassium fertilizer, was administered at a rate of 150 kg/ha. Once, 45 days following the date of planting, the whole quantity of potassium was applied.

A local seed company provided imported RR Celesta F1 hybrid seeds. RR seeds were split into two halves, with the first being steeped for 12 hours while the second part was immersed in distilled water for a similar amount of time and provided as a control therapy, in a mixture of BSE as a plant growth stimulant obtained from US Agent for Farming Development Organization. Then, on October 8, 2017, for the first season, and on October 10, 2018, for the second season, soaked seeds were immediately seeded onto the field at each side of 3 m long by 0.8 m wide rows in hills having a capability for three seeds per hill and 12 cm spacing between hills in the row. Then, BSE at rates of 2, 3, or 5 ml/l was foliar sprayed for 20 DAS at the stage of the third true leaf on RR plants. The sprinkling mixtures of BSE included a few drops of Tween-30 as a emulsifier ingredient.

Pigment content of leaves

Moran (1982) reported that the contents of photosynthetic pigments, such as chlorophyll a, b, a + b, and carotenoids, were assessed in crisp leaves tested at 30 days following planting date. 500 mg of fresh leaf disks were submerged in 15 ml of N, N-dimethylformamide for the whole night. The resulting extracts were analyzed for carotenoids, chlorophyll b, and chlorophyll a at 450, 620, and 645 nm in a UV/VIS spectrophotometer. N, N-di-methylformamide was employed as blank nutritional benefits of roots and leaves.

The nutritional value of the leaves and branches of the RR plant was also determined using samples of leaves and roots that had previously been used to estimate leaf or root yield. After washing the samples of leaves and roots to get rid of any surface-attached dirt and powder, the specimens were stretched out on tissue paper and left to air dry for one hour

Table 2.2 The commercial brown sea algae extract's chemical make-up.

Constituent	Estimate	Constituent	Estimate
Sodium	2%	Various hormones	12 mg/l
Potassium	19.6%	Glycine	0.04 mg/l
Calcium	0.18%	Algamic acid	700 ppm
Magnesium	0.51%	Auxin	15-18%
Iron	0.07%	-	22%
Sulphur	3.1%	Cytokinin	-

at room temperature. The roots were then sliced into slices before drying. Specimens were then baked in an oven at 65 °C until their weight remained consistent. The following nutritional value analyses were performed on the oven-baked specimens of leaves and branches that had been finely crushed in a stainless-steel grinder to pass a 3 mm screen dimension. Using the Folin-Ciocalteau reagent, as explained by [11], the total phenolic constituent of an absolute ethanolic extract was measured. A UV/VIS spectrophotometer was utilized to compute the reaction mixture consumption at 725 nm. Cumulative phenolic constituent was computed utilizing gallic acid as the usual reference and revealed as milligrams of gallic acid identical to per gram of dry weight tissue (mg GAE/g DW). Utilizing the aluminum chloride colorimetric technique published by [12], the total flavonoid content of an absolute ethanolic extract was measured. The absorption early in the morning, spraying was done using a manual pressure sprayer was calculated at 420 nm in a UV/VIS spectrophotometer [13] and spraying the fluid over the plant's leaves. In accordance with the advice of the UK Minister of Agriculture and Land Reformation, all farming managements for the production of RR plants under Trent River circumstances in the Staffordshire area of daily irrigation, fertilization, and management of weeds, diseases, and pests were carried out.

Characteristics of plant vegetative development

Forty days following the planting date, an arbitrary specimen of 14 RR plants was collected from every implemented sub-plot. To avoid damaging the roots, the plants were gently pulled from the ground. Next, the extra dirt was carefully scraped from around the roots. After that, a lot of tap water was used to wash the roots and remove any remaining dirt that had attached to them. The gathered plants were then utilized to quantify several plant vegetation development characteristics, such as plant length, leaf count per plant, branch diameter, and the fresh and dried weights of the leaves and branches.

An arbitrary specimen of 14 RR plants from every implemented sub-plot were picked at the harvesting stage, 60 days following the planting date, and cleaned as previously mentioned. Following harvest, plants' leaves and branches were separated. The average fresh weight of leaves and roots per plant was measured. The mean fresh weight of leaves or branches per plant was multiplied by the mean quantity of plants per implemented sub-plot to get the average production of leaves and branches per implemented sub-plot. The production of leaves or branches was then estimated as tons/ha. In terms of milligrams quercetin equal to per gram of dry weight tissue (mg QCE/g DW), the cumulative flavonoid constituent was represented. Utilizing a UV/VIS spectrophotometer, the anthocyanin concentration of a 70% methanol extract was spectrophotometrically examined. According to [14], the reaction mixture's absorbance was measured at 520 nm utilizing a blank of 70% methanol. The amount of anthocyanins was represented as mg of cyanidin chloride equivalent (mg CYE/g DW) per gram of dry weight tissue.

2.2.1 Content of Minerals

Wet digestion was performed on 200 mg of fine powdered, dried leaf and root specimens using a solution of H_2SO_4 98% and H_2O_2 30%. Using a Kjeldahl technique and a UDK 139 Semi-Automatic Kjeldahl Distillation Unit from VELP Scientific, Inc., in California, total nitrogen in acid digested solution was calculated. The amount of phosphorus was evaluated utilizing a spectrophotometer according to the adjusted colorimetric methods reported by [15]. As per [16] description, K was also calculated utilizing the flame photometer technique. Additionally, the elements Fe, Zn, and Mn were found utilizing the Chapman and Pratt (1982) technique utilizing an Analyst-200 from PerkinElmer, Inc., in Massachusetts, USA.

Statistic evaluation

The SPSS [17] software's Two-way-ANOVA [18] technique was used to statistically analyze the acquired data. The average of three values is used to convey values Standard deviation (SD) [19] in measurements. In accordance with the techniques described by [20], the least significant differences (LSD) test was utilized to examine the major differences between treatment means at a p0.05 level of significance.

2.3 Results and Analysis

Characteristics of plant vegetative development

In both the 2017 and 2018 growing seasons, SE-soaked seeds significantly outperformed water-soaked seeds in terms of plant length, quantity of leaves per plant, branch diameter, and fresh and dry weights of leaves and branches.

The benefits of the natural constituent of SE, such as polysaccharides, alginates, polyamines, amino acids, nutrients, naturally occurring cytokinins, auxins and auxin-like constituents, and gibberellins, may be responsible for the supremacy of SE-soaked seeds to water-soaked seed care on vegetation growing parameters. According to studies by [21], these components have an impact on cellular metabolic processes, which improve seed germination, seedling development, and accelerate plant vegetation growth.

There was a favorable rise in the vegetation growth parameters by expanding the implementation rate of SE with regard to the impact of foliar sprinkling of BSE 20 DAS on vegetation growth parameters of RR. The obtained findings showed that when foliar sprinkled with SE at the rates of 5 ml/l succeeded by 3 ml/l without a major difference between both of them, RR plants showed the maximum compelling estimates of every evaluated parameter of vegetation growing in all seasons. While the control therapy had the lowest results the results obtained are in accordance with those of [22] on celeriac plants, on tomato, on pepper, and on cowpea plants. They noted a connection between applying SE through foliar spray and increased plant vegetation growth.

The observed outcomes can be attributable to the fact that the implementation of SEs was likely to expand leaf cumulative chlorophyll constituent, which in turn had an impact on the capability and proficiency of the photosynthetic technique and the obtainability and soaking up of nutrients. These unquestionably have a good impact on plant vegetative development. Additionally, the availability of natural phytohormones and nutrients may explain why SE has the potential to improve plant vegetative development.

In the communication between pre-sowing seed soaking and leaf implementation of SE techniques, RR plant vegetative growth parameters showed significant differences in both the 2017 and 2018 growing seasons, with the exception of leaves and roots dry weight. It is clear that RR plants handled with SE as pre-sowing seed soaking at a rate of 3 ml/l for 14 h conformed by foliar sprinkled once having SE at a rate of 5 ml/l 20 DAS showed the maximum compelling (p0.05) estimates of every computed parameters of plant vegetation growth in all study seasons, succeeded lightly by the nursing of 3 ml/l. output of roots and leaves

Data demonstrated that using BSE as a seed soak significantly increased the yields of RR plants' leaves and roots throughout both research seasons. After being treated with seaweed-soaked seeds, the higher during both research seasons, yields of leaves and roots were higher than those of water-soaked seeds.

Because of the existence of phytohormones and nutritional constituents, this finding may be linked to the improvement of plant vegetation development after seed soaking in BSE. There is little doubt that such improvement will impact plant output.

RR plants produced significantly more leaves and roots after receiving foliar applications of BSE in comparison to controls in both seasons. The greatest estimates of leaves and roots productions were obtained by leaf sprinkling with the maximum rate (5 ml/l) of SE, followed by the rate of 3 ml/l insignificantly (p0.05). The control therapy, however, had the lowest results. These outcomes were comparable in both the 2017 and 2018 seasons. SE applied topically at a rate of 5 ml/l increased the yields of leaves and roots in the first and second seasons, respectively, by around 43.0%, 45.8%, and 19.8%, 23.1% above the control treatment. These improvements in the yields of leaves and roots might be directly related to the rise in the vegetation development parameters.

[23] are all in agreement with the findings that were attained. In comparison to the control treatment, they found that leaf application of SEs considerably boosted plant production. On the other hand, [24] came to the conclusion that foliar spraying with BSE had no discernible impact on the production of cabbage and broccoli plants.

BSE application increases the production of leaves and branches in part due to the existence of phytohormones, primarily cytokinins. The partitioning and mobilization of organ metabolites and nutrients is significantly influenced by cytokinins. Furthermore, the enhancement of plant vegetation development features with leaf sprinkling of BSE might be attributed to the findings found here. Better vegetation growth emerged as enhanced and having a higher yield.

In both the 2017 and 2018 growing seasons, the interaction significantly altered the RR plant's yields of leaves and branches. It's interesting to observe that RR plants handled with SE as pre-sowing seed soaked and then foliar sprinkled a single time with SE at 5 ml/l produced the greatest values of leaves and branches produces pigment content of leaves. Seeds treated with seaweed instead of water had more chlorophyll a, chlorophyll b, chlorophyll a + b, and carotenoids in their leaves. In both growth seasons, seaweed-soaked seeds had greater significant (p0.05) estimates of the RR leaf pigment concentrations than water-soaked seeds.

According to [25], soaking mung bean seeds in SE led to an enhancement in the leaf pigment levels when contrasted to control seeds. These results are consistent with their findings. Furthermore, according to [26], priming seeds in SEs improved the photosynthetic pigments concentration in radish plant leaves.

The spraying of BSE at a rate of 5 ml/l produced the greatest major quantities of chlorophyll a, chlorophyll b, chlorophyll a + b, and carotenoids in the leaves of RR plants when compared to control and other treatments. The control treatment produced the lowest results. Both growth seasons showed comparable tendencies.

According to [27], the findings were in agreement; they showed that adding SEs increased the quantity of total chlorophyll in the leaves of the treated plants, which had an impact on the effectiveness and capacity of the photosynthetic process. Effect of SE utilized as a pre-sowing seed-soaking solution and foliar treatment once 20 DAS at various rates and their relation on RR plant productions for the 2017 and 2018 growing seasons, as well as the amount of leaf pigment in the leaves.

According to [28], 10% application of seaweed extract improved the photosynthetic pigments in the leaves of Vigna radiata plants. However, [29] found that as compared to control treatment, SE at 2.6% increased leaf photosynthetic pigments in cluster bean seedlings. According to [30] the effects of SE treatment on leaf pigment contents may include a delay in leaf senescence, a reduction in the minimization of leaf pigments, or an improvement in the biosynthesis of leaf pigments. These effects may be caused by the occurrence of natural phytohormones in SE, mainly cytokinins.

Regarding the interactivity handling, substantial variations in leaf pigment concentrations were seen in both seasons, with the exception of chlorophyll a and carotenoids in the first season and second season, respectively. In the growing seasons of 2017 and 2018, plants whose seeds had been soaked in seaweed and those whose leaves had been sprayed with seaweed at the greatest rate (5 ml/l) had the greatest levels of RR leaf pigment concentrations. On the other hand, the plants receiving the control treatment had the lowest values.

During the two succeeding growing seasons, BSE-soaked seeds produced more phytochemical substances, such as cumulative phenolics, cumulative flavonoids, and cumulative anthocyanins, in the leaves and branches of RR plants than water-soaked seeds.

It is clear that in the first and second seasons, respectively, radish leaves seemed to have a greater flavonoid content (14.5 and 12.3 mg QCE/g DW), virtually double that of the roots (6.3 and 6.9 mg QCE/g DW). In a similar vein, it was discovered that the phenolic and anthocyanin levels of RR roots were greater than those of the leaves (28.7, 29.4 mg GAE/g DW and 2.42, 2.31 mg CYE/g DW, respectively). The observed findings are consistent with [31] who claimed that BSE used as a pre-sowing seed soaking for 14 hours at a rate of 2% led to the greatest estimates of biochemical constituents of radish plants.

Regarding the impact of foliar application of BSE on the levels of phenolics, flavonoids, and anthocyanins in the leaves and branches of RR plants, it was blatantly obvious that the implementation of the largest rate of SE (5 ml/l) resulted in mostly the largest estimates of leaves flavonoids and branches phenolics and anthocyanins concentrations in the two seasons. The least concentrations of phenolics and anthocyanins in the roots and leaves were produced by control treatment. The outcomes are consistent with those of [32]. They concluded that the implementation of brown seaweed extract might boost the aggregation of phytochemical components and antioxidant activity, hence improving the nutritional amount of managed plants.

According to several studies, these findings can be credited to BSE's huge concentrations of phenolic compounds and antioxidants, which may affect the metabolic processes of treated plants. This gathering of phytochemical compounds and increased antioxidant activity enhanced the treated plants' nutritional value. Additionally, according to [33], the physiologically active and potent antioxidant compounds tended to concentrate in Brassicaceae family crops.

The accumulation of phytochemical substances, such as cumulative phenolics, cumulative flavonoids, and cumulative anthocyanins, in the leaves and branches of RR plants in all seasons was mostly influenced differently by the interactivity. According to general consensus, RR effect of SE applied as a pre-sowing seed-soaking solution and as a foliar treatment a single time 20 DAS at various rates and their interactivity on cumulative phenolics, flavonoids and anthocyanin constituents in RR plants' leaves and roots throughout seasons 2021 and 2022.

Plants handled with SE as a pre-sowing soaked seed succeeded by foliar spray at a rate of 5 ml/l 20 DAS exhibited the greatest leaf flavonoid and root phenolic and anthocyanin levels content of minerals. In all seasons, the seaweed-soaked seeds treatment greatly outperformed the water-soaked seeds treatment in terms of mineral constituents in the RR leaves and branches. The highest mineral contents were found in the leaves and roots of RR when SE was foliar sprinkled at a rate of 5 ml/l. This treatment was mostly followed by treatments of 2 ml/l, 3 ml/l, and control management that produced the least mineral constituents.

As shown, the RR leaves appeared to have more nitrogen than the roots. However, compared to leaves, RR roots contain larger amounts of phosphorous, potassium, and iron. But it was discovered that the amounts of Zn and Mn in the leaves and roots were quite comparable. SE seed-soaking and foliar-sprinkling treatments led to the same outcomes in all seasons. The results of [34] are in accordance with the outcomes obtained. They claimed that as compared to untreated plants, plants treated with SEs exhibited greater mineral levels. The ability of plant roots to absorb water and nutrients from the soil was improved by the use of SEs. Furthermore, () found that foliar spraying SEs had a substantial impact on the estimates of nitrogen, phosphorous, and potassium in the leaves and roots of treated plants. The observed findings are most likely the consequence of several phytohormones present in SE, which enhanced the development of vigorous plant root systems and had an impact on the ability of plant roots to absorb water and nutrients from the soil. The amounts of minerals in the foliar-sprayed SE (Table 2.2) also contribute to this.

Regarding the interaction, it was shown that the mineral constituents of RR leaves and roots in both growth seasons had significantly different impacts. Except for the Fe and Zn concentrations of the roots throughout the two seasons, leaves contain no Fe. In general, it could be said that SE administered as a pre-sowing seed soaking for 14 h and then as foliar spraying once every 20 days at a rate of 5 ml/l may be used to demonstrate the mineral constituents in the leaves and branches of RR plants.

2.4 Conclusion

Sargassum vulgare BSE could be used as a plant-growing stimulant at a rate of 5 ml/l as a pre-sowing seed soaking for 14 h, along with foliar spraying once every 20 DAS, to improve plant vegetation growth, production, and nourishment value of RR plants. Additionally, it is considered to be a promising, practical, and excellent agricultural technique for crop productivity in sustainable and organic agriculture.

References

1. A. Dutta, P. K. Kumar, A. De, P. Kumar, S. Dwivedi and J. Harshith, "Ascribing Machine Learning Classifiers to diagnose the attacks of Alternaria solani on Leaves of Solanum tuberosum," *2023 2nd International Conference on Computational Systems and Communication (ICCSC)*, Thiruvananthapuram, India, 2023, pp. 1-6, doi: 10.1109/ICCSC56913.2023.10142978.
2. Q. Lv, G. Chen, Z. Yang, W. Zhong and C. Y. -C. Chen, "Meta Learning with Graph Attention Networks for Low-Data Drug Discovery," in *IEEE Transactions on Neural Networks and Learning Systems*, doi: 10.1109/TNNLS.2023.3250324.
3. Sahu, P., Kashaw, S. K. (2023). Polymeric Nanocomposites-Based Agricultural Delivery: Recent Developments, Challenges, and Perspectives. In: Fernandez-Luqueno, F., Patra, J.K. (Eds.) *Agricultural and Environmental Nanotechnology. Interdisciplinary Biotechnological Advances*. Springer, Singapore. https://doi.org/10.1007/978-981-19-5454-2_10
4. K. Newman, M. Endres, W. Weimer and B. Johnson, "From Organizations to Individuals: Psychoactive Substance Use by Professional Programmers," *2023 IEEE/ACM 45th International Conference on Software Engineering (ICSE)*, Melbourne, Australia, 2023, pp. 665-677, doi: 10.1109/ICSE48619.2023.00065.
5. X. Lyu, J. Zhou, P. Ren and A. C. Frery, "Classification with Unbalanced Samples by Self-Sampling and Semicorrelated Co-Training—An Application to Algal Bloom Detection," in *IEEE Transactions on Geoscience and Remote Sensing*, vol. 61, pp. 1-12, 2023, Art no. 4206612, doi: 10.1109/TGRS.2023.3299312.
6. Neupane, J.; Maja, J. M.; Miller, G.; Marshall, M.; Cutulle, M.; Greene, J.; Luo, J.; Barnes, E. The Next Generation of Cotton Defoliation Sprayer. *Agri-Engineering* 2023, *5*, 441-459. https://doi.org/10.3390/agriengineering5010029
7. M. A. Z. Abidin, M. N. Mahyuddin and M. A. A. M. Zainuri, "Optimal Efficient Energy Production by PV Module Tilt-Orientation Prediction without Compromising Crop-Light Demands in Agrivoltaic Systems," in *IEEE Access*, vol. 11, pp. 71557-71572, 2023, doi: 10.1109/ACCESS.2023.3293850.
8. Singh, R., Kumar, S. (2023). Nanofertilizers and Nanopesticides: Key to Healthier and Safe Food Products. In: *Nanotechnology Advancement in Agro-Food Industry*. Springer, Singapore. https://doi.org/10.1007/978-981-99-5045-4_5

9. P. T. Khanh, T. H. Ngọc, and S. Pramanik, Future of Smart Agriculture Techniques and Applications, in *Advanced Technologies and AI-Equipped IoT Applications in High Tech Agriculture*, Ed., A. Khang, IGI Global, 2023.
10. V. Vidya Chellam, V. Veeraiah, A. Khanna, T. H. Sheikh, S. Pramanik and D. Dhabliya, A Machine Vision-based Approach for Tuberculosis Identification in Chest X-Rays Images of Patients, *ICICC*, 2023, Springer, doi.org/10.1007/978-981-99-3315-0_3
11. Ahmad Royani, Muhammad Hanafi, Heddy Julistiono, and Azwar Manaf, The total phenolic and flavonoid contents of Aloe vera and Morinda citrifolia extracts as antibacterial material against Pseudomonas aeruginosa, *Materials Today: Proceedings*, Vol. 72, Part 6, 2023, pp. 2796-2802, ISSN 2214-7853, https://doi.org/10.1016/j.matpr.2022.06.466.
12. Y. Zhao *et al.*, "Development of Flexible Electronic Biosensors for Healthcare Engineering," in *IEEE Sensors Journal*, doi: 10.1109/JSEN.2023.3287291.
13. S. Praveenkumar, V. Veeraiah, S. Pramanik, S. M. Basha, A. V. Lira Neto, V. H. C. De Albuquerque and A. Gupta, Prediction of Patients' Incurable Diseases Utilizing Deep Learning Approaches, ICICC 2023, Springer, doi.org/10.1007/978-981-99-3315-0_4
14. Q. -D. Pham, N. Holle and J. Wilde, "Assessment methods for the characterization of flux material systems toward water absorption," *2023 IEEE 73rd Electronic Components and Technology Conference (ECTC)*, Orlando, FL, USA, 2023, pp. 2017-2023, doi: 10.1109/ECTC51909.2023.00346.
15. R. Nandeshwar, P. Mandal and S. Tallur, "Portable and Low-Cost Colorimetric Sensor for Detection of Urea in Milk Samples," in *IEEE Sensors Journal*, vol. 23, no. 14, pp. 16287-16292, July 15, 2023, doi: 10.1109/JSEN.2023.3282810.
16. S. Pramanik, S Joardar, OP Jena, AJ Obaid, "An Analysis of the Operations and Confrontations of Using Green IT in Sustainable Farming", Al-Kadhum 2nd International Conference on Modern Applications of Information and Communication Technology, *AIP Conference Proceedings* 2591, 040020 (2023). https://doi.org/10.1063/5.0119513
17. A. Roy and S. Pramanik, A Review of the Hydrogen Fuel Path to Emission Reduction in the Surface Transport Industry, *International Journal of Hydrogen Energy*, 2023.
18. R. Jayasingh, J. Kumar R.J.S, D. B. Telagathoti, K. M. Sagayam and S. Pramanik, "Speckle noise removal by SORAMA segmentation in Digital Image Processing to facilitate precise robotic surgery", *International Journal of Reliable and Quality E-Healthcare*, vol. 11, issue 1, doi: 10.4018/IJRQEH.295083, 2022.
19. V. Jain, M. Rastogi, J. V. N. Ramesh, A. Chauhan, P. Agarwal, S. Pramanik, A. Gupta, FinTech and Artificial Intelligence in Relationship Banking and Computer Technology, in *AI, IoT, and Blockchain Breakthroughs in E-Governance*, Eds., K. Saini, A. Mummoorthy, R. Chandrika and N.S. Gowri Ganesh, IGI Global, 2023.

20. S. Käss, S. Strahringer and M. Westner, "Practitioners' Perceptions on the Adoption of Low Code Development Platforms," in *IEEE Access*, vol. 11, pp. 29009-29034, 2023, doi: 10.1109/ACCESS.2023.3258539.
21. M. Busschaert, F. H. Vermeire and S. Waldherr, "Modeling Cell Size Distribution with Heterogeneous Flux Balance Analysis," in *IEEE Control Systems Letters*, vol. 7, pp. 1903-1908, 2023, doi: 10.1109/LCSYS.2023.3282699.
22. M. Kang, X. Wang, H. Wang, J. Hua, P. d. Reffye and F. -Y. Wang, "The Development of AgriVerse: Past, Present, and Future," in *IEEE Transactions on Systems, Man, and Cybernetics: Systems*, vol. 53, no. 6, pp. 3718-3727, June 2023, doi: 10.1109/TSMC.2022.3230830.
23. G. Roggiolani, M. Sodano, T. Guadagnino, F. Magistri, J. Behley and C. Stachniss, "Hierarchical Approach for Joint Semantic, Plant Instance, and Leaf Instance Segmentation in the Agricultural Domain," *2023 IEEE International Conference on Robotics and Automation (ICRA)*, London, United Kingdom, 2023, pp. 9601-9607, doi: 10.1109/ICRA48891.2023.10160918.
24. A. Kumar and B. Bhowmik, "Rice Cultivation and Its Disease Classification in Precision Agriculture," *2023 International Conference on Artificial Intelligence and Smart Communication (AISC)*, Greater Noida, India, 2023, pp. 200-205, doi: 10.1109/AISC56616.2023.10085072.
25. K. Ruangwong, N. Tephiruk, T. Rongsangchaicharean, W. Kanokbannakorn and S. Srisonphan, "Electrical Discharge Plasma for Seed Priming of Green Oak Lettuce (Lactuca sativa)," *2023 International Electrical Engineering Congress (iEECON)*, Krabi, Thailand, 2023, pp. 337-341, doi: 10.1109/iEECON56657.2023.10126704.
26. S.-J. Hong *et al.*, "Application of X-Ray Imaging and Convolutional Neural Networks in the Prediction of Tomato Seed Viability," in *IEEE Access*, vol. 11, pp. 38061-38071, 2023, doi: 10.1109/ACCESS.2023.3265998.
27. P. Das, R. Pegu, S. S. Bhattacharya and P. Nath, "Fluorescence-Based Accurate Estimation of Chlorophyll in Tea Leaves Using Smartphone," in *IEEE Sensors Journal*, vol. 23, no. 13, pp. 14864-14871, 1 July 1, 2023, doi: 10.1109/JSEN.2023.3275879.
28. T. N. Vipina Vinod, Jyothis Mathew, E. K. Radhakrishnan, Chapter 14 - Nano-based biofertilizers for horticulture, Eds.: Musa Seymen, Ertan S Kurtar, Ceknas Erdinc, Ajay Kumar, In *Developments in Applied Microbiology and Biotechnology, Sustainable Horticulture*, Academic Press, 2022, pp. 437-459, ISBN 9780323918619, https://doi.org/10.1016/B978-0-323-91861-9.00015-X.
29. P. Du, W. Zhong, X. Peng, L. Li and Z. Li, "Self-Healing Control for Wastewater Treatment Process Based on Variable-Gain State Observer," in *IEEE Transactions on Industrial Informatics*, vol. 19, no. 10, pp. 10412-10424, Oct. 2023, doi: 10.1109/TII.2023.3240937.
30. Y. Li and S. Liang, "Evaluation of Reflectance and Canopy Scattering Coefficient Based Vegetation Indices to Reduce the Impacts of Canopy Structure and Soil in Estimating Leaf and Canopy Chlorophyll Contents," in

IEEE Transactions on Geoscience and Remote Sensing, vol. 61, pp. 1-15, 2023, Art no. 4403015, doi: 10.1109/TGRS.2023.3266500.

31. K. Ruangwong, N. Tephiruk, T. Rongsangchaicharean, W. Kanokbannakorn and S. Srisonphan, "Electrical Discharge Plasma for Seed Priming of Green Oak Lettuce (Lactuca sativa)," *2023 International Electrical Engineering Congress (iEECON)*, Krabi, Thailand, 2023, pp. 337-341, doi: 10.1109/iEECON56657.2023.10126704.
32. K. Ruangwong, N. Tephiruk, T. Rongsangchaicharean, W. Kanokbannakorn and S. Srisonphan, "Electrical Discharge Plasma for Seed Priming of Green Oak Lettuce (Lactuca sativa)," *2023 International Electrical Engineering Congress (iEECON)*, Krabi, Thailand, 2023, pp. 337-341, doi: 10.1109/iEECON56657.2023.10126704.
33. D. Zhang and P. Li, "Study on Extraction Technology of Purple Cabbage Anthocyanin Pigment based on Artificial Neural Network and Interactive System," *2022 3rd International Conference on Smart Electronics and Communication (ICOSEC)*, Trichy, India, 2022, pp. 1065-1068, doi: 10.1109/ICOSEC54921.2022.9951880.
34. K. Ramaraj, G. Amiya, P. R. Murugan, V. Govindaraj, M. Vasudevan and A. Thiyagarajan, "Sensors for Bone Mineral Density Measurement to Identify the Level of Osteoporosis: A Study," *2022 4th International Conference on Smart Systems and Inventive Technology (ICSSIT)*, Tirunelveli, India, 2022, pp. 326-333, doi: 10.1109/ICSSIT53264.2022.9716536.

3

Mathematical Modeling of Carbon Sequestration in Agricultural Soils: Implications for Climate Change Mitigation Policies

Kailash Malode[1], Brijpal Singh Rajawat[2], Amar Shankar S.[3], Ravindra Kumar[4], Deepti Khubalkar[5] and Sabyasachi Pramanik[6]*

[1]Department of Soil Science, College of Agriculture, Parul University, Limda, Dist- Vadodara, Gujarat, India
[2]College of Agriculture Sciences, Teerthanker Mahaveer University, Moradabad, Uttar Pradesh, India
[3]Department of Food Technology, Faculty of Engineering and Technology, JAIN (Deemed-to-be University), Karnataka, India
[4]Department of Mathematics, Vivekananda Global University, Jaipur, Rajasthan, India
[5]Symbiosis Law School, Nagpur Campus, Symbiosis International (Deemed University), Pune, India
[6]Department of Computer Science & Engineering, Haldia Institute of Technology, Haldia, West Bengal, India

Abstract
This research used Excel-dependent techniques specified with information gathered via closed-ended questionnaire surveys, paired with a reference trial (RT), to estimate carbon footprint (CF) and evaluate reduction approaches for a few tropical crops. The majority of the estimations obtained via organized interviews were comparable to those determined through RT and existing research. Cumulative average emission intensities varied from 25 to 300 kg CO2eq•Mg1, with cassava in the RT (24 kg CO2eq•Mg1) and maize among the interrogated agriculturists in

**Corresponding author*: sabyalnt@gmail.com
Brijpal Singh Rajawat: Orcid: https://orcid.org/0009-0002-6003-6374
Amar Shankar S.: Orcid: https://orcid.org/0000-0002-5625-6959
Ravindra Kumar: Orcid: https://orcid.org/0009-0002-2776-4855

Sabyasachi Pramanik, Niranjanamurthy M., Ankur Gupta and Ahmed J. Obaid (eds.) Mathematical Modeling in Agriculture, (33–50) © 2024 Scrivener Publishing LLC

the Liverpool province (290 kg CO2eq•Mg1) having the highest values. In sum, the manufacture and use of fertilizers were responsible for 80% of all greenhouse gas (GHG) emissions. Emissions from conveyance were close to farm. values. A field's discharges may be cut by 15% to 30% by using mitigating measures. The technique adopted in this work is a practical and straightforward tool for evaluating beforehand the effects of rules and actions on carbon footprint in agricultural settings. Additionally, it may be utilized by many stakeholders for a variety of goals, along with but not restricted to: labeling products with GHG emission estimates on the market, making decisions to control discharges in the farming field, and giving agriculturists the ability to bargain costs and remunerations for ecological conservation with measurable data.

Keywords: Maize emissions, agriculture, sweet potato fertilizers

3.1 Introduction

About 6 Gigatons (1 Gt = 109 M Tons) of CO_2 equivalent (CO_2eq) from agricultural output are responsible for 12% to 15% of all human GHG emissions. Without appropriate policies, it is anticipated that the main GHGs generated by farming actions—N_2O discharges from 40% to 55% and CH4 by 70% by 2040—will rise due to the continued growth of the population, which is projected to increase by about 35% globally by 2050 (IPCC, 2013). The problem therefore is to figure out how to accomplish sustainability in addition to assuring a sustainable utilization of resources, such as nutrients, water, energy and others. Currently, the food business, rule, and survey are concerned about the sustainability of food production. The carbon footprint is suggested as a suitable measure of the sustainability of agro-ecosystems. The CF represents the volume of GHG released into the air as a result of human activity, such as agricultural production processes. Therefore, emissions are calculated as either the total amount of GHG in kilogram CO2 eq•ha1 (CFha) or the intensity of greenhouse gas in kg CO2 eq Mg1 (CFi). It is reasonable to predict that greenhouse gas emissions from food producing and usage will continue to increase given the exponential rise of the global population. The potential impact of Latin American (LAC) nations' commercially significant crops on climate change (CC) has been examined using gross estimates of CF in those crops. Additionally, trade regulations connected to the environment that are anticipated to be implemented in the near future and that would require CF labeling for all farming exports have begun to worry decision-makers. In LAC, this is crucial for crops like cassava, maize, and sweet potatoes that contribute to GHG emissions and have seen a substantial increase in area planted.

Nations where there aren't crop-based carbon footprint estimates are required to advise decision-takers about the ecological effect of newer rule actions for economically significant crops. Public environmental [1] regulations in the UK encourage collaboration with the corporate sector to develop long-term plans aimed at reducing greenhouse gas sources or enhancing sinks and adapting agro-ecosystems. The use of quantitative methodologies for CF estimation and determining the possible effects of suggested mitigation solutions is necessary for the application of these ecological rules. There is currently no such use of quantitative techniques.

Despite having recommendations for national greenhouse gas inventories, estimation programs lack globally accepted standards and there are no methodologies to find carbon footprint for a particular kind of user. Nevertheless, there is a little advancement in that route as the document supplies a cadre of tools to find CF from farming and forestry actions, e.g., CALM, IFSC, ClimAgri®, C-Plan, USAID FCC, Carbon Farming Calculator [2], AFOLU-Calculator and EX-ACT. Nevertheless, the accuracy of estimations is constrained by the scarcity of context-based reference data in poor nations. Particularly for developing nations, accessible and convenient models with simple input data collection from agriculturists will aid in producing CF estimations. In order to improve the accuracy of CF estimations while minimizing the complexity of the computations, approaches where in situ data are integrated with current empirical models are clearly needed.

This study's overarching goals were to examine calculable tools to identify the sources of GHG [3] emissions related to the yield of significant farming crops in the UK (maize, and sweet potato) and to start conversations with agriculturists about how these emissions may be minimized, utilizing simulation approximations. The definite goals were to: 1) assess the viability of utilizing reference trials to filter possible outliers in replies from agriculturists' evaluations for parameterizing simulation models based on excel for the relative evaluation of CF in maize, sweet potato, and cassava; 2) compute CF created by the application of agricultural [4-7] implementation in mechanized maize, sweet potato, and cassava crops, and 3) find and suggest reduction possibilities in the savanna.

3.2 Resources and Techniques

Of the UK's GDP, 7.3% [8] is attributed to agriculture, with the coastal area accounting for 46% of this total. The Department of London's coastal regions, meanwhile, have a big role to play.

The major significant national market is in the capital city of London, which is situated on the coast and is home to nearly a third of the country's inhabitants. Agriculture is performed in Lima's outskirts (represented in this research by Manchester) and its adjacent regions to the north and south in order to feed this megacity. Latitude and longitude of the implementation site, where the reference trial at London (inside the city borders) was performed, are 12°22′42.5″S and 81°36′41.3″W, respectively. Liverpool province's latitude and longitude are 12° S and 63° W, while London suburbs are 16° S and 68° W, respectively. There is hardly any annual precipitation along the shore.

The Central Coast is mostly composed of Fluvisols, Regosols, and Leptosols. The Soil, Tree, Water and Fertilizer Laboratory of the University of London in London examined the soils of a variety of agricultural units in each area. The soils of Manchester were subjected to similar analyses. Manchester's soil was determined to be loamy/sandy, Liverpool's soil was determined to be sandy and silty [9], and London suburb soil was determined to be sandy and loamy (Table 3.1). The Carter and Gregorich technique was utilized to find the quantity of specimens and sampling strategy per field since every factor is necessary as input to model GHG.

3.2.1 Reference Trial

Abiding by the suggestions made to agriculturists along the coastal area, a trial was set up at the implementation farm of the International Potato Organization in London, UK, to evaluate the usage of farming inputs and administration techniques. In order to exclude outliers from the data collected from producer interviews conducted in several Liverpool and London suburb provinces along London's central coastal area, this data was utilized as a standardization. From soil preparation to post-harvest (where relevant), agronomy [10] operations and input were meticulously documented and utilized as model I/P to calculate CF for maize, sweet potato, and cassava plants.

On loamy/sandy soils, the implementation was carried out between January 2022 and August 2023. Cassava was harvested in September, sweet potatoes in July, and maize in April. For each crop, six 150 m2 (1510 m) plots were created. Amanda, Robert, and Justin were the cultivars of sweet potato, cassava, and maize that were utilized, respectively. Foremost plantation density for maize plots was 70,600 plants/ha, or 0.90 m between furrows and 0.30 m between plants. Compared to cassava, which had 10,000 plants per ha, sweet potatoes had a density of 42,000 plants per ha (0.80 m between furrows and plants, respectively). Crops were hand harvested. The

Table 3.1 Places in Manchester, UK, with their average annual temperature and soil characteristics. Source: UNALM's Laboratory for Soil, Plant, Water, and Fertilizer surveys.

Area	Quality	Quality grade	Organic carbon (%)	Nitrogen (%)	Organic matter (%)	pH	Density (cm^{-2})	Yearly mean (°C)
Reddish Vale	Silt	Granular	0.60	0.05	0.74	8.09	1.33	19.84
Heaton Park	Sand	Medium	0.87	0.31	0.23	8.4	1.84	19.87
Old Trafford	Loam	Granular	0.43	0.41	0.45	8.12	1.74	19.88
Castlefield	Sand	Granular	0.33	0.07	0.74	8.33	1.61	19.15
Tatton Park	Sand	Granular	0.71	0.08	0.34	8.05	1.45	20.27
Cockfields	Loam	Granular	1.23	0.23	0.87	8.07	1.12 1.34	20.33
Grange Farm	Sand	Medium	0.88	0.12	0.33	7.87	1.84	21.41
Lancaster Park	Loam	Granular	0.64	0.07	0.74	8.22	1.54	19.78
Greater Manchester	Sand	Medium	0.37	0.21	0.64	8.15	1.75	19.74
Wigan	Loam	Granular	0.98	0.13	0.22	8.64	1.33	19.66

commercial production (Mg•ha1) required to find GHG was calculated using average yields per plot.

3.2.2 Interviews with Agriculturists in London Suburb and Liverpool

3.2.2.1 Overall Explanation of the Sampled Region and Organized Interviews

Between November 2022 and January 2023, 100 structured interviews with mechanized maize, sweet potato, and cassava agriculturists from Hertfordshire and Essex were conducted, having a calculated sample error for a limited census of 11.2%. Forty-seven farmers in all from San José de Pativilca, Arguay, La Campia de Supe, and 53 people in Derry, Omagh, and Strabane in the province of London, as well as the southwestern Fermanag in the province of London, were questioned. Each farm's position was tracked utilizing Global Positioning System [11-15] and shown in Google Earth. In particular, the Central Market in Lima coordinates was utilized to calculate the distance between the locations where the products were grown and the locations where they were sold. Closed-ended questions were included in the interview questionnaire to probe all aspects of crop production, like soil construction, seed administration, the general use of agrochemicals, commercial productions, and the last-mile transportation of agricultural goods. Local extension agents' expertise and the official database were used to supplement the data on the planted region, cumulative output, and average production. In Omagh and Strabane, the cropping area per farmer as extended by the 1972 London Reform is 5 ha and 4 ha, respectively. They reflect the agricultural frameworks of London's central coastal area, where 51%, 69%, and 6% of the nation's cumulative output of corn, sweet potatoes, and cassava, respectively, are harvested.

3.2.3 Online Tools for Calculating CF

Two open-access, Excel-dependent statistical models were utilized to calculate the coefficient of variation (CF): the CCAFS Mitigation Options was used.

Cool Farm Tool v.2 - beta 3 and Wp7DeU1LFeU. In accordance with international standards, both models employ the data harmonization, calibration, and validation processes. If the user is focused in calculating entire farm GHG discharges and in exploring possible reduction methods, these models complement one another. Using both models together, emissions related to sources of GHG emissions that estimate Carbon Footprint in the maize, sweet potato, and cassava crops using the CCAFS-MOT and CFT models.

The majority of agricultural tasks might be explained. That is, regardless of the model used, emissions for each activity were included. As a result, utilizing information from all the interviews for every model, predicted field greenhouse gas estimates, per site, included the average and positional average for each emission source. Since both models estimated the emissions related to fertilizer production and application, for these two sources we provided both the values found by every model and the average of the computations of multiple models, which can portray truth more perfectly than each model estimates. The CCAFS-MOT model combines a number of empirical models that have been shown effective in farming rain-fed crops, rice, pasture, and animals. It calculates the GHG emissions at every step of production, taking into account land usage, the usage and properties of agrochemicals, and the synthesis and application of synthetic fertilizers. The multivariate empirical model was utilized to find the discharges (CO2, N2O, and NO) related with the application and synthesis of synthetic and organic fertilizer. According to [16] the N loss caused by ammonia volatilization was assessed.

The emissions related to the manufacture of fertilizer (kg CO2eq•ha1) and its application (kg CO2eq•ha1 and kg CO2eq•ha1) were calculated using both models (CCAFS-MOT and CFT). The non-parametric WilcoXon test [17], commonly called the Mann-Whitney test, was used to compare the estimates. This test makes no assumptions on the distribution and evaluates the pair wise comparisons of various crop/site circumstances in a hierarchical sequence. According to the calculations by source of discharges by crop and locations, nine pairs of data were created. The wilcoX. test() function from the R [18-22] statistics package was utilized for the pair of two-sample instances. Having a test of variance and the post hoc Duncan test, emissions from locations and crops were compared. The CFT model itself is composed of smaller models that provide estimates.

The rate at which N was applied, the pH of the soil, the texture of the soil, the climate, and the kind of vegetation were the controlling variables for the nitrous oxide and nitric oxide discharges of farming lands by yielding frameworks. The worksheets below were utilized: 1. information of a general kind (country, location, year, crop, and production per unit area, climatic conditions, and average yearly temperature); 2. Crop management includes taking care of the harvests, soil, nourishments, amount and technique of fertilizer treatments, and the quantity of pesticide implementations (including fungicides and herbicides).

In order to calculate nitrogen usage efficiency (NUE), [23] modified the tool's default nitrogen content per crop and ideal synthetic nitrogen application rate for a crop. In contrast, soil mining occurs when N withdrawals outnumber N inputs. It was also taken into account how applying manure affected the N input. According to [24], the addition of animal dung increases soil carbon stocks. The quantity of residue a crop produces may also be estimated using formulae from IPCC (2006). The model employed tillage parameters to predict yearly soil organic carbon repository owing to switching from conventional tillage to no-till or decreased in order to evaluate the potential soil C sequestration in crops. The model also takes into account ways to reduce the environmental effect of production systems without reducing food yields, using less energy from fossil fuels, and protecting non-renewable energy [25] sources. In order to estimate the possible mitigation for the three crops, the 16 mitigation solutions were reviewed with farmers and those that they felt would be viable to adopt in their farms were utilized.

3.3 Results

3.3.1 Agricultural Data as Model I/P

3.3.1.1 Case Study

The seven documented mechanical soil cultivation operations for maize comprised a plough (1), two harrow passes (2), a furrow for planting, and three hills for adding fertilizer comparable to maize; cassava required comparable machinery for land preparedness and fertilizer addition. In sweet potato, only motorized soil tillage techniques were used. Low levels of Nitrogen (2.03%) and Phosphorous (5.1 ppm) were found in the soil analysis, but significant levels of K (102 ppm) were found. It is low, say local soil specialists. Low-level thresholds for Nitrogen, Phosphorous, and Potassium were introduced; they are as follows: 2.3%, 6.0 ppm, and 125 to 310 ppm, respectively. Fertilizers were thus treated in accordance, using formulas of 20-20-20 and 40-60-20 urea plus Phosphorous and Potassium sources. Furthermore, KCl was employed as a source of K2O. Applications of high Nitrogen (>150 kg ha), medium Phosphorous (60 to 70 kg P_2O_5 ha), and moderate Potassium (35 kg K2O ha) nutrient levels were applied to maize. Medium amounts of Nitrogen, Phosphorous, and Potassium were applied to sweet potato and cassava. Based on the in-situ diagnosis established after the corresponding census thresholds per phenological phases for every crop, pest control was used.

Five applications of pesticides were performed to suppress Spodoptera frugiperda and Heliothis zea pests in maize. Ten pesticide treatments were done to control the foliar pests Prodenia eridania and aphids in sweet potato. Tetranychus sp. in cassava, Erinnyis ello as well as Trialeurodes species. Bugs were discovered. Applications of six insecticides and acaricides were prepared for regulation. A 30 HP electric pump was used for gravity irrigation, with a conversion ratio of 1 HP = 0.81 kWh, to pump water from a well. A total of 166, 66, and 185 kWh were consumed for irrigation water delivery during the course of the maize (6625 m3), sweet potato (2150 m3), and cassava (7558 m3) crop cycles, respectively. Additionally, it was calculated that each crop cycle per hectare required 1395, 901 or 1465 MJ of direct energy (used by equipment).

Estimates of yield were recorded in order to determine GHG intensity. The average commercialized production per crop was 12.40 Mg ha 1 for maize, 4.1 Mg ha 1 for sweet potato storage roots, and 12.5 Mg ha 1 for dry foliage (equal to 11.34 Mg•ha; biomass, 40.2Mg•ha; and 11.34 Mg•ha (fresh weight: for cassava storage roots, 38.86 Mg•ha1)). Refreshing and dry weights were recorded in the reference experiment using the same units as commercial yields. Farmers describe the commercial yields of maize as dry weight rather than fresh weight for sweet potato and cassava. The following two harvests are also sold in fresh weight form.

3.3.1.2 From Discussions with Farmers

The average number of soil tillage operations conducted by corn farmers in both provinces was eight: two plough passes, two harrow passes, the furrow, and three hills. In London suburb, maize was cultivated by hand while the crops were fully older as opposed to 50% of maize farmers in Liverpool who used mechanized planting and harvesting. In the fields, whole cobs were sun-dried. According to agricultural purchasers, shelling was done in this province when grains had a moisture content of 14%. Because of the moisture in the grains, they may be preserved well and are resistant to a variety of pests and illnesses. A drying system was used to automate harvesting, and moisture levels were monitored on a regular basis. In Liverpool, there were ten traditional agricultural implementations for sweet potatoes, while there were five in London suburb. In response, eight traditional agricultural implementations for cassava in Liverpool and London suburb were documented. The three crops were treated with urea by every farmer surveyed.

Different sources changed based on the crop and location. Farmers of maize reported using 20-20-20 and NH4NO3 (NH4)2HPO4 in both locations. Farmers of sweet potatoes in Liverpool used $NH4NO3$, KCl, and K2SO4 as K2O sources in their fertilizers. Agriculturists moreover mixed Ca (H2PO4)2, and 0-45-0 in London suburb. Ca(H2PO4)2 and 0-46-0 were added to urea by farmers in Liverpool while growing cassava, whereas in London suburb, also applied on top of the cassava were (NH4)2HPO4, 20-20-20, and 0-46-0. In both sites, maize received an average nitrogen application of >120 kg/ha, compared to >95 kg/ha and > 125 kg/ha for sweet potato and cassava, respectively. Lower (55 kg ha 1) to higher (>95 kg ha 1) phosphorous and potassium dosages were found for maize in both locations, whereas medium levels were observed for sweet potato and cassava.

Farmers in both provinces reported seeing S. frugiperda,H. The life process of maize includes the pests zea, aphid, and tar spot complex. Farmers reported using chemicals to manage pests and illnesses in Liverpool. Eleven pesticide applications reported in London suburb were contrasted with 14 applications. The principal pests for sweet potatoes in Liverpool and London suburb were bacterial infections, aphids. With 15 pesticide treatments in Liverpool and 10 in London suburb, chemical control was carried out. Cassava growers in both provinces reported

Fungicides and insecticides were applied ten times to control E. Hello, Trialeurodes spp. and Tetranychus sp. Oidium species, too. According to estimations from Liverpool and London suburb, the growing of maize required 1461 and 838 MJ of energy per hectare, respectively 918 and 705 MJ for sweet potatoes and 779 and 838 MJ for cassava, respectively.

Modules for irrigation: 2500 m3 for sweet potatoes, 5000 m3 for maize, and the cassava crop provided irrigation water and gravity water for the maize, sweet potato, and cassava (users are organized in irrigation sub-districts). The local boards' irrigation managers were consulted, and they assessed the energy expenses for maize, sweet potatoes, and cassava (both provinces) to be 142, 75, and 200 kWh, respectively.

Regarding commercial yields, Liverpool reported an average of 11.5 Mg/ha for maize, compared to London suburb's 10.75 Mg/ha. Using samples from senescent plants, commercial yields of sweet potato and cassava roots were measured in fresh weight and approximated in collaboration with purchasers. No information was obtained because farmers did not weigh the fresh leaves they used as feed. Average sweet potato fresh storage root productions were found to be approximately 25 Mg•ha1 in Liverpool and London suburb. In Liverpool and London suburb, the fresh yields from cassava storage roots were 20 Mg/ha and 25 Mg/ha, respectively. It is important to note that producers favored XB-8010 hybrids when it came to maize varieties. The Jonathan and Jewel and the Milagroso and LimeoMorado (all purple variants), were the most popular sweet potato cultivars. The favorite types of cassava were Seorita and AmarillaCriolla.

3.3.2 Farms' Estimated GHG Emissions

Of the total greenhouse gas discharges, 77% were attributable to average emissions related to fertilizer production and application, across all locations and crop types, and using the mean of both models. The primary emission source was the application of fertilizer (39% on average). The estimated total GHG emissions in the tested provinces (kg CO2eq•ha1) were greater than in the Manchester reference trial (p 0.02). For cassava, it was especially higher (thrice higher in London suburb and five times higher in Liverpool). The projected yield in Manchester was similarly greater (38.86 Mg•ha1), but the estimated yields for the other two crops were statistically comparable. In other words, the predicted GHG intensity and announced productions for maize and sweet potatoes were comparable among regions. Due to the large quantities of nitrogen applied to the maize crops in Manchester, Liverpool and London suburb as well as the sweet potato crops in Liverpool and London suburb, the estimated excess was linked to N2O discharges (180, 209, 425, 460, and 180 kg CO2eq•ha1, respectively). In Liverpool and London suburb, every cassava grower reported that 10% of the stems were set aside for vegetal products, and the other 90% were burnt, creating a source of CH4 and N2O discharges via burning. The procedure was used as a field cleaning technique after harvest. The justification for this is that diseases and pests that are present in plant leaves may be directly controlled. The discharge factors for CH4 and N2O from the burning of farming waste also increase as a result of this.

Remainders were collected from [26] and the IPCC (2006). According to [27] the IPCC (2006) recommendations' formulae may be used to estimate the amount of straw from agricultural leftovers. Additionally, 20% of the sweet potato and cassava farmers surveyed in London suburb admitted to adding 8 Mg/ha of chicken feces to the soil. This practice results in CH4 emissions, which are then converted by direct oxidation into anaerobic digestion after the aerobic phase. It was noted that the fertilizing component, in the sweet potato and cassava production structures in London suburb, constituted, on average, 26.55% and 18.25% of the relative economic value of the overall cost, respectively. Our modeling findings show that by putting these reduction solutions in place, expenses might go down while yields stay the same. Nevertheless, the sum of these emissions represented less than 0.5% of the predicted agricultural emissions overall.

The locations have a problem with emissions brought on by the transit of goods beyond Lima. In each of the examined provinces, the majority of the maize was carried up to 63 kilometers. 90% of the sweet potato and cassava crops from Liverpool and London suburb are carried to the significant wholesale market in London, which are 165 to 220 kilometers away from Liverpool and 180 to 190 kilometers away from London suburb. The remaining production is eaten at nearby hotels. The intensity or CFi for maize after accounting for Liverpool's transportation emissions was 306 kg CO2eq•Mg1. 221 and 258 kg CO2eq•Mg1 for sweet potato and cassava, respectively. In London suburb, these crops' respective emission intensities were 187 and 167 kg CO2eq•Mg1. Then we included the greenhouse gas (GHG) emissions from transporting the fruit to Lima or the neighborhood market. Since local marketplaces are where maize is often sold, London suburb's transport offering to overall greenhouse gas emissions was up to 18%. In Liverpool and London suburb, discharge values in transport were higher as those for farming operations for sweet potato and cassava.

3.3.3 Effects of Mitigating Measures

The primary mitigation strategy that farmers consistently found to be simple to apply across crops and settings was the usage of organic fertilizer. Reduced tillage was then used, which may also be a crucial mitigating strategy for the three crops. Contrarily, Manchester's maize benefited significantly from the incorporation of garbage.

3.4 Discussion

Greenhouse gas discharges from maize, sweet potatoes, and cassava are estimated at 4.1. The caliber of the input that is introduced into the models affects how reliable CF estimations are. We performed a reference experiment in circumstances as close as possible to the fields on the coastal areas where the data were acquired from agriculturists via structured interviews in order to generate a measure against the data obtained via interviews can be evaluated. Various inorganic fertilizers (Nitrogen, Potassium, and Phosphorous) used to fertilize maize were reported to be within or similar to the amounts used in the reference experiment. The proclaimed range for K in London suburb was an outlier, as farmers claimed to have applied up to three times the amount utilized in Manchester. Agriculturists in each of the examined provinces claimed to have applied more P, but equivalent amounts of N and K, to sweet potatoes. In contrast to the other two crops, farmer reports of N, P, and K applications to cassava varied more from those to the reference trials. Farmers in Liverpool claimed to have applied two to four times as nitrogen (N) as farmers in Manchester, although reported values for London suburb are only slightly higher than amounts applied in Manchester. Nevertheless, the announced P levels in each of the studied provinces were lower than those in Manchester. Additionally, agriculturists in Liverpool claimed to have used less K (in cassava and sweet potatoes) than had been used in the reference study. However, farmers in London suburb claimed to have used up to four times as much K as those in Manchester.

The kind of fertilizers used was said to be responsible for the greatest estimated average discharges for maize fields in Manchester. Utilizing a significant quantity of fossil fuels is required for the manufacturing of applied fertilizers, which vary by country and production technology. With a few exceptions, the values reported by farmers were generally consistent with those used in the reference experiment in Manchester, despite the significant variability characteristic of data obtained via interviews. It was notable that CFT continually predicted greater discharges for fertilizer implementations (p 0.01, in kg) even though we complied with the advice that the average of the values of multiple models can show the truth more perfectly and consistently than predictions depending on each model where the fertilizer is produced, and statistics show that 99% of nitrogen fertilizer imports are made in the US.

Through nitrification, denitrification [28], and volatilization—processes that happen naturally in soils—the implementation of these fertilizers hikes the discharges of N2O, NO, and NH3. Both models performed similarly well ($p > 0.05$) for estimating GHG emission magnitudes. Because the applied N fertilizer surpassed the entire crop need, the CCAFS-MOT model predicted that the NUE was very low (70%) having a danger of substantial N losses and GHG discharges (representing 26% and 19% of the cumulative CF of corn in London suburb and Liverpool, respectively). However, there were no GHG emissions when N fertilizer was used to meet the overall crop requirement, or when intake and production of N were balanced. Liverpool's maize fields had the greatest yields and used up to 345 kg of nitrogen (1.7 times more urea and different N_2 fertilizers than in Manchester), which affected how total CF was calculated. These findings are in line with those from fields in Canada that have been planted with maize for 20 consecutive years. Additionally, higher Nitrogen levels up to 350 kg (four times greater than in Manchester) are applied in Liverpool's cassava production systems. As a result, substantial (500 kg CO2 eq•ha1 as N2O) discharges were calculated.

In Manchester, Liverpool, and London suburb, fertilizer applications for corn were responsible for 44%, 39%, and 34%, respectively, of the total emissions. The traditional maize production system's computed CF at all locations reflects the extent to which agronomic techniques depending on heavy and ongoing usage of agrochemicals, particularly N_2 fertilization, are used. Even the greatest CFi in maize that was obtained for Liverpool was equivalent to estimate announced for Uruguay; 350 kg CO2eq•Mg1 in terms of GHG intensity. In terms of sweet potato, GHG intensity ranged from 81 to 98 kg CO2eq•Mg1 across all locations. The Philippines saw similar intensities (Flores et al., 2016, reported up to 95 kg CO2eq•Mg1) as well. Cassava, on the other hand, varied from 28 to 130 kg CO2eq•Mg1. These estimates are comparable to those for smallholder farmers in Nigeria reported in the literature (reporting 105 kg CO2eq•Mg1).

3.4.1 Evaluating the Possible Effects of Mitigating Measures

To address climate change, effective policy and sustainability management depend on the computation of CF per agricultural system. Farmers indicated that the following mitigation strategies for maize, sweet potatoes, and cassava can be effortlessly executed: (i) adding farming wastage to the soil to improve fertility; and (ii) applying organic manure that has been processed beforehand to be utilized to improve or to supply soil nutrients and improve water retention (sweet potato and cassava crops in London suburb). Its execution depends on cutting back on the fossil fuels used to make synthetic fertilizers, which accounted for the majority of greenhouse gas emissions (GHG), implementing a single semi-heavy harrow pass of minimal tillage, and using inexpensive nitrification inhibitors, which are readily available in Peru and restraint NH4 from being converted into NO2 and then NO3. A limited potential for emission reduction was assessed using the CCAFS-MOT (a largest average CF minimization in cassava of 225 kg CO2eq•ha1 kg year-1).

In Manchester, the capacity to minimize cumulative discharges per crop varies depending on the option chosen. For instance, the average CF for maize is decreased from 1700 to 1634 kg CO2eq•ha1 kg year-1 as a result of minimized tillage or the limited elimination of weeds in the field. According to [29] this approach has reduced energy use, erosion risk, soil compaction, and cost of land preparation by up to 19%. When utilizing nitrification-inhibiting fertilizers, minimized tillage, organic fertilizers, and farming waste, CF may be minimized by 50%. If the coverage of maize plant waste per mechanized harvest is established, CF minimization will be up to 1200 kg CO2eq•ha1 kg year-1. The average total cost of using organic fertilizers in cassava and sweet potatoes, as measured by relative economic value.

In the future, more nitrogen fertilizers will be used and more manure will be produced, particularly in developing nations. Ex-ante evaluations of the choices' prospective effects are possible. The models utilized in this research represent a helpful and simple usable tool for CF detection under agricultural circumstances, that may be utilized by various stakeholders for various reasons, including but not confined to: labeling products provided in the market with GHG discharge values, making predictions to modify the discharges in the farming domain, and giving farmers access to quantitative information to negotiate costs and incentives for environmental preservation.

3.5 Conclusions

The in-person, organized interviews revealed that the coast of Peru is dominated by mechanized agriculture methods that rely on petrochemical fertilizers and pesticides. The technique used in this research, which included the farmers' interviews, open Excel-dependent models, and a reference experiment to weed out probable outliers from elicited data (calculated to be less than 5% of the data), produced findings that were consistent with those reported in the literature. The majority of the estimations made using structured interviews matched those determined in the control trials. Input data must be confirmed using a better accurate model input prediction where differences were statistically noteworthy, such as with cassava in Liverpool, where predicted discharge intensity was up to six times greater than the reference trial. This would improve the accuracy of estimations. According to estimates, fertilizers used along the Peruvian coast are the primary human-caused source of N2O emissions. In cooperation with the producers, these emissions may be decreased by technical interventions if modeling estimates for reduction measures approximate reality. These scenarios should be tested in the real world while taking into account the climate, soil, and crops. Producing as near to the final market as is practical is the only approach to decrease transportation emissions. To encourage researchers in developing nations to provide comparable values utilizing similar methods which may be integrated in debates on farming CF and to give input for policy debate and development, the case in Peru was sufficiently described at length.

References

1. S. Pramanik, S. Joardar, O. P. Jena, A. J. Obaid, "An Analysis of the Operations and Confrontations of Using Green IT in Sustainable Farming", Al-Kadhum 2nd International Conference on Modern Applications of Information and Communication Technology, *AIP Conference Proceedings* 2591, 040020 (2023). https://doi.org/10.1063/5.0119513
2. A. F. Soofi and S. D. Manshadi, "Carbon-Aware Operation of Resilient Vertical Farms in Active Distribution Networks," in *IEEE Transactions on Smart Grid*, doi: 10.1109/TSG.2023.3284752.
3. A. Roy and S. Pramanik, A Review of the Hydrogen Fuel Path to Emission Reduction in the Surface Transport Industry, *International Journal of Hydrogen Energy*, 2023.
4. S. Fölster and A. Litwin, "Hidden Consequences of Innovation—Economic Activity Beyond GDP," in *IEEE Technology and Society Magazine*, vol. 42, no. 1, pp. 37-44, March 2023, doi: 10.1109/MTS.2023.3241323.
5. P. T. Khanh, T. H. Ngọc, and S. Pramanik, Future of Smart Agriculture Techniques and Applications, in *Advanced Technologies and AI-Equipped IoT Applications in High Tech Agriculture*, Ed., A. Khang, IGI Global, 2023.
6. T. H. Ngọc, P. T. Khanh and S. Pramanik, Smart Agriculture using a Soil Monitoring System, in *Advanced Technologies and AI-Equipped IoT Applications in High Tech Agriculture*, Ed., A. Khang, IGI Global, 2023.
7. S. Pramanik, Intelligent Farming Utilizing a Soil Tracking Device, in *Convergence of Cloud Computing, AI and Agricultural Science*, Eds., A. K. Sharma, N. Chanderwal and R. Khan, IGI Global, 2023.
8. V. Veeraiah, V. Talukdar, Manikandan K., S. B. Talukdar, V. D. Solavande, S. Pramanik, A. Gupta, Machine Learning Frameworks in Carpooling, in *Handbook of Research on AI and Machine Learning Applications in Customer Support and Analytics*, Eds., Md S. Hossain, R. C. Ho and G. Trajkovski, IGI Global, 2023.
9. S. Praveenkumar, V. Veeraiah, S. Pramanik, S. M. Basha, A. V. Lira Neto, V. H. C. De Albuquerque and A. Gupta, Prediction of Patients' Incurable Diseases Utilizing Deep Learning Approaches, *ICICC* 2023, Springer, doi.org/10.1007/978-981-99-3315-0_4
10. S. Pramanik and S. Bandyopadhyay, Identifying Disease and Diagnosis in Females using Machine Learning, in *Encyclopedia of Data Science and Machine Learning*, Ed. John Wang, IGI Global, 2023, doi: 10.4018/978-1-7998-9220-5.ch187.
11. R. Bansal, B. Jenipher, V. Nisha, Jain, Makhan R., Dilip, Kumbhkar, S. Pramanik, S. Roy and A. Gupta, "Big Data Architecture for Network Security", in *Cyber Security and Network Security*, Wiley, 2022, https://doi.org/10.1002/9781119812555.ch11
12. V. Vidya Chellam, V. Veeraiah, A. Khanna, T. H. Sheikh, S. Pramanik and D. Dhabliya, A Machine Vision-based Approach for Tuberculosis

Identification in Chest X-Rays Images of Patients, *ICICC* 2023, Springer, doi.org/10.1007/978-981-99-3315-0_3

13. S. Pramanik, Niranjanamurthy M. and S. N. Panda, Using Green Energy Prediction in Data Centers for Scheduling Service Jobs, *ICRITCSA 2022*, Bengaluru, 2023.
14. S. Pramanik, "Carpooling Solutions using Machine Learning Tools", in *Handbook of Research on Evolving Designs and Innovation in ICT and Intelligent Systems for Real-World Applications*, K. K. Sarma, N. Saikia and M. Sharma, IGI Global, 2022, doi: 10.4018/978-1-7998-9795-8.ch002.
15. V. Veeraiah, D. J. Shiju, J. V. N. Ramesh, Ganesh K. R., S. Pramanik, D. Pandey, A, Gupta, Healthcare Cloud Services in Image Processing, *Handbook of Research on Thrust Technologies' Effect on Image Processing*, IGI Global, 2023.
16. S. N. Ciolofan, M. Drăgoicea and D.-C. Popeangă, "Enhanced cyber-physical system with semantic technologies and machine learning to support smart farming," *2023 24th International Conference on Control Systems and Computer Science (CSCS)*, Bucharest, Romania, 2023, pp. 56-60, doi: 10.1109/CSCS59211.2023.00018.
17. H. Myriam *et al.*, "Advanced Meta-Heuristic Algorithm Based on Particle Swarm and Al-Biruni Earth Radius Optimization Methods for Oral Cancer Detection," in *IEEE Access*, vol. 11, pp. 23681-23700, 2023, doi: 10.1109/ACCESS.2023.3253430.
18. Y. Meslie, W. Enbeyle, B. K. Pandey, S. Pramanik, D. Pandey, P. Dadeech, A. Belay, A.Saini, "Machine Intelligence-based Trend Analysis of COVID-19 for Total Daily Confirmed Cases in Asia and Africa", in *Methodologies and Applications of Computational Statistics for Machine Learning*, D. Samanta, R. R. Althar, S. Pramanik and S. Dutta, Eds., IGI Global, 2021, pp. 164-185, doi: 10.4018/978-1-7998-7701-1.ch009
19. A. Mandal, S. Dutta, S. Pramanik, "Machine Intelligence of Pi from Geometrical Figures with Variable Parameters Using SCILab", in *Methodologies and Applications of Computational Statistics for Machine Learning*, D. Samanta, R. R. Althar, S. Pramanik and S. Dutta, Eds., IGI Global, 2021, pp. 38-63, doi: 10.4018/978-1-7998-7701-1.ch003
20. A. Bhattacharya, A. Ghosal, A. J. Obaid, S. Krit, V. K. Shukla, K. Mandal and S. Pramanik, "Unsupervised Summarization Approach with Computational Statistics of Microblog Data", *in Methodologies and Applications of Computational Statistics for Machine Learning*, D. Samanta, R. R. Althar, S. Pramanik and S. Dutta, Eds., IGI Global, 2021, pp. 23-37, doi: 10.4018/978-1-7998-7701-1.ch002
21. Dhamodaran S, S. Ahamad, J.V.N. Ramesh, G. Muthugurunathan, Manikandan K,S. Pramanik, D.Pandey, Food Quality Assessment using Image Processing Technique, in *Handbook of Research on* Thrust Technologies' Effect on Image Processing, IGI Global, 2023.
22. Dhamodaran S. S. Ahamad, J.V.N. Ramesh, S. Sathappan, A. Namdev, R. R. Kanse, S. Pramanik, Fire Detection System Utilizing an Aggregate

Technique in UAV and Cloud Computing, in *Handbook of Research on Thrust Technologies' Effect on Image Processing*, IGI Global, 2023.
23. S. Sarkar, A. Gokhale, M. Parashar and K. Saha, "Efficient Signal Processing for Low-Cost Magnetometry Using Nitrogen Vacancy Center in Diamond," in *IEEE Transactions on Instrumentation and Measurement*, vol. 72, pp. 1-12, 2023, Art no. 7001812, doi: 10.1109/TIM.2023.3234086.
24. A. R. C., A. P. M., R. K. C., and A. Mohanarathinam, "Clinical Waste Storage System with Contamination Prevention Mechanism using UV and Arduino Microcontroller," *2023 International Conference on Inventive Computation Technologies (ICICT)*, Lalitpur, Nepal, 2023, pp. 1585-1591, doi: 10.1109/ICICT57646.2023.10134053.
25. S. Pramanik, A. J. Obaid, N M., and S.K. Bandyopadhyay, "Applications of Big Data in Clinical Applications", Al-Kadhum 2nd International Conference on Modern Applications of Information and Communication Technology, *AIP Conference Proceedings* 2591, 030086 (2023), https://doi.org/10.1063/5.0119414
26. K. I. Shahin, S. Lazarova-Molnar and P. Niloofar, "Multi-objective Decision Support Tool for Sustainable Livestock Farming," *2023 8th International Conference on Smart and Sustainable Technologies (SpliTech)*, Split/Bol, Croatia, 2023, pp. 1-6, doi: 10.23919/SpliTech58164.2023.10192938.
27. Y. Xiang, G. Qing, P. Zeng, Z. Tan, W. Xu and J. Liu, "Electricity-carbon Nexus: Concept, Framework and Practice," *2023 Panda Forum on Power and Energy (PandaFPE)*, Chengdu, China, 2023, pp. 979-986, doi: 10.1109/PandaFPE57779.2023.10140249.
28. V. Veeraiah, D. J. Shiju, J.V.N. Ramesh, Ganesh Kumar R. S. Pramanik, A. Gupta, D. Pandey, Healthcare Cloud Services in Image Processing, in *Handbook of Research on Thrust Technologies' Effect on Image Processing*, IGI Global, 2023
29. A. K. Sangaiah, A. Javadpour, F. Ja'fari, H. Zavieh and S. M. Khaniabadi, "SALA-IoT: Self-Reduced Internet of Things with Learning Automaton Sleep Scheduling Algorithm," in *IEEE Sensors Journal*, doi: 10.1109/JSEN.2023.3242759.

4

Optimizing Livestock Feed Formulation for Sustainable Agriculture: A Mathematical Modeling Approach

Rutul Patel[1], Upasana[2], Ashutosh Pattanaik[3], Deepak Kumar[4], Ahmar Afaq[5] and Soma Bag[6*]

[1]Department of Agronomy, College of Agriculture, Parul University, Limda, Dist- Vadodara, Gujarat, India
[2]College of Agriculture Sciences, Teerthanker Mahaveer University, Moradabad, Uttar Pradesh, India
[3]Department of Mechanical Engineering, Faculty of Engineering and Technology, JAIN (Deemed-to-be University), Karnataka, India
[4]Department of Mathematics, Vivekananda Global University, Jaipur, Rajasthan, India
[5]Symbiosis Law School, Nagpur Campus, Symbiosis International (Deemed University), Pune, India
[6]Asadtala Nivedita Kanya Vidyamath, Nandigram, West Bengal, India

Abstract
In order to manage sow herds, many sow models that have been published in the literature and using various mathematical techniques are surveyed in this study. A broad taxonomy of models was considered, including simulation and optimization. The latter were comprised of the Markov decision models and optimal control as its two main exemplary models, together with linear programming and dynamic programming. We examined basic characteristics and the modeling underpinnings of herd management models in the first stage, and then we looked at various features of sow herd models that have been previously published. The primary variables, the source of the parameters, validation, output, and intended

**Corresponding author*: bagsoma2016@gmail.com
Upasana: ORCID: https://orcid.org/0009-0006-4093-7166
Ashutosh Pattanaik: ORCID: https://orcid.org/0000-0003-2453-1343
Deepak Kumar: ORCID: https://orcid.org/0009-0005-7455-8124

Sabyasachi Pramanik, Niranjanamurthy M., Ankur Gupta and Ahmed J. Obaid (eds.) Mathematical Modeling in Agriculture, (51–80) © 2024 Scrivener Publishing LLC

usage all get special consideration. The majority of these models was created as teaching and research aids. In reality, there hasn't been a rise in the number of decision support tools that include such complicated models in real-world settings to match the improvement in our capacity to describe complex systems. It was determined that there was a need for fresh ideas for dealing with variables that weren't time homogeneous and transitory circumstances. For pragmatic reasons, the integration of variability-risk elements and multicriteria decision approaches was moreover intriguing. New management herd models, particularly those that take into account more than one herd at once, are expected as a result of actual developments in the pig industry.

***Keywords*:** Sow management, modeling, herd management

4.1 Introduction

Within the European Union (EU), swine production has undergone significant change during the last ten years and is anticipated to undergo further transformation. The assessment and comparison of the management practices that result in sow farms have come under increased scrutiny as a consequence of the recent EU expansion, rules pertaining to the welfare of pigs, competition, and the ongoing development of the census. In addition, the capacity of swine farmers to improve productivity efficiency rather than farm size will determine how successful they are in the future. In order to maximize management options or to consider new ones, sow herd management models might be crucial. However, it seems that a critical review of the current plans is sufficient for adjusting them or adding newer ideas for future improvements.

Herd management is the method by which a farm manager achieves certain objectives, represented as a quantity of output, by using a matching quantity of production inputs. Knowing the key relationships between these parameters and their impact on the system's ultimate productivity is crucial to be able to combine them in the best possible manner. System simplifications are often made for getting useful herd models while preserving the core of the original framework. For extracting useful information from a challenging scenario which can at first tend to be disordered, the task for the livestock modeler is to depict what is vital in the framework. In the study and development of livestock, mathematical models that depict the yield behavior of a livestock herd are employed. Sow herd models in particular, as well as livestock herd frameworks in general, are crucial tools for analyzing various herd management techniques. A "sow herd model" is defined as a simulation of a breeding herd of pigs.

Over time, animals were occasionally replaced. Researchers can better understand and control actual farm behavior via the use of herd models, followed by swine specialists and farm managers. Researchers now have access to improvements in computer, database, and problem-solving software that make it easier and more detailed to describe agricultural systems. As an illustration, programming models from the 1980s engaged with thousands of states, but by the 2020s, the amount had increased to some millions. Additionally, methodological advancements have been made to accurately depict the system, and as a consequence, solutions or workarounds for issues with complicated models have been found. However, research models are often highly sophisticated in relation to the system they represent, making them less useful for everyday usage at this time. Although improving decision-making should be the final goal of model development in sow herd management, experts, advisors, and even farmers seldom employ models. In reality, the demand for computerized decision support tools that include such complicated models among farm managers has not increased in tandem with advances in our capacity to describe complex systems.

This work reviews current sow herd frameworks that depict the fruitful and reproductive behavior of a herd of breeding sows over time, as well as the mathematical underpinnings of such models. As a result, additional management elements are not taken into account when using such herd models to regulate sow reproduction and replacement. The objective of this study is to identify strong and weak areas that affect how practical use-suitable a model is. This evaluation aims to contribute to the creation of more useful and trustworthy technologies for on-farm sow decision assistance.

4.2 Managing Swine Herds Using Modeling

4.2.1 System of a Sow Herd

The sow herd framework may be seen as a collection of several interconnected components, or breeding animals, that respond as a unit to external stimuli. Modern piglets are produced using intensive techniques that have been shown to be more successful than older ones. Piglet yield, the supply of foods, breeding, sow replacement, and faeces destruction are key issues involved in this activity and can have a considerable influence on framework performance.

The key characteristics of the sow farm operation must be modeled in the modeling of sow herd management. Thus, a sow's life cycle typically begins while it is bought or raised and brought to a farm after a suggested quarantine (Figure 4.1). While mating gilts for the first time, factors including weight, age, and observed heats should be taken into consideration. In general, when heat is discovered, gilts and sows should be prepared for mating. There are several breeding techniques for gilts, like mating them in the second or third heat. The two primary methods of reproduction are natural mating and artificial insemination. In order to arrange farm activities more simply, sows are often managed in herd batches; however, individual management is still an option, especially on smaller farms. Following mating, gilts and sows are monitored to find and verify pregnancies. Sows may be mated again or rejected as unproductive after conception has aborted because they go into heat once more in the later oestrus cycle. Instead, the gestational period may result in farrowing or an abortion if fertilization is successful. Piglets stay with the sow for a few days after birth throughout the breastfeeding phase till weaning. Fostering is another option, and it may result in a sow's lactation lasting less time. Other lactation/weaning methods, for as the Isowean system, are feasible. These systems are less common, more sophisticated, more refined.

When a litter is weaned, it is either sold or sent to a nursery facility. The two common methods for ending a reproductive cycle are weaning and abortion. Because of this, a female pig serves as the herd's fundamental production unit. Herd production, as calculated by the quantity of piglets weaned and sold or by the kilograms of live weight sold, is closely linked to the technique of reproduction. Although fabricated insemination is greatly employed for breeding rigorous livestock, boars might be overlooked during business point of view despite the fact that male animals are as crucial to the reproduction technique. Boars play a more significant part when

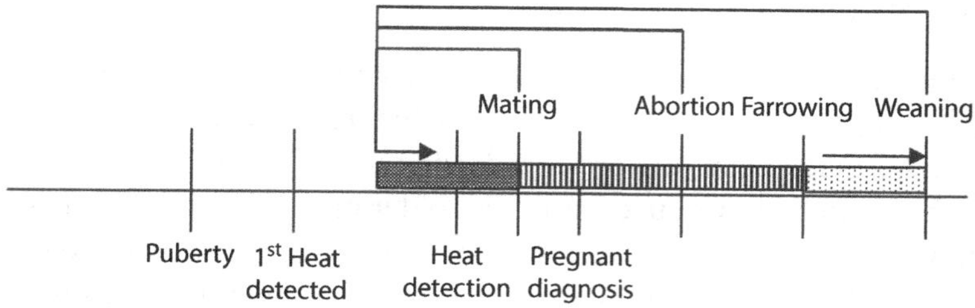

Figure 4.1 The main occurrences and cyclical pattern in sow reproduction.

discussing quality issues, such as genetic features that might influence crucial choices in breeding herds.

4.2.2 Major Statistical Techniques Used in Modeling Cattle Herds

4.2.2.1 Literature Review on Herd Modeling for Cattle

A statistical model [1] is a description of a framework in terms of quantitative and logical connections that makes the trade-off between accuracy and tractability. Although several statistical models of animals were developed, swine production has gotten very little care. In a basic introduction of animal herd modeling, for instance, [2] shows a more detailed analysis of dairy and sow flock frameworks with an emphasis on replacement and reproduction administration.

The female animal is the same in all livestock herd models as a production unit. Most of the time, the flock is shown by either a single female or a collection of autonomous individuals that are scattered in a similar manner. As a result, the sowing lifetime is divided into several reproductive stages that are limited by occurrences as indicated in Figure 4.1 in order to deal with the discrete event aspect of the reproduction technique. The impact of herd structure and dynamics on the computation [3] must be taken into account when assessing production efficiency [4].

Every herd management strategy makes an effort to somehow capture the dynamics of the herd. In intensive livestock herds, where productivity is not constant over time and is mostly influenced by breeding animals' ages or parities, this is essential to modeling [5] performance. The dynamics of the herd are thus generated from the animals' transition between states and are influenced by both intentional and involuntary culling. The total quantity of cycles completed and the total output of every herd unit define the final production. Until 1995, reproduction, mortality, and production were defined as the three primary factors influencing performance in cattle systems in this manner.

Herd models may be divided into simulation and optimization models methodologically. While optimization [6] models provide an objective function of predicted utility or profit which is maximized in consideration of production options, pricing, and resource availability, simulation models, in general, are more adapted to connecting with the complexity and variety of livestock yield. When optimization models are only solved to identify the best course of action or results, simulation models are designed to obtain knowledge into the cattle framework.

4.2.2.2 Models for Simulation

In terms of the starting condition, time horizon, discount rate, administrative techniques, and stochasticity [7], simulation models are adaptable. They may be categorized using several criteria, including (i) random components, discrete event vs. continuous, (ii) simulation time step, (iii) incorporation of time as a variable, and (iv) static vs. dynamic. The average result and its dispersion over time may be studied via simulation, whether it is deterministic or stochastic. However, the overwhelming variety of potential possibilities is the frequent downside. Deterministic simulation is considered to mean that the similar set of inputs yields the similar outcome; hence, it is unable to produce estimates of the variability associated to the outcome. Occurrence based simulation seems to be preferable to continuous time models since the reproduction process is significant in herd management. Suitable probability distributions and pseudorandom number generators govern the simulation of discrete occurrences. To convey that not all of the parameters are found out randomly, nearly every simulation framework is characterized as somewhat stochastic [8]. Additionally, fewer stochastic frameworks include pseudorandom number generators for certain continuously distributed variables, such as live weight fluctuations, litter size, and production of milk. Since random numbers are being used, numerous runs are required to get a trustworthy confidence interval of the herd's mean findings. As a result, uncertainty is tackled by first declaring an a priori distribution and then undertaking risk evaluations, often in a similar static framework. Still, distributions for every parameter are not always specified. The model is seen as static when time is not a variable; otherwise, it is regarded as dynamic.

4.2.2.3 Models for Optimization

The two most common optimization techniques used in livestock modeling are dynamic programming and linear programming. The latter includes both continuous time or optimum control frameworks and discrete time models in the form of Markov decision processes. They all have the characteristic of being designed to provide the best solution to a well-defined issue. Because time is often ignored or has no impact on them (as in the case of so-called stationary models), optimization models are frequently static models. Static models evaluate a hypothetical huge population in which the impact of the produced alteration has stabilized the herd layout; this is the supposed steady-state static frameworks abstract from the situation after an alteration. Herd structure at equilibrium produces a variety of outputs that are important for comparing or evaluating systems.

Although linear programming is the most used optimization approach in agriculture, a model for sow herd management has not yet been suggested. Instead, [9] noted that there are instances of them in the production of dairy products and cattle. When using optimum control models, something distinct happens. For example, [10] presented a framework of continuing biological growing in pigs to highlight dynamic features of pig production in contrast to static techniques, which applied a similar technique to address the feeding and promotion issue; still, neither of them had sow herd supervision as their primary goal. Again, discrete time models are preferable over the alternatives because sow reproduction is a discrete event process. However, the dimensionality issue might arise from large state spaces produced by the decretisation process. This issue, which is sometimes intractable, benefits from both computer power and mathematical advancements in that the dimension of solvable frameworks is now greater. As an example [11] suggested a technique depending on hierarchical Markov techniques, which was fruitfully used to resolve extremely huge experiments with numerous states.

4.2.2.4 *The Integration of Simulation and Optimization*

A wide range of mathematical models are composed of optimization and simulation approaches. Both optimizing and simulating problems may be solved using different mathematical approaches.

These models include the Markov ones. Optimization models are used to understand typical Markov decision processes. However, when a stationary policy is fixed, a Markov chain serves as the model for the system. Here, considering the transition matrix to be irreducible and aperiodic, a singular stationary distribution corresponds to this matrix. As a result, the Markov chain technique considers the probabilistic aspect of herd dynamics and allows for the execution of stochastic and deterministic simulations. Due to its simplicity, shorter computation time, and consideration of just expectations obtained from the stationary distribution, the final simulation is the one that is most often utilized. An excellent way for assessing management practices is to compare the outcomes of herds at steady-state, although it is also feasible in stochastic simulation if each animal is simulated using the similar modeling strategy. In such circumstance, it is possible to estimate the mean and variance of the distribution of predicted outcomes as well as cope with temporary events.

4.3 Models of a Sow Herd

4.3.1 Chosen Models

Simulation was the approach mostly employed to depict sow herds (10 vs. 4) among the 14 publications relating to matching sow herd models that were taken into consideration. While all of these models were interested in simulating a sow herd and had used single sow behavior as a guide, not every one of them had similar objectives. To demonstrate several mathematical ways to sow farms, they were examined. Only one of them additionally took genetic factors into account. The majority of them were capable of assessing the impact of alterations in reproduction or replacement. Table 1 presents a summary of their many traits.

A classification criterion includes the purpose for which sow herd frameworks were created. So, the authors discovered that the majority of them were developed for research reasons and that their major goal was to appropriately depict farm dynamics. The models provided by [12] were the only ones that were specifically intended to be employed in field settings. They suggested using certain agricultural data to run the framework, using actual farm data; [13] accomplished this. Later, [14] backed the similar contention for which herd-based criteria are necessary to assist decision-making at the stage of the particular farm.

Herd dynamics were all discrete in time since the optimization models portrayed them as transitions between several (reproductive) states. The transient nature of such transitions was one way they varied from one another in terms of how time was represented. Weekly transitions were used by [15], which was a logical decision driven by the typical scheduling of agricultural operations by weeks and with analytical benefits associated with constant time transition matrixes. But because of the ambiguity generated by this pattern every (reproductive) state has to be dependent on a weekly basis; while [16] looked at natural gaps between transitions, [17] focused on transitions via parities.

According to [18], comparable models are dynamic programming frameworks with a functional formulation that is maximized and is based on herd dynamics. In order to tackle an innovative challenge, [19] investigated the Markov decision process incorporated in a semi-Markov decision framework. As a result, they realized computation savings and a better accurate state characterization. In [20], a methodical advancement first created at [21] that took use of the transition matrix's structure was used. Depending on the division of the transition matrix into several sub-processes (i.e., sub-models), they provided a hierarchical model. In addition to better handling of big models, the benefit was the way the issue was structured. All of the writers took into account incentives, time-homogeneous transition problems, and deterministic managing practices. Here, even though it wasn't stated clearly, they guaranteed the stochastic technique's ergodicity and its convergence to a steady-state distribution. The predicted mean reward per unit of time served as the common optimization criteria since the optimization process was thus tied to this steady-state distribution. Only [22] approximated the solution to the optimization issue via subsequent iterations; all other writers used an exact approach to tackle the problem.

According to a predetermined management philosophy, the simulation techniques portrayed the sows in the flock. A deterministic model has to employ categories of animals as the simulation unit to cope with the discrete processes like conception, sex of babies, and demise. For this reason, Allen and Tess (1983a) mathematically modeled and connected a number of subsystems using rather straightforward connections. Later, as computing power increased, more complicated simulation models could be created. Only [23] examined a synchronized simulation of the flock, and as a result, the authors were capable of describing batch operation. Sows were simulated sequentially, assuming animal independence. [24] thought of Markov chain models to mimic herd dynamics; instead of simulating individual sows, they were interested in steady-state herd distribution. Week-by-week simulations of the system were conducted by [25], while [26] developed an event-directed framework. The rest of the models used a day as a measure of time. Production facilities were only taken into consideration. Tess *et al.* (1983a) provided a more thorough accounting of the development process and nutritional needs, whereas [27] focused on genetic features and breeding selection.

4.3.2 Input Criteria

Depending on the kind of model we were referring to, the input parameters would vary. Optimization models often had a far reduced establishment than simulation frameworks. Since these models' goals were more wide and flexible, input parameters for simulation models were often thought to be greater.

4.3.2.1 Parameters Used as Inputs in Optimization Models

The optimization models were dependent on sow herd dynamics by using a partition in states of the sow's lifetime as shown in Figure 4.1. By taking into account parity-based characteristics (probability of survival, discount rate, marginal profit per parity, length, maximum amount of parities permitted, departure from average parity-specific litter size), [28] developed the more generic division. The other optimization models' parameter considerations were generally rather comparable. These factors might be divided into transition probabilities, stage and state variables, and economic inputs. The methodologies used for parameter estimates and the total number of states came into sharp contrast. For gestation, nursing, the time from weaning to the first breeding, and the time between breeding, state variables were used. Only [29] specifically considered the certain-state time gap (e.g., breastfeeding duration, gestation period), which is primarily why the final number of states varied. There are now more states to more accurately depict the production's variation and shifts in level of output. In this context, repeatability of litter size was taken into consideration by [30]. Less information from actual farms was used from the literature to analyze model behavior. Using various techniques, [31] reported particular farm characteristics (such as everyday feed consumption, litter dimension, and transition likelihood) that were calculated from actual farm data. To demonstrate model operation, [32] retrieved parameter values from the literature or used acceptable amounts. In general, writers did not take into consideration the unique variability of average parameters (such as gestation length, lactation duration, oestrus interval, etc.).

4.3.2.2 Parameters Used as Inputs in Simulation Techniques

Simulation techniques had random parameters with a defined distribution rather than a fixed amount. The organic production characteristics, which comprised conception rate, live piglet birth rate, stage-specific death rates, gestation period, weaning to initial oestrus gap, oestrus cycle duration, and growth rate per state, were relatively comparable in all models. The model's structure, design, and aim all had an impact on how these factors were considered and evaluated. Empirical distributions were examined by Marsh and Singh (1986a). In order to get values for litter dimension, mortality rates, and weaning to first oestrus gap, [33] used empirical distributions of Hawaii's sow farms. However, he also used random distributions for various factors, such as gestation and oestrus cycle duration. Generally, normal univariate distributions for continuous variables or actual uniform distributions for state transitions were utilized to generate random input parameters. However, alternative distributions—such as log-normal by [34] and exponential by [35]—were used by a number of writers to depict the weaning to oestrus period. To determine the oestrus cycle, gestation period, age at puberty, weight at puberty, and litter size, [36] used the normal distribution. If individual behavior is represented, real uniforms are frequently used as the factor to produce various distributions; otherwise, the rate is applied directly to the flock. It was the instance, for example, when events were depicted as successful conception and unanticipated fatalities. Injuries and issues with reproduction were the most common weeding out causes. For instance, [37] examined culling depending on parity limit and demise; however, different writers were more explicative and detailed infertility and extra causes for killing.

[38] tackled the system from the perspective of a nutritionist by basing their different models on the growth process and feeding needs. While [39] developed a stochastic model, [40] carried out their work in a deterministic manner. As a result, [41] explicitly regarded as communications between nutrition and reproduction factors, although other models generally reduced the needs for feeding. For instance, [42] considered the everyday food consumption of pigs in a nursery by age at weaning and chronological age; [43] suggested a multiple regression approach set up in the literature to compute the everyday food intake for the lactation time.

[44] incorporated certain parameters from earlier simulation techniques, as did different writers, and added a far specific explanation of the ovulation and growth techniques using a set of equations. Although they were taken into account by various writers, they did not accurately reflect the availability of amenities. While [45] specified a highest quantity of farrowings per week as a reference for room demands, [46] considered floor needs and imposed a limitation. Breeding, gestation, farrowing, nurseries, and growing-finishing spaces were spatially dispersed among several structures, according to [47]. Only [48] took into account sow and boar selection indices in addition to other parity-dependent variables.

Last but not least, we should point out that not every simulation approaches contained economic inputs, although optimization techniques did.

4.3.3 Results from the Models

The models' outputs were matched to their intended use. There existed more outputs in the simulation techniques than in the optimization models. The goal of the optimization models was to find an optimum (a largest or at least). [49] discovered the best replacement strategy for sows. The ideal replacement policy, together with a related mating strategy based on sow quality and a description of the technical and financial outcomes, were supplied by [50]. Following that, a sensitivity study or a post-optimum analysis is carried out, depending on the author. As a result, Huirne *et al.* (1993) only computed a few performance indices, while [51] provided technical indexes and a sensitivity analysis of various factors. Curiously, the only study to provide CPU time for the optimization procedure as an outcome was Huirne et al.

There were many different simulation model outcomes, many of which were heavily influenced by their design. A large number of outputs were then supplied by Marsh (1986), which divided them into seven types: census, attainment indicators, reproductive attainment, monthly graphics, cash flow survey, earnings document, and livestock value. He had previously utilized them in a dairy model, and they were based on commercial information systems. [52] provided various herd dynamics-related outcomes. [53] calculated data about herd dynamics in more detail. They estimated yearly revenues, expenses, and rates of yield for economic survey besides various prices and costs. Similar calculations were made by [54] using technical and profitable factors obtained from the distribution of swine throughout equilibrium states. The cost of replacement gilts, the price of piglets, the slaughter value of culled sows, the quantity of litters per sow per year, and the percentage of fertilizations were the most crucial factors. [55] computed differently and uniquely for each farm while also deriving from the distribution of sows across states at equilibrium. They also examined technical and economic factors. Additionally, they included several graphics pertaining to seed distribution throughout states. In their models, [56] took animal development into account; as a result, they displayed plots of sows' body weight. While [56] included data characterizing the flow of animals between phases of life cycle in the flock, mean sow age every day, and simulated amount of animals every day, [57] analyzed growth curves, performance indices, and certain rates of biological effectiveness. In terms of calculating outputs, [58] were more specific. In order to determine the means of various production characteristics, [59] estimated the litter dimension at birth, pigs born/sow/year, piglets weaned/sow/year, conception rate, and kg of pig sold per kg of feed. [60] computed sow days/pig, sows/year, litters/sow, and sows/pigs by simulation year.

4.3.4 The Models' Validation

Not all of the models under examination underwent validation. For instance, the optimization models were not tested; they simply took average values into account and were equal to deterministic approaches that dealt with well-defined situations. The major goal of the optimization models was to demonstrate mathematical approaches to tackle certain issues.

As an example, [61] provided a novel method for modeling sow herds called hierarchical Markov decision models, which was improved upon from traditional Markov decision processes. These papers were not intended for validation. Instead of doing a formal validation, writers like [62] examined the impact of altering certain key parameters to better understand the behavior of the model.

In contrast, a number of validation techniques were used to simulation models. As none of the parameters were familiar in operation nor appropriate data for validation were obtainable, authors presenting simulation models concurred that it is challenging to accomplish a thorough validation. As an alternative, various writers have described the model in detail without using any further tests to verify it. In other instances, the typical approach was to carry out verification dependent on a thorough model description and ensure that the technique was correctly functioning at various times during its life cycle, inclusive of the ultimate cumulative of inputs and outputs. [63] compared average findings from the literature with those from the simulation as proof. While [64] contrasted pig weaning weights (at birth and at 20 kg) with those mentioned in the review work, [65] evaluated various outputs such as lactation weight pattern, total body composition, litter weight during birth and at weaning, feed/gain ratios, and milk production. After presenting a model behavior research based on sensitivity analysis, [66] contrasted the overall findings with those acquired from a management information system. Previous articles did not offer any statistical analysis; it was only accomplished by [67], which claimed that because their simulation technique was focused on the sows' reproductive cycle, the validation process should concentrate on the reproductive events the model anticipated. To determine if the noticed and implemented specimens of the farrowing to initial oestrus gap came from similar distribution, they employed a non-parametric test called the Kolmogorov-Smirnov test. Comparing the sow herd distribution across states generated by the approach with the true distribution noticed, [68] utilized a non-parametric test called the Chi-square test to determine if both distributions were obtained from the same.

4.3.5 Opportunities for Implementation and Integration

Typically, at least in the first stage, sow herd management techniques were created by the authors. Only a handful of the approaches explicitly stated their intention to be used on farms. The majority of the models were created for research or teaching objectives. These factors may help to explain why user interfaces for farmers or advisors were not sufficiently detailed.

The most widely used programming languages for software implementation were procedural or object-oriented languages. For instance, [69] programmed their model in ANSI C, [70] used Pascal, ExtendTM, a C-based language, and [71] used Java. As an alternative, simulation-specific programming languages like SLAM II (Simulation Language for Alternative Modeling) or GPSS (General Purpose System Simulator) were used by [72] to create the simulation models. The models' implementation was not discussed in the remaining publications.

Such models' incorporation as modules into already-existing information systems had a significant impact on how they were used on farms. For instance, PORKchop suggested that pertinent data from VAMPP and PigCHAMP may have been transferred. PigORACL was developed as a PigCHAMP module. According to [73], TACTSys was a management information system for tactical decision support that integrated many models. The model of [74] was supposed to be placed within as a module of the BD-Porc system (2000), a MIS which holds the official database of Spanish pig industry; however, it has not been done yet. Regardless of the introduction of computerized herd models, generally they have not yet been entirely effective when used as stand-alone apps.

4.3.6 Management of Risk

Risk is the term for uncertainty that results from farmer activities because of unforeseen events like price changes and biological reactions to various agricultural methods. A huge quantity of independent scrutiny must be made on the target random variable in order to provide statistically meaningful findings from a stochastic model, which is what simulation models are. In order to evaluate the mean income and get the 93% confidence interval of the annual average income, [75] ran the model for 12 successive years, collecting a specimen each year. In a totally random design, [76] reproduced each option three times and contrasted them using an ANOVA.

Although it was taken into account implicitly for the majority of the model's parameters, optimization models overlooked the uncertainty connected to their outcomes. Because they were just expectations, the outcomes produced by optimization models were aimed at risk-neutral decision makers.

4.3.7 Additional Submissions and Literature Review

When describing the models' features and capabilities wasn't sufficient, several writers provided succinct examples of their application. For instance

[77] examined various management techniques for lactations lasting three and six weeks. [78] simulated a number of possibilities (increased prolificacy, split weaning, increased litter size, and lower mortality), comparing them. To determine their effect on model performances, [79] conducted a study of sensitivity for the key production factors.

In general, the majority of the models under evaluation were used in subsequent publications that provide more specific illustrations of prospective uses. For instance, [80]'s methodology for studying facility occupancy was published in [81]. Similar to [82], and others, the same issue was studied using the models developed by [83], respectively. Additionally, it provided instances of implementations in real-world settings. For instance, [84] utilized empirical data to the model developed by [85], while researchers developed a DSS for use on farms. The Huirne prototype used to analyze the economic factors in replacement was also utilized in several publications to assess replacement options. For the purposes of calculating litters/sow/year, pigs weaned/sow/year, profit/sow, and profit/herd, [86] adapted the model of [87]. Later, they used their model to analyze the results of various replacement and interim programs in order to determine the combination that would work best. Using the model they had previously established, [88] conducted a profitable evaluation of managing options for reproduction and replacement in sow flocks. Review models have sometimes been included into larger systems that have been detailed in other places.

4.4 Discussion

Reproduction was highlighted in the sow herd models under examination as the primary factor affecting herd output. Although [89] claimed that these kinds of models show a huge diversity in structure, numerous commonalities in the modeling technique at a mathematical level have been discovered. The adoption of discrete models, rather than continuous ones, to account for the discontinuous aspect of the reproduction technique was one such shared characteristic. Additionally, primary decision-making factors were replacement, lactation, and breeding approaches, individually or combined. Dynamic models were developed in each instance as a result of the variability in sow performance throughout the course of the sow's lifetime and the makeup of the herd.

There was considerable agreement about the factors to consider while explaining the herd system, but there were observable disparities in the manner, level of information, and emphasis employed in their explanation. For instance, elements characterizing the reproductive process were covered more heavily than those relating to food, growth, economics, facilities, or genetics. [90] had subsequently seen this finding as a significant flaw for the practical application of existing livestock models because it misrepresents the economic efficiency of cattle production. However, contrary to what [91] believed, recognized herd management parameters cannot both fully explain all noticed feasibility nor be substituted by theories or educated approximations. As a result, assumptions and simplifications must be appropriate for the model's intended purpose.

The majority of animal features were thought to be independent and uniformly distributed, allowing for using a modeling method, for instance, aggregating states of separate animals. As a result, it was challenging to simulate animal relationships. As an example, batch management and housing facilities have been modeled more simply in simulation than in optimization model. On the other hand, the involvement of enormous unpredictability in models for on-farm decision assistance may cause confusion and lower end-user approval.

[92] attempted to simulate effects that were difficult to understand using empirical distributions, such as seasonal impacts on fertility. To accurately depict each farm behavior, input factors relevant to the farm are required. Only [93] employed farm-specific input parameters, as can be shown, while the majority of the models used values derived from broad databases or literature. Since a particular farm is not represented, the latter is ideal for testing the model as well as for academic and research motives but not for providing advise at the herd level. However, [94] noted that while input parameters were thought to be time homogeneous in all studied models, fresh data are sometimes obtained on farms, making it fair to update input parameters or revise hypotheses. In this regard, [95] suggested the investigation of true herd formation over time to gauge the stability of the framework, while [96] employed a dynamic linear model (DLM) to improve litter dimension assumption based on prior data.

The starting state of the farm was not taken into account since the simulation and optimization techniques focused on steady-state studies. As [97] pointed out, this is highly helpful for comparing management tactics at equilibrium. The productive route a farm would take from its current management approach to the newer one suggested by equilibrium by any of the models is not shown or appreciated, however. Considerable practical ramifications may result from the convergence from the initial condition to the newer steady-state. This indicates that there aren't any temporary models or analyses connecting the theoretical, academic, and practical worlds.

Existing models' outputs, including simulation techniques, only offer mean values without calculating the related confidence intervals or measuring variability, which makes it impossible to statistically compare substitutes or a compatible risk analysis. Moreover, despite the fact that the investigated optimization models included probabilities in an effort to reflect the dynamic-stochastic character of the system, their operational formulation and resolution was dependent on analogous static-deterministic techniques that in fact gave the remedy. [98] had previously made note of the necessity for distinct criteria to indicate variability-risk characteristics rather than the typical mean criteria addressed in the research on Markov decision processes in order to adequately capture the varied components of a decision maker. In this regard, [99] use of the variance of cumulative rewards might be considered a replacement. The use of multicriteria techniques in place of the prevalent mono-criterion approaches in these kinds of models was supported by other writers.

To achieve acceptability and credibility for practical usage, the models must be validated. Numerous researchers had claimed that they needed the necessary data to carry out the validation, but [100] suggested various approaches of validation, in spite of situations where data is lacking, putting an emphasis on statistical approaches which produce repeatable, objective, and quantitative data about the excellence of the simulation techniques. For example, [101] recognized that herd form was important in all models and key to calculating outputs in several of them. The so-called ideal herd structure and its tight relationship to the culling rate is another recurring problem in actual herd management. In order to validate this form of model, [101] suggested the herd distribution at equilibrium. Additionally, with regard to Markov simulation models, the mean output of the anticipated outputs in stochastic simulation must converge to that in deterministic simulation; [102] exploited this idea to validate several examples of sophisticated simulation techniques.

The real-life utilization of evaluated frameworks on farms was hindered in the 1980s by the simplifications necessary for a sufficient solution procedure by computers. According to [103], the use of mathematical models may significantly influence the farmer's decision-making behavior anytime models run interactively or results are promptly accessible. The quantity of possible implementations and the use of DSS increased as a result of methodological and computational advancements (cf. Section 3.7). Because their goal was to be employed in field circumstances rather than the actual system, as [104] had noted, this feature stimulated the aggregation of such various models into current MISs, albeit with little accomplishment. The knowledge and abilities needed to describe the outcomes of these frameworks and how the end-users—farmers, swine experts, or extension service advisers—were involved at various phases of development. According to [105], farmers are more concerned with making the appropriate large choices and significant tactical modifications than real decision models are.

The interface of practical applications based on complicated models that are used at the farm level should be clear, easy to understand, and capable of avoiding errors or unexpected results. The incorporation of expert systems, as suggested by several writers, would be one way to meet these requirements; however, this would result in software with complex user interfaces but no practical applications. The emergence of the Internet, e-commerce, and online facilities, however, is seen to alter or supplement this tendency. For instance, several MISs are now beginning to facilitate help through the Internet. The employment of these automations to help decision techniques, however, might be stymied if the counterweight is inadequate.

Finally, since producers are vertically aggregated in larger businesses, cooperatives, or organizations, which mean that farmers contract their output under various circumstances, the prognosis for the swine sector is shifting. Therefore, this contract is often a rental of agricultural equipment and labor. Additionally, each business often employs a team of advisors who serve as actual pig managers (farmers must follow their recommendations in order to avoid being penalized in accordance with the contract). Goals and objectives are therefore predetermined by businesses, and really autonomous farmers in charge of a single sow farm are predicted to become quite few. The number of automated data-collecting devices is growing at the same time as the amount of data that has to be processed and the requirement for their aggregation into current systems. In this situation, even identical decision issues need several farms and a vast quantity of data. To deal with this novel practical circumstance, they can introduce fresh modeling advancements. As a result, it is anticipated that fresh additions to the literature in this area will emerge soon.

4.5 Conclusions

A sow herd model cannot be constructed in a single, idealized manner. It depends on the objective, but assumptions and simplifications must be appropriate for the model's intended use. Such models have been created in large numbers effectively as teaching and research tools. They may be used to investigate presumptions and hypotheses, which are useful for learning but not for giving advice. There has been a lot of work done in the methodological field, and it has been shown that even highly complicated models can be solved; nevertheless, more work has to be done before sow herd models may be effectively utilized to assist decision tasks. Today, a respectable quantity of data from actual farms is accessible for validation as well as for the inference of particular parameters characterizing the behavior of each unique farm. Specific farm factors, as well as the utilization of actual farm data for validation, have just lately been incorporated. The validation of these models is still an issue, however. To establish confidence and ensure their broad usage in practical settings, they must be trustworthy instruments.

It is determined that short time horizon decision models or transient models are required in order to react to the changing pig production environment. The proposals offered thus far are thought of as steady-state scenarios that are not as prevalent in actual farms as would be preferred. Similar results are shown when the parameters are assumed to be time-homogeneous, which is expected in a dynamic environment where estimates must be updated often.

The utilization of predicted total return per unit of time can be woefully inadequate to characterize the issue from the perspective of the farm manager; thus it is also advisable to revise the optimality criteria. In this regard, it would appear appropriate for future ideas to include variability-risk characteristics of the situation or other multicriteria techniques.

The interface has also contributed to the failure of the usage of these models by farmers and their integration into current management information systems. However, if these models are able to provide pertinent responses for the users, it is anticipated that the Internet will have a significant influence on their development and usage for on-farm decision assistance. Finally, the sector's changing organizational structure, which includes larger corporations and/or organizations and fewer individual farmers, causes new decision-making issues to emerge and shifts the decision-making center away from the farm. In order to address these issues, new models that consider a pool of farms rather than a single autonomous farm must be created.

References

1. A. D. Andrushia, T. M. Neebha, A. T. Patricia, K. M. Sagayam, S. Pramanik, Capsule Network based Disease Classification for VitisVinifera Leaves, *Neural Computing and Applications*, doi: 10.1007/s00521-023-09058-y, 2023.
2. M. Molapo, C. Tu, D. Du Plessis and S. Du, "Management and Monitoring of Livestock in the Farm Using Deep Learning," *2023 International Conference*

on Artificial Intelligence, Big Data, Computing and Data Communication Systems (icABCD), Durban, South Africa, 2023, pp. 1-6, doi: 10.1109/icABCD59051.2023.10220556.

3. A. Roy and S. Pramanik, A Review of the Hydrogen Fuel Path to Emission Reduction in the Surface Transport Industry, *International Journal of Hydrogen Energy*, https://doi.org/10.1016/j.ijhydene.2023.07.010, 2023.
4. S. Pramanik, An Adaptive Image Steganography Approach depending on Integer Wavelet Transform and Genetic Algorithm, *Multimedia Tools and Applications*, 2023, https://doi.org/10.1007/s11042-023-14505-y
5. R. Jayasingh, J. Kumar R. J. S., D. B. Telagathoti, K. M. Sagayam and S. Pramanik, "Speckle noise removal by SORAMA segmentation in Digital Image Processing to facilitate precise robotic surgery", *International Journal of Reliable and Quality E-Healthcare*, vol. 11, issue 1, doi: 10.4018/IJRQEH.295083, 2022.
6. S. Pramanik and S. Suresh Raja, "A Secured Image Steganography using Genetic Algorithm", *Advances in Mathematics: Scientific Journal*, vol. 9, issue 7, pp. 4533-4541, 2020. doi: https://doi.org/10.37418/amsj.9.7.22
7. S. Pramanik, R. P. Singh and R. Ghosh, "Application of Bi-orthogonal Wavelet Transform and Genetic Algorithm in Image Steganography", *Multimedia Tools and Applications*, https://doi.org/10.1007/s11042-020-08676-2020, 2020.
8. G. R. Ashisha, Mary X. Anitha, T. George, Martin Sagayam, Unai Fernandez-Gamiz, Hatıra Günerhan, M. N. Uddin and S. Pramanik, Analysis of Diabetes Disease Using Machine Learning Techniques: A Review, *Journal of Information Technology Management*, 2023.
9. K. P. Kumar, H. Murthy, V. J. Pillai, B. R. Prathap, M. Moses and Y. Urs, "Data Analysis and Machine Learning Observation on Production Losses in the Food Processing Industry," *2023 IEEE International Conference on Contemporary Computing and Communications (InC4)*, Bangalore, India, 2023, pp. 1-4, doi: 10.1109/InC457730.2023.10263147.
10. D. -Y. Jeong, S. -K. Jo, I. -B. Lee, H. Shin and J. -G. Kim, "Digital Twin Application: Making a Virtual Pig House Toward Digital Livestock Farming," in *IEEE Access*, vol. 11, pp. 121592-121602, 2023, doi: 10.1109/ACCESS.2023.3313618.
11. A. Heidari, M. A. J. Jamali, N. J. Navimipour and S. Akbarpour, "A QoS-Aware Technique for Computation Offloading in IoT-Edge Platforms Using a Convolutional Neural Network and Markov Decision Process," in *IT Professional*, vol. 25, no. 1, pp. 24-39, Jan.-Feb. 2023, doi: 10.1109/MITP.2022.3217886.
12. S. Pramanik, K. M. Sagayam, and O. P. Jena, Machine Learning Frameworks in Cancer Detection, *ICCSRE* 2021, Morocco, 2021.
13. S. Pramanik, M. G. Galety, D. Samanta, N. P Joseph, Data Mining Approaches for Decision Support Systems, *3rd International Conference on Emerging Technologies in Data Mining and Information Security*, 2022.

14. R. Bansal, A. J. Obaid, A. Gupta, R. Singh, S. Pramanik, "Impact of Big Data on Digital Transformation in 5G Era", *2nd International Conference on Physics and Applied Sciences (ICPAS 2021)*, doi:10.1088/1742-6596/1963/1/012170, 2021.
15. V. Veeraiah, K. O., Thejaswini, R. Dilip, S. K. Jain, A. Sahu, S. Pramanik, and A. Gupta, The Suggested Use of Big Data in Medical Analytics by Fortis Healthcare Hospital, Adoption and Use of Technology Tools and Services by Economically Disadvantaged Communities: Implications for Growth and Sustainability, IGI Global, 2023.
16. S. Ahamad, S. Janani, V. Talukdar, T. Sharma, A. Sahu, S. Pramanik, A. Gupta, Apollo Hospital's Proposed Use of Big Data Healthcare Analytics, in *Big Data Analytics Techniques for Market Intelligence*, IGI Global, 2023.
17. S. Pramanik, Intelligent Farming Utilizing a Soil Tracking Device, in *Convergence of Cloud Computing, AI and Agricultural Science*, Eds., A. K. Sharma, N. Chanderwal and R. Khan, IGI Global, 2023.
18. P. T. Khanh, T. H. Ngọc, and S. Pramanik, Future of Smart Agriculture Techniques and Applications, in *Advanced Technologies and AI-Equipped IoT Applications in High Tech Agriculture*, Ed., A. Khang, IGI Global, 2023.
19. T. H. Ngọc, P. T. Khanh and S. Pramanik, Smart Agriculture using a Soil Monitoring System, in *Advanced Technologies and AI-Equipped IoT Applications in High Tech Agriculture*, Ed., A. Khang, IGI Global, 2023.
20. Reepu, S. Kumar, M. G. Chaudhary, K. G. Gupta, S. Pramanik and A. Gupta, Information Security and Privacy in IoT, in *Handbook of Research in Advancements in AI and IoT Convergence Technologies*, Eds., J. Zhao, V. V. Kumar, R. Natarajan and T. R. Mahesh, IGI Global, 2023.
21. V. Veeraiah, V. Talukdar, Manikandan K., S. B. Talukdar, V. D. Solavande, S. Pramanik, A. Gupta, Machine Learning Frameworks in Carpooling, in *Handbook of Research on AI and Machine Learning Applications in Customer Support and Analytics*, Eds., Md S. Hossain, R. C. Ho and G. Trajkovski, IGI Global, 2023.
22. V. Jain, M. Rastogi, J. V. N. Ramesh, A. Chauhan, P. Agarwal, S. Pramanik, A. Gupta, FinTech and Artificial Intelligence in Relationship Banking and Computer Technology, in *AI, IoT, and Blockchain Breakthroughs in E-Governance*, Eds., K. Saini, A. Mummoorthy, R. Chandrika and N. S. Gowri Ganesh, IGI Global, 2023.
23. S. Pramanik and S. Bandyopadhyay, Identifying Disease and Diagnosis in Females Using Machine Learning, in *Encyclopedia of Data Science and Machine Learning*, Ed. John Wang, IGI Global, 2023, doi: 10.4018/978-1-7998-9220-5.ch187.
24. S. Pramanik, "An Effective Secured Privacy-Protecting Data Aggregation Method in IoT", in *Achieving Full Realization and Mitigating the Challenges of the Internet of Things*, Eds., M. O. Odhiambo and W. Mwashita, IGI Global, 2022, doi: 10.4018/978-1-7998-9312-7.ch008

25. S. Pramanik and S. Bandyopadhyay, Analysis of Big Data, in *Encyclopedia of Data Science and Machine Learning*, Eds. John Wang, IGI Global, 2023, doi: 10.4018/978-1-7998-9220-5.ch006
26. R. Bansal, B. Jenipher, V. Nisha, Jain, Makhan R., Dilip, Kumbhkar, S. Pramanik, S. Roy and A. Gupta, "Big Data Architecture for Network Security", in *Cyber Security and Network Security*, Wiley, 2022, https://doi.org/10.1002/9781119812555.ch11
27. D. Pradhan, P. K. Sahu, N. S. Goje, H. Myo, M. M.Ghonge, M., Tun, R, Rajeswari and S. Pramanik, "Security, Privacy, Risk, and Safety toward 5G Green Network (5G-GN)", in *Cyber Security and Network Security*, Wiley, 2022, doi: 10.1002/9781119812555.ch9
28. R. Anand, J. Singh, D. Pandey, B. K. Pandey, V. K. Nassa and S. Pramanik, "Modern Technique for Interactive Communication in LEACH-Based Ad Hoc Wireless Sensor Network", in *Software Defined Networking for Ad Hoc Networks*, M. M. Ghonge, S. Pramanik and A. D. Potgantwar, Eds., Springer, 2022, https://doi.org/10.1007/978-3-030-91149-2_3
29. S. Choudhary, V. Narayan, M. Faiz and S. Pramanik, "Fuzzy Approach-Based Stable Energy-Efficient AODV Routing Protocol in Mobile Ad hoc Networks", in *Software Defined Networking for Ad Hoc Networks*, M. M. Ghonge, S. Pramanik and A. D. Potgantwar, Eds., Springer, 2022, https://doi.org/10.1007/978-3-030-91149-2_6
30. K. Dushyant, G. Muskan, A. Gupta and S. Pramanik, "Utilizing Machine Learning and Deep Learning in Cyber security: An Innovative Approach", in *Cyber security and Digital Forensics*, M. M. Ghonge, S. Pramanik, R. Mangrulkar, D. N. Le, Eds., Wiley, 2022, https://doi.org/10.1002/9781119795667.ch12
31. B. K. Pandey, D. Pandey, S. Wairya, G. Agarwal, P. Dadeech, S. R. Dogiwal and S. Pramanik, "Application of Integrated Steganography and Image Compressing Techniques for Confidential Information Transmission", in *Cyber Security and Network Security*, https://doi.org/10.1002/9781119812555.ch8, Eds., Wiley, 2022.
32. A. Gupta, A. Verma and S. Pramanik, Security Aspects in Advanced Image Processing Techniques for COVID-19, in *An Interdisciplinary Approach to Modern Network Security*, S. Pramanik, A. Sharma, S. Bhatia and D. N. Le, Eds, CRC Press, 2022.
33. A. Ahmadi, S. Taheri, R. Ghorbani, V. Vahidinasab and B. Mohammadi-ivatloo, "Decomposition-Based Stacked Bagging Boosting Ensemble for Dynamic Line Rating Forecasting," in *IEEE Transactions on Power Delivery*, vol. 38, no. 5, pp. 2987-2997, Oct. 2023, doi: 10.1109/TPWRD.2023.3267511.
34. F. A. Almalki and H. Aljahdalia, "Optimizing Propagation Predictions for Last-mile Connectivity in a CubeSat Perspective," *2023 3rd International Conference on Computing and Information Technology (ICCIT)*, Tabuk, Saudi Arabia, 2023, pp. 550-555, doi: 10.1109/ICCIT58132.2023.10273920.

35. R. V. Aravind and P. Balasubramaniam, "An Exponential Stabilization Analysis for Switched Interval Type-2 Fuzzy Sampled-Data Control Systems," in *IEEE Transactions on Fuzzy Systems*, vol. 31, no. 10, pp. 3652-3663, Oct. 2023, doi: 10.1109/TFUZZ.2023.3262657.
36. W. Lee, Y. Ham, T. -W. Ban and O. Jo, "Analysis of Growth Performance in Swine Based on Machine Learning," in *IEEE Access*, vol. 7, pp. 161716-161724, 2019, doi: 10.1109/ACCESS.2019.2951522.
37. K. J. Dsouza and S. Arumugam Muthukumaraswamy, "A Robotics-Based Surveillance System for Livestock Wellbeing and Early Disease Detection in Poultry Farms," *2023 International Conference on Computational Intelligence and Knowledge Economy (ICCIKE)*, Dubai, United Arab Emirates, 2023, pp. 221-225, doi: 10.1109/ICCIKE58312.2023.10131735.
38. A. Y. Majid, R. F. Nurmansyah, M. L. A. Pratama, H. Susanti and N. Prihatiningrum, "IoT-Based Cat Feeding and Monitoring System," *2023 8th International Conference on Instrumentation, Control, and Automation (ICA)*, Jakarta, Indonesia, 2023, pp. 160-165, doi: 10.1109/ICA58538.2023.10273126.
39. B. Sun *et al.*, "A Dynamics-Learning Multirate Estimation Approach for the Feeding Condition Perception of Complex Industry Processes," in *IEEE Transactions on Cybernetics*, doi: 10.1109/TCYB.2023.3263571.
40. D. Yang *et al.*, "DetFed: Dynamic Resource Scheduling for Deterministic Federated Learning over Time-sensitive Networks," in *IEEE Transactions on Mobile Computing*, doi: 10.1109/TMC.2023.3303017.
41. S. Naseem *et al.*, "DeepFert: An Intelligent Fertility Rate Prediction Approach for Men Based on Deep Learning Neural Networks," in *IEEE Access*, vol. 11, pp. 75006-75022, 2023, doi: 10.1109/ACCESS.2023.3290554.
42. C. Navaneethan and S. T. Prasath, "Analysis of the Impacts of Consuming Red and Processed Meat on Colorectal Cancer and the Role of Machine Learning in Clinical Diagnostics," *2023 International Conference on Intelligent and Innovative Technologies in Computing, Electrical and Electronics (IITCEE)*, Bengaluru, India, 2023, pp. 855-861, doi: 10.1109/IITCEE57236.2023.10091060.
43. F. Torres-Cruz et al., "Comparative Analysis of High-Performance Computing Systems and Machine Learning in Enhancing Cyber Infrastructure: A Multiple Regression Analysis Approach," *2022 2nd International Conference on Innovative Practices in Technology and Management (ICIPTM)*, Gautam Buddha Nagar, India, 2022, pp. 69-73, doi: 10.1109/ICIPTM54933.2022.9753839.
44. K. Mahajan and P. Mane, "Follicle Detection of Polycystic Ovarian Syndrome (Pcos) Using Yolo," *2023 9th International Conference on Advanced Computing and Communication Systems (ICACCS)*, Coimbatore, India, 2023, pp. 1550-1553, doi: 10.1109/ICACCS57279.2023.10112821.
45. Y.-B. Lin, W.-E. Chen and T. C.-Y. Chang, "Moving from Cloud to Fog/Edge: The Smart Agriculture Experience," in *IEEE Communications Magazine*, doi: 10.1109/MCOM.001.2200633.

46. D. -Y. Jeong, S. -K. Jo, I. -B. Lee, H. Shin and J. -G. Kim, "Digital Twin Application: Making a Virtual Pig House Toward Digital Livestock Farming," in *IEEE Access*, vol. 11, pp. 121592-121602, 2023, doi: 10.1109/ACCESS.2023.3313618.
47. Z. Pan, H. Chen, W. Zhong, A. Wang and C. Zheng, "A CNN-Based Animal Behavior Recognition Algorithm for Wearable Devices," in *IEEE Sensors Journal*, vol. 23, no. 5, pp. 5156-5164, 1 March 1, 2023, doi: 10.1109/JSEN.2023.3239015.
48. A. Hussain, S. Ali, M.-I. Joo and H. C. Kim, "A Deep Learning Approach for Detecting and Classifying Cat Activity to Monitor and Improve Cat's Well-being Using Accelerometer, Gyroscope, and Magnetometer," in *IEEE Sensors Journal*, doi: 10.1109/JSEN.2023.3324665.
49. J. Song, D. Gunduz and W. Choi, "Optimal Scheduling Policy for Minimizing Age of Information with a Relay," in *IEEE Internet of Things Journal*, doi: 10.1109/JIOT.2023.3308113.
50. A. J. Patil, S. J. Patil, S. D. Patil, M. Dharme, R. K. Jarial and R. N. Sharma, "Random Forest Models and Ensemble Learning-Based dynamic Sowing Methodology," *2023 11th International Conference on Internet of Everything, Microwave Engineering, Communication and Networks (IEMECON)*, Jaipur, India, 2023, pp. 1-6, doi: 10.1109/IEMECON56962.2023.10092347.
51. T. Keerthi, A. Kumari, M. C. Chinnaiah, M. Parvathi, C. D. Reddy and D. Sumana, "An Efficient Image Analysis Approach to Evaluate the Quality of Cotton Seeds," *2023 4th International Conference for Emerging Technology (INCET)*, Belgaum, India, 2023, pp. 1-7, doi: 10.1109/INCET57972.2023.10170358.
52. S. K. K. R, G. T, B. S. Jebaraj, V. S, V. M and K. C, "Design and Fabrication of Multi-functioned Seed Sowing Machine," *2023 Second International Conference on Electronics and Renewable Systems (ICEARS)*, Tuticorin, India, 2023, pp. 151-155, doi: 10.1109/ICEARS56392.2023.10085567.
53. A. A. Khan, V. P. Meena, A. V. Waghmare, U. K. Yadav and V. P. Singh, "Elephant Herding Optimization Assisted Reduction of Level-Up DC-DC Converter," *2023 IEEE 3rd International Conference on Sustainable Energy and Future Electric Transportation (SEFET)*, Bhubaneswar, India, 2023, pp. 1-6, doi: 10.1109/SeFeT57834.2023.10244929.
54. P. Suebsaiprom, T. Satiramatekul and A. Engkaninan, "Swine Body Temperature Monitoring System," *2023 International Electrical Engineering Congress (iEECON)*, Krabi, Thailand, 2023, pp. 188-192, doi: 10.1109/iEECON56657.2023.10127042.
55. G. Zhou, J. Xu, W. Chen, X. Li, J. Li and L. Wang, "Deep Feature Enhancement Method for Land Cover With Irregular and Sparse Spatial Distribution Features: A Case Study on Open-Pit Mining," in *IEEE Transactions on Geoscience and Remote Sensing*, vol. 61, pp. 1-20, 2023, Art no. 4401220, doi: 10.1109/TGRS.2023.3241331.

56. H. Hwang *et al.*, "System Configuration and Navigation of a Guide Dog Robot: Toward Animal Guide Dog-Level Guiding Work," *2023 IEEE International Conference on Robotics and Automation (ICRA)*, London, United Kingdom, 2023, pp. 9778-9784, doi: 10.1109/ICRA48891.2023.10160573.
57. S. Fu *et al.*, "Robust vascular segmentation for raw complex images of laser speckle contrast based on weakly supervised learning," in *IEEE Transactions on Medical Imaging*, doi: 10.1109/TMI.2023.3287200.
58. T. Govindan *et al.*, "Design and Analysis of a Conformal MIMO Ingestible Bolus Sensor Antenna for Wireless Capsule Endoscopy for Animal Husbandry," in *IEEE Sensors Journal*, doi: 10.1109/JSEN.2023.3323658.
59. B. Natarajan, R. Elakkiya, R. Bhuvaneswari, K. Saleem, D. Chaudhary and S. H. Samsudeen, "Creating Alert Messages Based on Wild Animal Activity Detection Using Hybrid Deep Neural Networks," in *IEEE Access*, vol. 11, pp. 67308-67321, 2023, doi: 10.1109/ACCESS.2023.3289586.
60. W. Lee, Y. Ham, T. -W. Ban and O. Jo, "Analysis of Growth Performance in Swine Based on Machine Learning," in *IEEE Access*, vol. 7, pp. 161716-161724, 2019, doi: 10.1109/ACCESS.2019.2951522.
61. V. Papaspyros, D. Burnier, R. Cherfan, G. Theraulaz, C. Sire and F. Mondada, "A Biohybrid Interaction Framework for the Integration of Robots in Animal Societies," in *IEEE Access*, vol. 11, pp. 67640-67659, 2023, doi: 10.1109/ACCESS.2023.3290960.
62. C. Kim, A. Jang, S. Kim, S. Cho, D. Lee and Y. -J. Kim, "A Wearable ECG Monitoring System Using Microneedle Electrodes for Small Animals," in *IEEE Sensors Journal*, vol. 23, no. 18, pp. 21873-21881, 15 Sept.15, 2023, doi: 10.1109/JSEN.2023.3300992.
63. R. Falque, T. Vidal-Calleja and A. Alempijevic, "Semantic Keypoint Extraction for Scanned Animals using Multi-Depth-Camera Systems," *2023 IEEE International Conference on Robotics and Automation (ICRA)*, London, United Kingdom, 2023, pp. 11794-11801, doi: 10.1109/ICRA48891.2023.10160307.
64. R. Saha *et al.*, "Toward a Smart Sensing System to Monitor Small Animal's Physical State via Multi-Frequency Resonator Array," in *IEEE Transactions on Biomedical Circuits and Systems*, vol. 17, no. 3, pp. 521-533, June 2023, doi: 10.1109/TBCAS.2023.3284823.
65. H. Mao *et al.*, "Deep-Learning-Based Flying Animals Migration Prediction With Weather Radar Network," in *IEEE Transactions on Geoscience and Remote Sensing*, vol. 61, pp. 1-13, 2023, Art no. 5101513, doi: 10.1109/TGRS.2023.3242315.
66. M.-T. Pham, H. Gangloff and S. Lefèvre, "Weakly Supervised Marine Animal Detection from Remote Sensing Images Using Vector-Quantized Variational Autoencoder," *IGARSS 2023 - 2023 IEEE International Geoscience and Remote Sensing Symposium*, Pasadena, CA, USA, 2023, pp. 5559-5562, doi: 10.1109/IGARSS52108.2023.10282672.

67. Wei J, Tang X, Liu J, Zhang Z. Detection of Pig Movement and Aggression Using Deep Learning Approaches. *Animals*, 2023; 13(19):3074. https://doi.org/10.3390/ani13193074
68. A. A. Rahman, M. R. Kabir, R. H. Ratul, F. A. Shamns, M. M. Nishat and F. Faisal, "An Efficient Analysis of EEG Signals to Perform Emotion Analysis," *2023 4th International Conference on Artificial Intelligence, Robotics and Control (AIRC)*, Cairo, Egypt, 2023, pp. 1-7, doi: 10.1109/AIRC57904.2023.10303179.
69. E. Kurian, P. Braione, D. Briola, D. D'Avino, M. Modonato and G. Denaro, "Automated Test Case Generation for Safety-Critical Software in Scade," *2023 IEEE/ACM 45th International Conference on Software Engineering: Software Engineering in Practice (ICSE-SEIP)*, Melbourne, Australia, 2023, pp. 483-494, doi: 10.1109/ICSE-SEIP58684.2023.00049.
70. E. Baidal-Bustamante, C. Mora and M. S. Alvarez-Alvarado, "STEAM Project-Based Learning Approach to Enhance Teaching-Learning Process in the Topic of Pascal's Principle," in *IEEE Transactions on Education*, doi: 10.1109/TE.2023.3283850.
71. R. Croft, M. A. Babar and M. M. Kholoosi, "Data Quality for Software Vulnerability Datasets," *2023 IEEE/ACM 45th International Conference on Software Engineering (ICSE)*, Melbourne, Australia, 2023, pp. 121-133, doi: 10.1109/ICSE48619.2023.00022.
72. X. Shi *et al.*, "Analysis of Ultra-Low Frequency Oscillation in Hydro-Dominant Power System and Suppression Strategy by GPSS," in *IEEE Transactions on Industry Applications*, vol. 59, no. 3, pp. 2796-2806, May-June 2023, doi: 10.1109/TIA.2023.3237658.
73. Y. Geng *et al.*, "Relational Message Passing for Fully Inductive Knowledge Graph Completion," *2023 IEEE 39th International Conference on Data Engineering (ICDE)*, Anaheim, CA, USA, 2023, pp. 1221-1233, doi: 10.1109/ICDE55515.2023.00098.
74. S. Jalil, S. Rafi, T. D. LaToza, K. Moran and W. Lam, "ChatGPT and Software Testing Education: Promises & Perils," *2023 IEEE International Conference on Software Testing, Verification and Validation Workshops (ICSTW)*, Dublin, Ireland, 2023, pp. 4130-4137, doi: 10.1109/ICSTW58534.2023.00078.
75. A. A. Alhussan *et al.*, "A Binary Waterwheel Plant Optimization Algorithm for Feature Selection," in *IEEE Access*, vol. 11, pp. 94227-94251, 2023, doi: 10.1109/ACCESS.2023.3312022.
76. X. Wang, R. -P. Liu, X. Wang and F. Bouffard, "A Comparative Study of Polynomial Chaos Expansion-Based Methods for Global Sensitivity Analysis in Power System Uncertainty Control," in *IEEE Transactions on Circuits and Systems II: Express Briefs*, doi: 10.1109/TCSII.2023.3295805.
77. B. Phadermrod and V. Attasena, "Improving Lactation Curve Prediction by Incorporating Weather and Cow Behaviour," *2023 Joint International Conference on Digital Arts, Media and Technology with ECTI Northern Section Conference on Electrical, Electronics, Computer and Telecommunications*

Engineering (ECTI DAMT & NCON), Phuket, Thailand, 2023, pp. 219-224, doi: 10.1109/ECTIDAMTNCON57770.2023.10139578.
78. M. K. Gourisaria, V. Singh, R. Chatterjee, S. K. Panda, M. R. Pradhan and B. Acharya, "PneuNetV1: A Deep Neural Network for Classification of Pneumothorax Using CXR Images," in *IEEE Access*, vol. 11, pp. 65028-65042, 2023, doi: 10.1109/ACCESS.2023.3289842.
79. Y. Zhang and N. M. Freris, "Adaptive Filter Pruning via Sensitivity Feedback," in *IEEE Transactions on Neural Networks and Learning Systems*, doi: 10.1109/TNNLS.2023.3246263.
80. A. Mirheli and L. Hajibabai, "Charging Network Design and Service Pricing for Electric Vehicles with User-Equilibrium Decisions," in *IEEE Transactions on Intelligent Transportation Systems*, vol. 24, no. 3, pp. 2888-2902, March 2023, doi: 10.1109/TITS.2022.3227888.
81. Y. Li and Z. Li, "Bi-Objective Optimization for Multi-Row Facility Layout Problem Integrating Automated Guided Vehicle Path," in *IEEE Access*, vol. 11, pp. 55954-55964, 2023, doi: 10.1109/ACCESS.2023.3281554.
82. S. A. Asad Rizvi, A. J. Pertzborn and Z. Lin, "Development of a Bias Compensating Q-Learning Controller for a Multi-Zone HVAC Facility," in *IEEE/CAA Journal of Automatica Sinica*, vol. 10, no. 8, pp. 1704-1715, August 2023, doi: 10.1109/JAS.2023.123624.
83. L. Hao and Y. Xu, "Semisupervised Learning-Based Occupancy Estimation for Real-Time Energy Management Using Ambient Data," in *IEEE Internet of Things Journal*, vol. 10, no. 20, pp. 18426-18437, 15 Oct.15, 2023, doi: 10.1109/JIOT.2023.3280361.
84. N. Manikandan, S. N. Tadiboina, M. S. Khan, V. P. R, R. Singh and K. K. Gupta, "Automation of Smart Home for the Wellbeing of Elders Using Empirical Big Data Analysis," *2023 3rd International Conference on Advance Computing and Innovative Technologies in Engineering (ICACITE)*, Greater Noida, India, 2023, pp. 1164-1168, doi: 10.1109/ICACITE57410.2023.10183103.
85. P. Wu, W. Wang, F. Chang, C. Liu and B. Wang, "DSS-Net: Dynamic Self-Supervised Network for Video Anomaly Detection," in *IEEE Transactions on Multimedia*, doi: 10.1109/TMM.2023.3292596.
86. E. M. Pechlivani, G. Gkogkos, N. Giakoumoglou, I. Hadjigeorgiou and D. Tzovaras, "Towards Sustainable Farming: A Robust Decision Support System's Architecture for Agriculture 4.0," *2023 24th International Conference on Digital Signal Processing (DSP)*, Rhodes (Rodos), Greece, 2023, pp. 1-5, doi: 10.1109/DSP58604.2023.10167922.
87. M. C. S. Irawan, A. Purnomo, A. N. Sanjaya, Meiryani, S. Ubud and F. I. Maulana, "Global Patent Landscape of Decision Support System in the Business: An Overview," *2023 International Conference on Information Management and Technology (ICIMTech)*, Malang, Indonesia, 2023, pp. 464-469, doi: 10.1109/ICIMTech59029.2023.10277829.
88. E. Sun, Z. Xiao, F. Yuan, Z. Wang, G. Ma and J. Liu, "Method of Classified Counting of Mixed Breeding Chickens Based on YOLOV5," *2023 42nd*

Chinese Control Conference (CCC), Tianjin, China, 2023, pp. 7644-7649, doi: 10.23919/CCC58697.2023.10240449.

89. F. -A. Croitoru, V. Hondru, R. T. Ionescu and M. Shah, "Diffusion Models in Vision: A Survey," in *IEEE Transactions on Pattern Analysis and Machine Intelligence*, vol. 45, no. 9, pp. 10850-10869, 1 Sept. 2023, doi: 10.1109/TPAMI.2023.3261988.
90. X. Huang, Z. Hu, Y. Qiao and S. Sukkarieh, "Deep Learning-Based Cow Tail Detection and Tracking for Precision Livestock Farming," in *IEEE/ASME Transactions on Mechatronics*, vol. 28, no. 3, pp. 1213-1221, June 2023, doi: 10.1109/TMECH.2022.3175377.
91. Y. Guo, W. Hong, J. Wu, X. Huang, Y. Qiao and H. Kong, "Vision-Based Cow Tracking and Feeding Monitoring for Autonomous Livestock Farming: The YOLOv5s-CA+DeepSORT-Vision Transformer," in *IEEE Robotics & Automation Magazine*, doi: 10.1109/MRA.2023.3310857.
92. J. Ramya and M. Sankar, "Crop Recommendation Using Supervised Learning Techniques," *2023 4th International Conference on Smart Electronics and Communication (ICOSEC)*, Trichy, India, 2023, pp. 1076-1084, doi: 10.1109/ICOSEC58147.2023.10276199.
93. Y. Oktarina, Z. Nawawi, B. Y. Suprapto and T. Dewi, "Digitized Smart Solar Powered Agriculture Implementation in Palembang, South Sumatra," *2023 10th International Conference on Electrical Engineering, Computer Science and Informatics (EECSI)*, Palembang, Indonesia, 2023, pp. 60-65, doi: 10.1109/EECSI59885.2023.10295805.
94. M. Satoh, A. Bhat, T. Usami and K. Tatsumi, "A Multi-Purpose Autonomous Mobile Robot as a Part of Agricultural Decision Support Systems," *2023 IEEE 19th International Conference on Automation Science and Engineering (CASE)*, Auckland, New Zealand, 2023, pp. 1-7, doi: 10.1109/CASE56687.2023.10260483.
95. R. E. Mohamed, R. Hunjet, S. Elsayed and H. Abbass, "Connectivity-Aware Particle Swarm Optimisation for Swarm Shepherding," in *IEEE Transactions on Emerging Topics in Computational Intelligence*, vol. 7, no. 3, pp. 661-683, June 2023, doi: 10.1109/TETCI.2022.3195178.
96. H. Zhao, Y. Ding, L. Meng, Z. Qin, F. Yang and A. Li, "Bayesian Multiple Linear Regression and New Modeling Paradigm for Structural Deflection Robust to Data Time Lag and Abnormal Signal," in *IEEE Sensors Journal*, vol. 23, no. 17, pp. 19635-19647, 1 Sept. 1, 2023, doi: 10.1109/JSEN.2023.3294912.
97. T. Pang, H. J. T. Suh, L. Yang and R. Tedrake, "Global Planning for Contact-Rich Manipulation via Local Smoothing of Quasi-Dynamic Contact Models," in *IEEE Transactions on Robotics*, doi: 10.1109/TRO.2023.3300230.
98. S. S. Shinde and D. Tarchi, "A Markov Decision Process Solution for Energy-Saving Network Selection and Computation Offloading in Vehicular Networks," in *IEEE Transactions on Vehicular Technology*, vol. 72, no. 9, pp. 12031-12046, Sept. 2023, doi: 10.1109/TVT.2023.3264504.
99. H. R. Chi *et al.*, "Multi-Criteria Dynamic Service Migration for Ultra-Large-Scale Edge Computing Networks," in *IEEE Transactions on Industrial*

Informatics, vol. 19, no. 11, pp. 11115-11127, Nov. 2023, doi: 10.1109/TII.2023.3244321.

100. D. DeAlcala, I. Serna, A. Morales, J. Fierrez and J. Ortega-Garcia, "Measuring Bias in AI Models: An Statistical Approach Introducing N-Sigma," 2023 *IEEE 47th Annual Computers, Software, and Applications Conference (COMPSAC)*, Torino, Italy, 2023, pp. 1167-1172, doi: 10.1109/COMPSAC57700.2023.00176.
101. L. S. Martinez-Rau, V. Adın, L. L. Giovanini, B. Oelmann and S. Bader, "Real-Time Acoustic Monitoring of Foraging Behavior of Grazing Cattle Using Low-Power Embedded Devices," *2023 IEEE Sensors Applications Symposium (SAS)*, Ottawa, ON, Canada, 2023, pp. 01-06, doi: 10.1109/SAS58821.2023.10254175.
102. Y. Lai, Y. Yang, Z. Ma, Q. Huang, S. Wang and Z. Luo, "Development of Simulation Prediction Techniques for Low Frequency Emissions," *2023 IEEE Applied Power Electronics Conference and Exposition (APEC)*, Orlando, FL, USA, 2023, pp. 2690-2696, doi: 10.1109/APEC43580.2023.10131625.
103. L. C. Voumik, R. Karthik, A. Ramamoorthy and A. Dutta, "A Study on Mathematics Modeling using Fuzzy Logic and Artificial Neural Network for Medical Decision Making System," *2023 International Conference on Computational Intelligence and Sustainable Engineering Solutions (CISES)*, Greater Noida, India, 2023, pp. 492-498, doi: 10.1109/CISES58720.2023.10183534.
104. X. -F. Qin and J. -X. Shen, "Mathematical Modeling of High-Speed PMSM Considering Rotor Eddy Current Reaction Effect," in *IEEE Transactions on Energy Conversion*, doi: 10.1109/TEC.2023.3288608.
105. F. -Y. Wang, J. Yang, X. Wang, J. Li and Q. -L. Han, "Chat with ChatGPT on Industry 5.0: Learning and Decision-Making for Intelligent Industries," in *IEEE/CAA Journal of Automatica Sinica*, vol. 10, no. 4, pp. 831-834, April 2023, doi: 10.1109/JAS.2023.123552.

5

Modeling the Economic Impact of Agricultural Regulations: A Case Study on Environmental Compliance Costs

Vikesh Rami[1], Sunil Kumar[2], Gautham Krishna[3], Abhinav[4], Sukhvinder Singh Dari[5] and Dharmesh Dhabliya[6]*

[1]Department of Agricultural Economics, College of Agriculture, Parul University, Limda, Dist- Vadodara, Gujarat, India
[2]College of Agriculture Sciences, Teerthanker Mahaveer University, Moradabad, Uttar Pradesh, India
[3]Department of Civil Engineering, Faculty of Engineering and Technology, JAIN (Deemed-to-be University), Karnataka, India
[4]Department of Agriculture, Vivekananda Global University, Jaipur, Rajasthan, India
[5]Symbiosis Law School, Nagpur Campus, Symbiosis International (Deemed University), Pune, India
[6]Department of Information Technology, Vishwakarma Institute of Information Technology, Pune, Maharashtra, India

Abstract
Some sizable sections of the primary Everglades which were directed and depleted for evolution are the Everglades Agricultural Area (EAA). The Everglades National Park and other downstream areas get nutrients, particularly phosphorus, from the agricultural fields in the EAA's drainage waters. Agriculturists in the EAA are taking part in a significant initiative to put farming administration techniques into place which reduce the amount of nutrients that are ascribed to their discharge waters. Studies conducted at the farm and basin scales revealed that the main source of supplied particulate phosphorus (PP) is biotic substance developing

**Corresponding author*: dharmesh.dhabliya@viit.ac.in
Sunil Kumar: ORCID: https://orcid.org/0009-0004-9975-5506
Gautham Krishna: ORCID: https://orcid.org/0000-0001-6257-3056
Abhinav: ORCID: https://orcid.org/0009-0009-8115-5808

Sabyasachi Pramanik, Niranjanamurthy M., Ankur Gupta and Ahmed J. Obaid (eds.) Mathematical Modeling in Agriculture, (81–102) © 2024 Scrivener Publishing LLC

in agricultural canals. It went against earlier theories that the main cause of PP was soil erosion. Different control strategies may be economically evaluated using computer modeling of PP transport. A statistical framework which properly captures the geographical and earthly variability of the biologically derived dissolved solids found in the EAA is necessary for such an endeavor. This study proposes a mathematical model with lumped parameters that represents the sourcing, mobilization, transportation, and demobilization of organic content on farms that contains phosphorus. The 560-ha sugarcane field in the EAA served as the model's calibration and validation site. It was then put to use in assessing a number of common management techniques.

***Keywords*:** Drainage, particulate phosphorus, water grade, everglades agricultural area

5.1 Introduction

Ecological pressure on the South Florida Everglades, a natural resource that is special in the US, was present for a good many decades. The Kissimmee River basin in the present, Lake Okeechobee, and the southernmost point of the Florida peninsula all formed the primary Everglades watershed (Figure 5.1). It had over 20,000 km^2 of highland and marsh area. The historical flow proceeded at relatively lower velocity from north to south over a riverbed which was often 85 to 112 km width, resulting from a practically flat land slope of 8 to 17 cm km^{-1}. The last release was made into the seas at Florida's southernmost point. Large sections of the ancient Everglades were directed and evacuated for expansion beginning in the late 1800s. There exist more over 2,452 km of significant canals and levees assisting business, communities, and farming. This area includes the Everglades Agricultural Area (EAA).

The EAA is situated south of Lake Okeechobee (Figure 5.1). This 300,000-ha tract's soils, which are mostly histosols, are underlain by marl and limestone. According to [1], 150,000 hectares are sowed with sugarcane, while 39,000 ha are sowed with sod, vegetables, and rice. Open farm ditches are used to regulate water tables and accomplish farm water management. Rainfall varies greatly by season and is usually heavy. To reduce the water levels in open ditches, big volume, low head axial flow pumps are used almost exclusively for drainage. Extreme changes in flow may occur in the drainage and irrigation networks, moving from no flow to maximum flow quickly. The Everglades National Park, the South Florida coastal estuaries, or the Water Conservation Areas (WCAs) are where the majority of drainage water from the EAA is finally dumped.

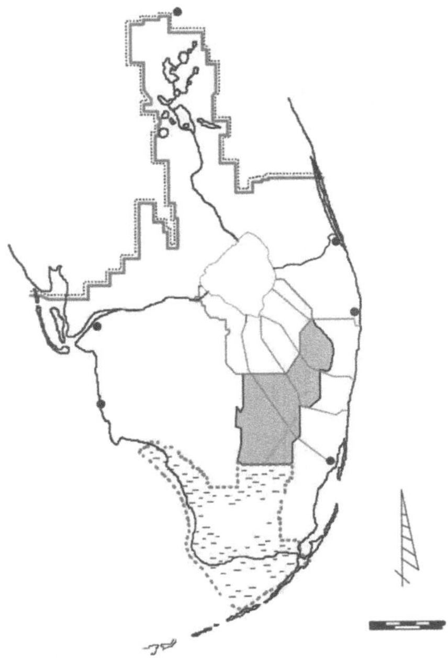

Figure 5.1 The Everglades Location illustrates the EAA in relation to the different geographical elements and South Florida's ecologically sensitive regions.

Agricultural drainage water contains more nutrients than the initial flows that the Everglades developed under. According to [2], one of the reasons for ecological alterations in the WCAs and the ENP is this enrichment, notably with regard to phosphorus (P). Best management practices (BMPs) must be used by local producers as part of the cleanup plans now in place, and many controlled wetlands must be built inside the EAA.

The model shown here is customized for the unique conditions of the agricultural canals in the EAA. It acknowledges time-variable biomass yield and P constituent features which were seen at the field scale and incorporate the idea of concurrent removal/erosion and settling of organically derived substance in a dynamic system. Its main use is to imitate PP discharge within a narrow range of pumping circumstances.

5.2 Mechanisms Study Time and Location

Eight farms are being studied by the University of Florida inside the EAA to gauge the effectiveness of BMPs. A 530-ha sugarcane field having two major canals and a pump station served as the study's setting (Figure 5.2).

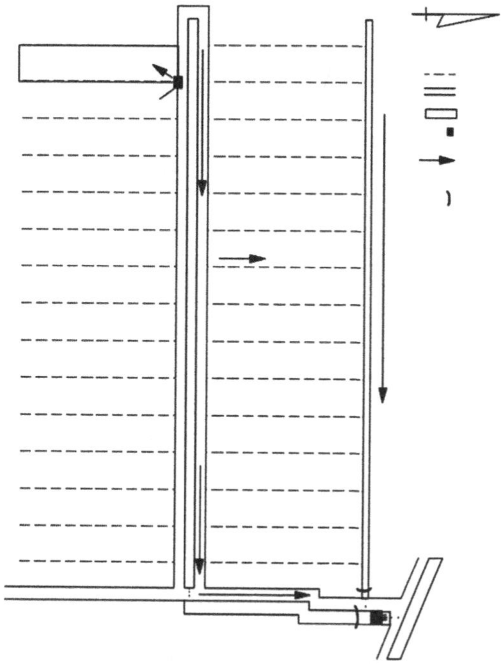

Figure 5.2 The overall flow paths, BMP structures, and the hydraulic system configuration.

This is an example of a medium-sized to big, actively maintained sugarcane farm with rather deep soils. The primary pump station at the field has single-speed electric pumps having on-off controls which set largest and least water levels in the intake sump. The property was selected as the target farm for model testing due to its consistency of functioning and straightforward design.

The bulk of discharge-pumping activities occur in South Florida during the rainy season, which lasts from April to October. In a year when the total amount of precipitation through October was within 6% of the 72-year average, six rainy season occurrences were sampled for model calibration. There are two types of phosphorus: insoluble forms (minerals and particulate organic) and soluble forms (orthophosphate and soluble organic). Particle transfer of P is also possible and has received less attention than the transport of soluble materials showed how much particulate phosphorus (PP) was exported. Two significant pumping episodes were created by tropical depression storms at the year's conclusion and sampled for model validation. The sampling was only possible for one water year due to project restrictions.

Additional research revealed that PP made from 25% to 75% of the entire amount of phosphorus (TP) exported from EAA farms and that PP export was typically the reason for increases in TP loading. The efficiency of conventional runoff management and sediment trap devices to lower PP loads has been examined in many EAA studies. These investigations either came to no conclusions or showed the inefficiency of the conventional approaches.

According to research on farm-scale PP transport by [3], the bulk of PP in the EAA comes through in-stream organic growth rather than soil erosion. They called this theory the "biological contribution mechanism" (BCM). According to their theory, a large portion of the released PP consists of currently sediment biomass, such as plankton, algae, and detritus from floating and bank-rooted water plants. If a weakly bound substance is released by turbulent shear pressures, floating or suspended plants may potentially contribute directly to the export of solids. The fact that exported particulate matter exhibited properties more like living plant material than field soils or fundamental canal accumulation at the farm scale substantially supported the BCM.

5.3 Sampling

ISCO automated water samples were used to collect pumping event discharge samples under the direction of Campbell Scientific CR10 data loggers. When the farm discharge pumps were running, the data loggers started recording pumping times and rates and started sampling the discharge water. After 40 mins of interval of pumping time, samples were taken at a height of 0.6 m beyond the base of the pump discharge sump, and they were combined during a 2-hr. pumping duration. These specimens were collected during the pumping process. Daily sample caddies were gathered during pumping and kept chilled until analysis.

5.4 Analysis, Both Physical and Chemical

Dry solids, volatile solids, and total suspended solids (TSS) were all measured using APHA Techniques 2540D, 2054B, and 2054E, respectively. According to ASTM Designations D4531-86 and D854-92, respectively, mass density and specific gravity were calculated. APHA Technique 4500-P E was used to measure the amount of soluble orthophosphate. The ignition/HCl digestion technique of Andersen was used to measure Particulate P.

Development of a Hydraulic Model

For their suitability for EAA transport modeling, a variety of hydraulic models for irrigation and drainage networks were examined. The European model, DUFLOW [4], was selected because it has many features with the EAA and has been widely used in Netherlands locales. The DUFLOW model has enough precision to simulate transients that happen during pump start-up and shutdown, and it can accurately depict flow in a network similar to EAA farms during the course of a pumping event. A subroutine for integrated water quality is also included. For this study, a groundwater flow component was introduced to the model.

1-D St. Venant [5] equations of continuity and motion that do not include terms for wind-induced flow—are the foundation of the DUFLOW program. They are as follows:

$$G\frac{\partial F}{\partial t}+\frac{\partial R}{\partial y}=0 \tag{5.1}$$

$$\frac{\partial R}{\partial t}+fB\frac{\partial F}{\partial y}+\frac{\partial}{\partial y}(\beta Ru)+\frac{f|R|R}{D^2BS}=0 \tag{5.2}$$

where,
R = flow (m^3 s^{-1})
f = acceleration due to gravity (m s^{-2})
B = cross-sectional area (m^2)
G = cross-sectional storage width (m)
F = water level relative to reference level (m)
β = flow non-uniformity correction factor
u = mean velocity over cross-sectional area (m s^{-1})
D = coefficient of Chezy (m$^{1/2}$ s^{-1})
S = hydraulic radius (m)
y = longitudinal distance (m)
t = time (s).

Gaussian elimination is used in computation to solve the equations using a 4-point implicit Preissmann technique. This approach permits the inclusion of any network or loop design. When there are internal and boundary conditions, such as weirs and gates, this strategy works effectively. The user enters the initial flows and levels, channel measurements, boundary conditions, and hydraulic parameters for interior structures.

A quasi-steady-state method was used to simulate groundwater drainage to the field ditches. It included an algorithm which used the Hooghoudt equation together with a user-specified water balance for evapotranspiration and rainfall.

5.5 Module for Water Quality

The one-dimensional transport equation serves as the foundation for the DUFLOW water quality module:

$$\frac{\partial(BD)}{\partial t} = -\frac{\partial(RD)}{\partial y} + \frac{\partial}{\partial y}(B_g E \frac{\partial D}{\partial y}) + K \tag{5.3}$$

where D = elemental aggregation (g m^{-3})
R = flow (m^3 s^{-1})
B = cross-sectional dimension (m^2)
B_g = cross–sectional flow dimension (m^2)
E = dispersion coefficient (m^2 s^{-1})
y = longitudinal distance (m) t = time (s)
K = yield or expenditure of the element per unit sectional length (g m^{-1} s^{-1}).

The transport parameter (S) that is the amount of the component moving across a cross-sectional dimension per unit of time (g s^{-1}), must be introduced in order to solve equation 5.3 numerically.

The production term (J) is described in the DUFLOW water quality subroutine. A user-created external model file is accepted by the computational package. Because this file is constructed such that the program reads its output as input at every time step, changing the quality subroutine doesn't require recompiling the whole program.

5.6 Particulate Phosphorus and Suspended Solids

According to fieldwork previously published, the system TSS may have many origins. The idea of erosion gains context when hydraulically dislodgable plant debris from floating plants is included. Erosion [6] now occurs at the surface of both the channel's silt and water. Uncertainty in source locations and amounts is introduced by the plant population's mobility, variability, and detrital P content. To create an explicit distributed-parameter PP delivery model, further sampling and testing are needed across many populations with different erosion and sedimentation features. The choice was made to create a physically dependent lumped-parameter transport framework, where erosion and sedimentation parameters will be fit to the available data as calibration parameters. This was motivated by the need to begin with a simple first approximation and the complexity of a distributed-parameter framework.

According to [7], flow experiments of a prototype organic sediment revealed that the organic sediments had few characteristics with cohesive sediments. Following the achievement of a critical shear stress on the sediment bed, cohesive sediment erosion is a function of shear stress. The re-deposition rate [8] is a function of concentration in the water column and the likelihood that a particle will be captured upon collision with the bed. Eroded material is vertically movable in the water column. In this study, general mathematical forms that describe cohesive sediment erosion and deposition were modified for use with organic sediments.

The Appendix displays the framework's derivation and assumptions. The ultimate appearance is composed of two linked linear differential equations which have the same three fundamental parameters: the sedimentation velocity, the erosion coefficient, and the critical velocity for erosion. A starting determination of the concentration of erodible and suspended constituent is necessary for the equations to be solved.

5.7 Calculation of PP

The field data revealed a significant negative association between the TSS [9] concentrations and P content. According to [10], the most plausible explanation for this is that most of the PP comes from biological constituent [11] that typically has a lesser specific gravity and a larger organic constituent as its feasibility rises. Lower TSS concentrations were discovered in the field under the normal benign hydraulic [12] conditions. The major feasible, light substance having the greatest P concentration was these suspended solids. The mean P concentration of the TSS was successfully diluted as hydraulic intensity rose because denser materials having lesser viability and P constituent were dissolved.

Additionally, it was shown that specimens collected prior to an event, namely in the first four hours, had a slightly greater P concentration than ones collected afterwards. The link between the P concentration of the dissolved particles and TSS for each of the six calibration events is shown

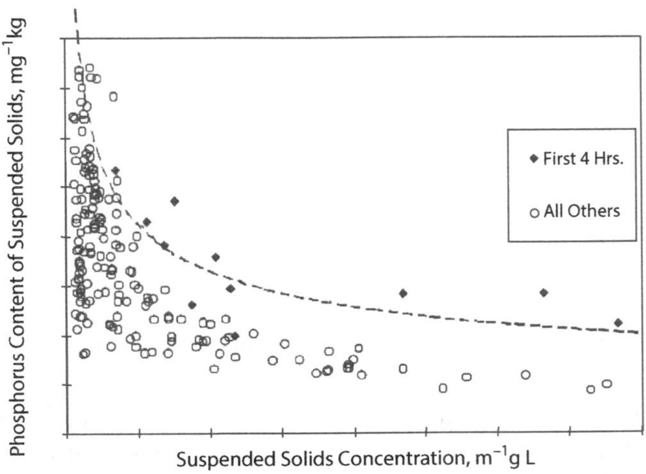

Figure 5.3 Comparison of the overall phosphorus constituent to the cumulative concentrations of suspended particles.

in Figure 5.3, with the early samples and the latter samples being correlated independently. The discharge PP concentrations were calculated in the model as a function of the released TSS concentrations using these relationships.

5.8 Model Caliphy

While the amount of P exported is associated by the formerly stated correlations, the bulk of released TSS is calculated instantly (as the product of flow × concentration). Because of this, the model's calibration events were based on cumulative TSS load. Cumulative PP loads were then computed using the correlations.

Early simulations revealed that TSS concentration changed quickly during pump start-up, shallow channel depth periods, and the discharge pump's frequent on-off cycles. The real data had a two-hour temporal resolution, but succeeding simulations were done with 1-5 minute time intervals to encompass these transients.

An analysis of the data revealed a tendency of rising exported solids with an expanding time between the conclusion of one event's pumping and the beginning of the next event's pumping. The connection between inter-event time and CEM(0) was assumed to be monotonically growing in order to start calibration. Finding the settings for the beginning condition

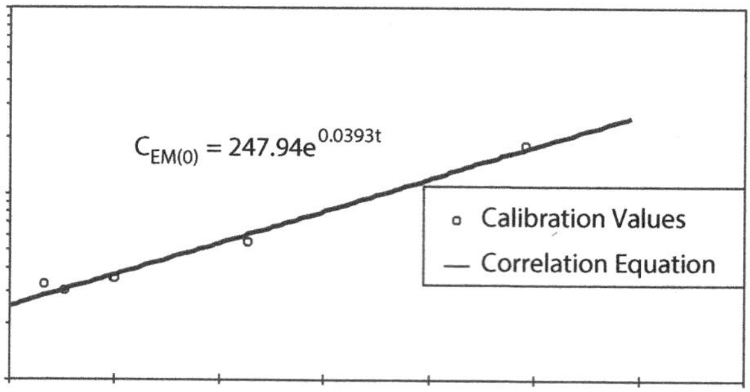

Figure 5.4 CEM(0) correlation as an equation for exponential growth.

(CEM(0)) and the three fundamental parameters (ε, Vc, and wsed) which would provide the greatest match of sum of TSS loads for every six consecutive calibration events was part of the calibration procedure. With the restriction that no single event difference exceed 5%, the calibration criteria used was the reduction of the difference between simulated and noticed cumulative TSS loads for the cumulative of the six calibration events. These best-fit parameters were utilized in the model validation process to hind cast the outcomes of the two tropical depression-based events that occurred in December of the research year.

Figure 5.4 illustrates an exponential growth equation that accurately characterized the calibration estimates of the uniform starting distribution of erodible mass (CEM(0)), with inter-event time serving as the independent variable.

Result and Discussion

Validation and Calibration of Models

The pumping profiles went through many stages as a result of level restrictions in the agricultural canal. High flow rates were seen at first, but they eventually decreased as the major canal level and the pump's suction head fell. If the canal got to the lower-level control depth, the pumping halted. Typical flow patterns and TSS concentrations are shown in Figure 5.5, specifically for Event 237. Events are identified by the day of the year that they usually begin. There are traits in the sample profile that apply to every occurrence. The first pumping cycle was characterized by a "U" shape in the TSS profiles. As the cycle continued, concentrations rose at first, fell down quickly, and then surged once again.

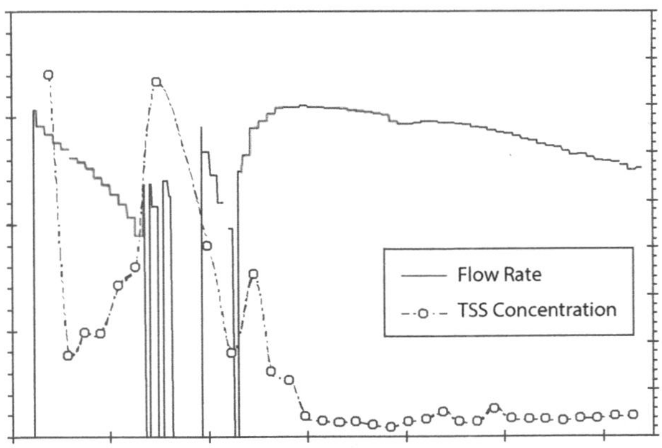

Figure 5.5 Event 237 flow and cumulative dissolved solids materials.

The TSS concentrations seemed to vary irregularly throughout the on-off intervals that followed the first pumping session.

The implemented and actual TSS profiles for Event 237 are shown in Figure 5.6. The stair step pattern results from the noticed profiles showing the sample concentration to be a constant across the sampling duration. The "U" form was accurately approximated by the model for each calibration event. Additionally, it rather accurately replicated the TSS variations throughout the on-off times. The model demonstrated that the durations and intervals between the on-off cycles were responsible for the apparently unpredictable TSS variations.

Figure 5.7 displays the cumulative PP export for Event 237 that was computed from the TSS export utilizing the PP-TSS correlation [13]. The parameter of most significance is the phosphorus load on the receiving stream; hence the PP export is shown in sum load form. For each calibration event, the model accurately predicted the form of the sum PP load curves.

The calibrated framework simulated a load of 24,800 kg, while the total TSS load for every six incidents was 25,894 kg. The framework predicted [14] a weight of 80.1 kg, whereas the actual summation of PP load was 88.7 kg. Cumulative TSS load was overestimated by 2.5%, while summation of PP load was underestimated by 8.2% in the calibrated model. For TSS and PP, respectively, the calibration period implementation efficiency (variance) was 82.4% and 74.5%.

The inter-event temporal correlation depicted in Figure 5.4 was used to determine the starting concentrations of erodible mass for model validation. Event 319 and Event 355 simulations were conducted using the

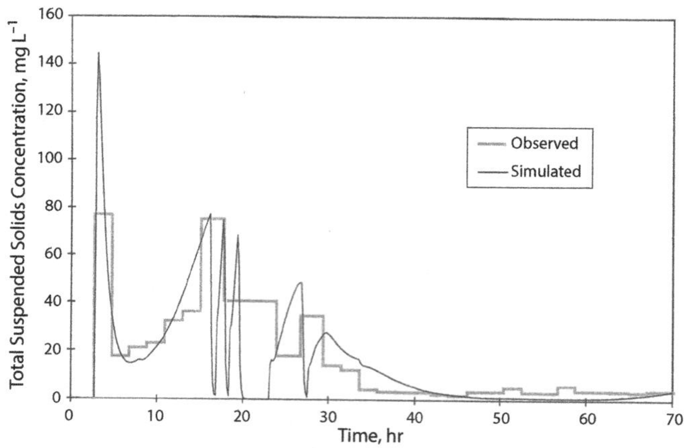

Figure 5.6 Implemented and noticed cumulative dissolved solids materials for Event 237.

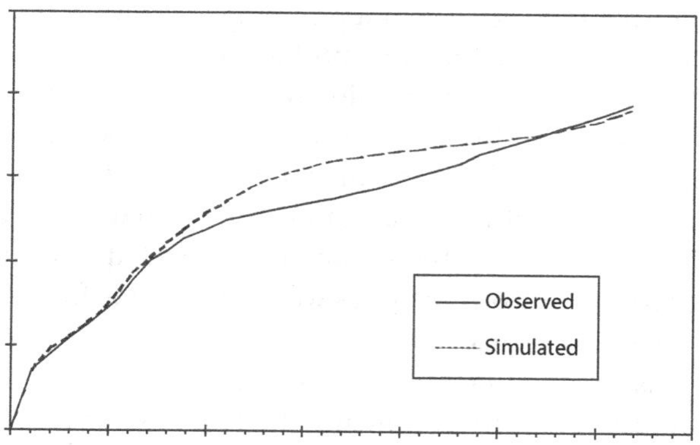

Figure 5.7 Summation of PP load for Event 237.

calibration parameters. The implemented and actual TSS assemblage for Event 355 is shown in Figure 5.8. Figure 5.9 displays the simulated and actual sum of PP loads for the similar occurrence. Identical outcomes were seen with Event 319. Although there were minor mismatches between the simulated and actual timing of delivery of the loading, as shown in Figure 5.9, the final loads for every event were accurately implemented. The erodible mass in the model is initially distributed uniformly. In the comparably small EAA farm canals under the tropical depression circumstances which followed Events 319 and 355, the actual distribution was probably not uniform. Gusty winds move floating aquatic herbs, and they can also create currents which are driven by the wind and move mobile sediments. The cumulative measured TSS load during the simulation periods of the two

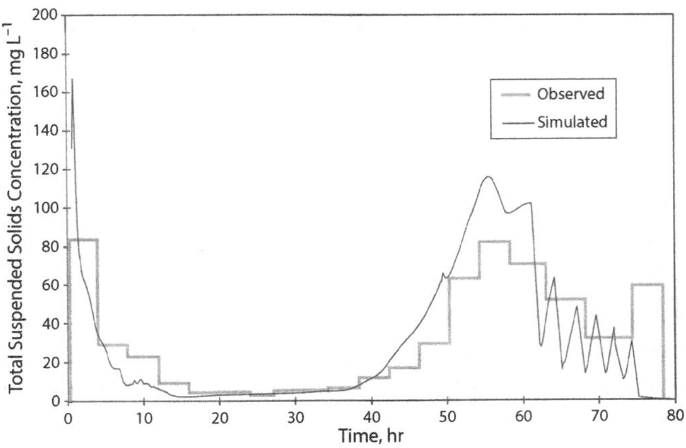

Figure 5.8 Implemented and noticed cumulative dissolved solids materials for Event 355.

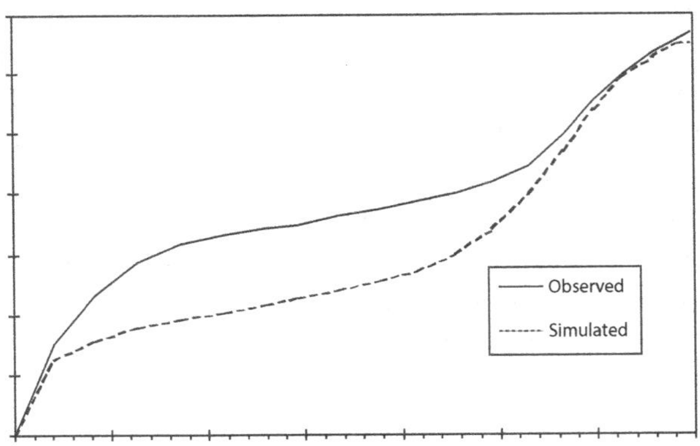

Figure 5.9 Sum of PP load for Event 355.

incidents was 25,230 kg, compared to a predicted load of 25,683 kg, a 2.9% overestimation of cumulative load. The 3.4% overestimation seen for the total TSS loads for the six calibration events was practically comparable.

The total measured PP load during the same time period was 110.9 kg, which was 11.3% more than the predicted load of 98.3 kg. This contrasts very well with the overall underestimation for the calibration period, which was 7.8%. For TSS and PP, respectively, the validation period simulation efficiency was 86.2% and 73.5%. These reflect the calibration findings similarly. For Events 319 and 355, the isolated PP implementation efficiency was 74.2% and 68.5%, respectively. The accuracy and variance explanation of the findings from the validation simulations were comparable to those from the calibration simulations, proving that the validation was effective.

5.9 Scientifications Described by the Model

Multiple mechanisms are superimposed to explain the form of the TSS curve. The amount of transportable biomass increases in the interim between events. In Figure 5.6, the flow dimension narrowing at the pump inlet results in a hike in velocity. Cumulative dissolved solid forms for Event 238, the area directly upstream of the pump, were simulated and observed. High initial concentrations result from the assembly of solids which have been collected in the area between episodes brought on by the velocity increase. The quick assembly and exhaustion of transported materials in the vicinity of the pumping station causes the first dropping leg of the U-shaped curve.

Particulate matter moves downstream in a wave-like form due to concurrent erosion and deposition upstream.

The implementation of TSS and erodible mass, respectively, for a few chosen upstream places in the east canal of the farm are shown in Figures 5.10 and 5.11. Every site experiences a maximum in total suspended solids (TSS) concentration, which becomes more pronounced the farther downstream the point, is. In a given volume of water flowing downstream, the concentration rises with time as long as cumulative erosion outpaces cumulative sedimentation. Because of intake from field ditches, the major canals' velocity increases with distance downstream, increasing the pushing power for erosion as position goes downstream. TSS concentration tends to rise as a result of several circumstances.

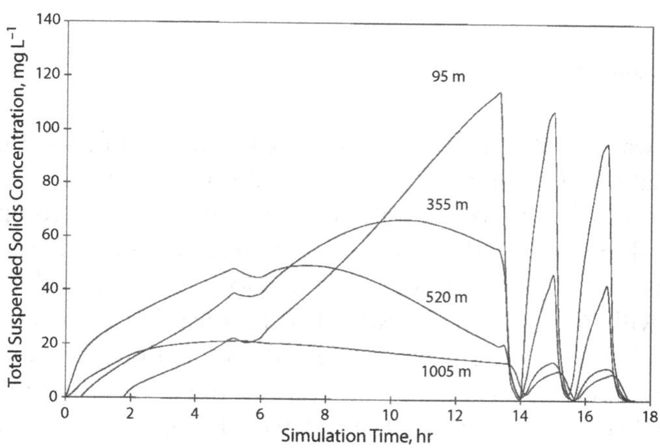

Figure 5.10 Dissolved solids implementation at particular positions for Event 237.

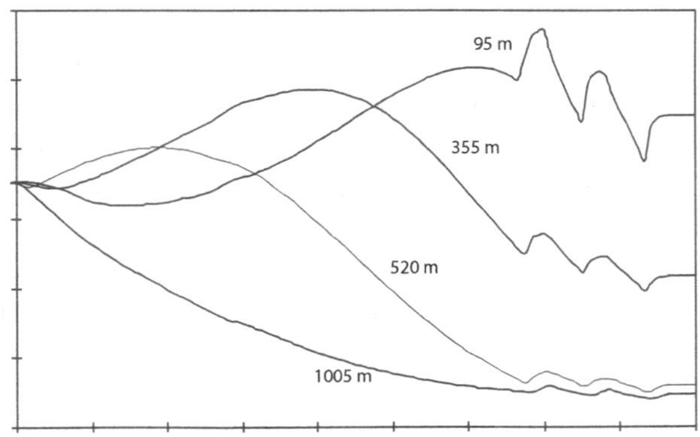

Figure 5.11 Erodible mass implementation at chosen positions for Event 237.

The erodible mass loss in the channel opposes these processes. This often lessens the suspension of solids. Figure 5.11 provides a vivid illustration of the process, showing how erosion outpaces sedimentation [15] at the simulated point furthest upstream, causing that region to lose erodible mass.

Because the net deposition of solids from upstream surpasses the erosion of solids at that place, the corroded mass increases first at the locations farther downstream. When the amount of sediment material available upstream decline then the amount of erosive mass begins to decline. The TSS concentration rises and then falls at each site as a result of the erodible mass increasing and then decreasing. The TSS concentration at the downstream site that was the farthest away increased steadily and did not peak prior to the pump being turned off. The U-shaped curve's [16] ascending leg is the result of this. The pump's on-off cycles were the cause of the TSS concentrations' sharp rises and falls after the first pumping time. When the pump began up, the particulate matter quickly re-suspended and settled once the pump was stopped.

The TSS concentration maxima were associated with solids re-suspension into the water column. The deposition of materials from the water column was correlated with the maxima in the concentrations of corroded mass. The model provided a qualitative explanation for the seemingly irregular values of the TSS concentrations that were noticed in pump cycling. It was shown that the pump run duration and the amount of time between pump cycles were factors of the discharge concentrations during cycling. This demonstrates the error-proneness of grab-sampling techniques when there are sudden fluctuations in flow.

Between events, the BCM encourages the buildup of transportable material in the farm channels. With a starting corroded mass of 1,920 g m^{-2}, Event 220 had an inter–event period of 49.1 days. Event 258 had a starting erodible mass of 315 g m^{-2} and an inter–event period of 5.5 days. For the first ten hours of pumping, the discharge rates from both events were almost the same. The implemented and measured TSS depositions for both events' first pumping phases are shown in Figure 5.12. This figure clearly shows the impact of the lengthy inter-process interval and the larger accumulation of corroded mass for Event 220; the increased mass level in Event 220 led to an increase in TSS released. Compared to Event 258 the TSS concentration U-shaped curve for Event 220 was much greater. Both legs of the curve showed a very noticeable difference. It seems that the model can be utilized to estimate the significant impact of changes in administrative practices at the site since it generated simulations that were appropriate for the calibrated location. Though the framework was only calibrated for the research location, the site's simulation findings may serve as a qualitative manual for managing other farms and could assist in determining the research goals for the future. Here are some instances of these applications.

5.10 Simulation of Sediment Trap

As was previously mentioned, tests conducted in the EAA on typical sediment traps revealed that they were inefficient in lowering TSS discharge. Dredging organic silt from a canal's lower reaches allowed for the construction of these traps without significantly altering the canal's depth.

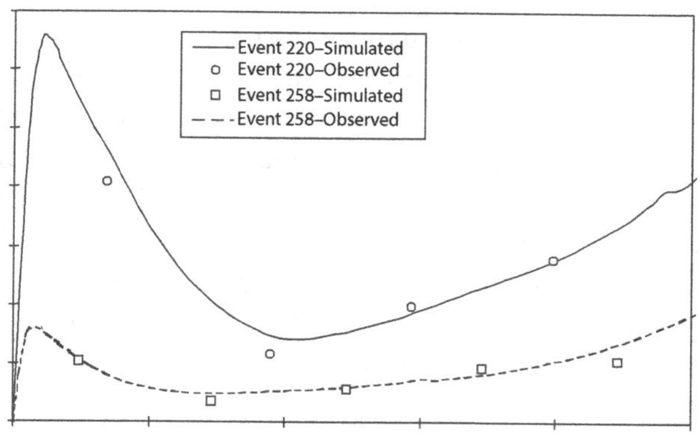

Figure 5.12 Implementation of Events 220 and 258.

At the endpoints of the dredged sections, walls were inserted into the canal bottom. In the segment 750 m upstream of the pump, it was mimicked by keeping the starting accumulation of erodible mass to 0. During Event 220, the implementation was run for the first 150,230 m^3 of release. In this simulation, 2,990 kg of TSS were released, while 10,608 kg were released within the same time frame during the real incident. Although it is just temporary, this seems to be a remarkable decrease. Figure 5.13 shows the implemented accumulation of erodible mass at certain points as the process goes on. The concentration at the initial zero mass sites in the sediment trap rapidly grew due to the assembly of mass from upstream. The mass concentration at the top termination of the trap was almost equivalent to the concentration in the segment directly upstream by the eleventh hour of the event. Therefore, the trap's only function was to momentarily impede the movement of erodible mass. The trap began to function during the twelfth hour, just as it would have if the original concentration of erodible mass had been much larger. The trap only offered a transient advantage in these conditions, and it vanished after a single or two pumping episodes.

Conventional deposition traps are intended to catch materials that fall or saltate as they travel along the foot of the channel bed. Rather than moving as a bed load, the deposits in this framework advance due to corrosion, suspension, and depositing. The ineffectiveness of deposit traps examined on EAA fields may be explained by this model of sediment behavior. Due to erodible mass motion shown in the implementation, there is no long-term impact until the uncovering of the trap is sufficiently intense to significantly reduce the water velocity in the channel. In the EAA, notable deepness enlargements are not usually cheaply possible. Already, canals have been excavated to or below the limerock-soil boundary. It would be necessary to blast and remove the underlying limerock in order to reach further depths. According to the model, sediment trapping requires a significant enough shift in the cross-sectional dimension of the channel to lower the velocity to the point where it doesn't get close to the critical velocity for transportation. Thereby, till the velocity is maintained less than critical, TSS sedimentation [17] will surpass erosion and sediment collecting will continue.

Examining equations 6 through 9 makes this clear, and a simulation is not needed to illustrate it. During the calibration procedure, the critical velocity was found to be 0.096 m s^{-1}. In Event 220, highest velocities were on the order of 0.36 m s^{-1}. This comparison demonstrates that ahead of the channel started to function as a deposition trap for light, wooly materials, the cross-sectional area would need to be increased by an estimated three to four times. Based on hydraulic considerations, it is advisable to redistribute the flow at the bigger channel's entrance in order to minimize jetting at the channel hub and make efficient utilization of the extra breadth.

5.11 Pumping Profile Modifications Simulation

Because there is a threshold velocity in transportation and erosion rates depend exponentially on channel velocity, reducing the maximum flow velocity must decrease the export of TSS. The study farm's pumping method involves operating the biggest pump at its highest pace while maintaining on-off level management [18]. This results in times when there is no flow followed by times when the flow rate is significant. Pumping at a lower constant rate is an option that will be used to discharge similar capacity in similar amount of time without turning the pump with time to time. In 40 hours, Event 230 released 150,000 m^3, or a mean pumping rate of around 0.8 m^3 sec^{-1}. This constant discharge rate was used to imitate the event rather than the real pumping profile. The simulated TSS load for both the initial and continuous discharge conditions is shown in Figure 14. According to the simulation [19], TSS export decreased for the similar dispensed volume during the similar time interval, from 10,608 kg to 1,023 kg.

5.12 Conclusion

The model that is being given here was created for use in semi-tropical hydraulic conveyance systems with light, flocculent, organic particulate matter that contains P. The research farm's TSS movement was effectively replicated by the model in both typical wet season and anomalous tropical depression scenarios. It is good news because it validates the fundamental structure and underlying guesses of the framework that consists of the following: simultaneous erosion and deposition; scale of erosion to mass existing, intra-event biomass generation and exponential erosion dependence on channel velocity. As of right now, the framework is limited by the temporal and spatial constraints of the calibration data set. The coefficients of the association between P content and TSS can change depending on cropping and water handling techniques, according to data from different EAA farms. The rates of inter-event bioaccumulation, shown in Figure 5.4, may vary from year to year due to climatic variations and adjustments made to management strategies to regulate in-canal plant growth (Vol. 44(5): 1117–1126 1125). There is not enough information available at this time to make assumptions about how effectively the corrosion and assemblage parameters established for the survey field will address other fields in the EAA that are comparable. The framework is still under development and needs more thorough calibration tests conducted over longer times and at more places in order to become a reliable forecasting tool. After that, the framework must be able to be translated to the watershed scale using the transport ideas that were utilized to construct it. The model's applicability will be further expanded by testing it on various field hydraulic and cropping patterns and by expanding the TSS P constituent correlation to include site-based parameters like cropping techniques, canal volume turnover rates, and the amount of floating macrophyte coverage. The model, even in its early stages of development, can be utilized qualitatively to assess control techniques and provide substitutes that merit further research. The examples used to illustrate the shortcomings of traditional sediment trap construction and the potential advantages of altering pumping techniques were specifically selected to highlight prevalent present-day procedures that could be insufficient or harmful to the decrease of P exports. The model has been used to simulate beneficial alternative water management strategies (Stuck, 1996), and work is now being done to assess these strategies in real-world settings.

References

1. P. Pal, R. P. Sharma, S. Tripathi, C. Kumar and D. Ramesh, "NSGA-III Based Heterogeneous Transmission Range Selection for Node Deployment in IEEE 802.15.4 Infrastructure for Sugarcane and Rice Crop Monitoring in a Humid Sub-Tropical Region," in *IEEE Transactions on Wireless Communications*, vol. 22, no. 6, pp. 3643-3656, June 2023, doi: 10.1109/TWC.2022.3220146.
2. M. Thomas *et al.*, "A Framework to Assess Remote Sensing Algorithms for Satellite-Based Flood Index Insurance," in *IEEE Journal of Selected Topics in Applied Earth Observations and Remote Sensing*, vol. 16, pp. 2589-2604, 2023, doi: 10.1109/JSTARS.2023.3244098.
3. N. Ye, J. P. Walker, Y. Gao, I. PopStefanija and J. Hills, "Comparison Between Thermal-Optical and L-Band Passive Microwave Soil Moisture Remote Sensing at Farm Scales: Towards UAV-Based Near-Surface Soil Moisture Mapping," in *IEEE Journal of Selected Topics in Applied Earth Observations and Remote Sensing*, vol. 17, pp. 633-642, 2024, doi: 10.1109/JSTARS.2023.3329015.
4. N. Houhat, L. Saengdee, T. Devaux, S. Callé, S. B. Gauchet and F. Vander Meulen, "Ultrasonic Instrumentation for Simultaneous Rheological and Ultrasonic Measurements on a Rheometer," *2023 IEEE International*

Ultrasonics Symposium (IUS), Montreal, QC, Canada, 2023, pp. 1-3, doi: 10.1109/IUS51837.2023.10308367.

5. H. Jie and W. Hong, "Water Hammer Suppression with Variable Terminal Time via PDE-Constrained Dynamic Optimization," *2023 42nd Chinese Control Conference (CCC)*, Tianjin, China, 2023, pp. 1648-1653, doi: 10.23919/CCC58697.2023.10240378.
6. Z. Zheng, W. Ren and C. Zhang, "Visualization and Characterization of Cumulative Arc Erosion Behavior Using 2-D Microscopic Images," in *IEEE Transactions on Components, Packaging and Manufacturing Technology*, vol. 13, no. 1, pp. 34-44, Jan. 2023, doi: 10.1109/TCPMT.2023.3234884.
7. C. Le, M. Wu, S. Cai, N. Liu and J. He, "Remote Sensing of Particulate Organic Carbon on the Western Antarctic Peninsula Shelf Using a Color Index-Based Algorithm," in *IEEE Transactions on Geoscience and Remote Sensing*, vol. 61, pp. 1-13, 2023, Art no. 4203113, doi: 10.1109/TGRS.2023.3256367.
8. V. Chulukhadze *et al.*, "Frequency Scaling Millimeter Wave Acoustic Resonators Using Ion Beam Trimmed Lithium Niobate," *2023 Joint Conference of the European Frequency and Time Forum and IEEE International Frequency Control Symposium (EFTF/IFCS)*, Toyama, Japan, 2023, pp. 1-4, doi: 10.1109/EFTF/IFCS57587.2023.10272038.
9. O. O. Elemile *et al.*, "Assessment of the Effects of Bathroom Effluents on the Soil in Halls of Residence, Landmark University, OMU-ARAN, Nigeria," *2023 International Conference on Science, Engineering and Business for Sustainable Development Goals (SEB-SDG)*, Omu-Aran, Nigeria, 2023, pp. 1-9, doi: 10.1109/SEB-SDG57117.2023.10124481.
10. A. K. G and R. Swarnalatha, "Simulation and Analysis of Ammonium-Based Aeration Control Strategies to Enhance Efficiency in Wastewater Treatment Plant," *2023 IEEE World Conference on Applied Intelligence and Computing (AIC)*, Sonbhadra, India, 2023, pp. 142-147, doi: 10.1109/AIC57670.2023.10263849.
11. F. Mirlou and L. Beker, "Wearable Electrochemical Sensors for Healthcare Monitoring: A Review of Current Developments and Future Prospects," in *IEEE Transactions on Molecular, Biological and Multi-Scale Communications*, vol. 9, no. 3, pp. 364-373, Sept. 2023, doi: 10.1109/TMBMC.2023.3304240.
12. A. Roy and S. Pramanik, A Review of the Hydrogen Fuel Path to Emission Reduction in the Surface Transport Industry, *International Journal of Hydrogen Energy*, https://doi.org/10.1016/j.ijhydene.2023.07.010, 2023
13. Ashisha, G. R., Anitha, Mary X., T. George, Martin, Sagayam, K, Unai Fernandez-Gamiz, Hatıra Günerhan, M. N. Uddin and S. Pramanik, Analysis of Diabetes disease using Machine Learning Techniques: A Review, *Journal of Information Technology Management*, 2023.
14. V. Jain, M. Rastogi, J. V. N. Ramesh, A. Chauhan, P. Agarwal, S. Pramanik, A. Gupta, FinTech and Artificial Intelligence in Relationship Banking and Computer Technology, in *AI, IoT, and Blockchain Breakthroughs in*

E-Governance, Eds., K. Saini, A. Mummoorthy, R. Chandrika and N. S. Gowri Ganesh, IGI Global, 2023.

15. P. T. Khanh, T. H. Ngọc, and S. Pramanik, Future of Smart Agriculture Techniques and Applications, in *Advanced Technologies and AI-Equipped IoT Applications in High Tech Agriculture*, Ed., A. Khang, IGI Global, 2023.
16. V. Veeraiah, V. Talukdar, Manikandan K., S. B. Talukdar, V. D. Solavande, S. Pramanik, A. Gupta, Machine Learning Frameworks in Carpooling, in *Handbook of Research on AI and Machine Learning Applications in Customer Support and Analytics*, Eds., Md S. Hossain, R. C. Ho and G. Trajkovski, IGI Global, 2023.
17. T. H. Ngọc, P. T. Khanh and S. Pramanik, Smart Agriculture using a Soil Monitoring System, in *Advanced Technologies and AI-Equipped IoT Applications in High Tech Agriculture*, Eds., A. Khang, IGI Global, 2023.
18. S. Ahamad, S. Janani, V. Talukdar, T. Sharma, A. Sahu, S. Pramanik, A. Gupta, Apollo Hospital's Proposed Use of Big Data Healthcare Analytics, Big Data Analytics Techniques for Market Intelligence, IGI Global, 2023.
19. Reepu, S. Kumar, M. G. Chaudhary, K. G. Gupta, S. Pramanik and A. Gupta, Information Security and Privacy in IoT, in *Handbook of Research in Advancements in AI and IoT Convergence Technologies*, Eds., J. Zhao, V. V. Kumar, R. Natarajan and T. R. Mahesh, IGI Global, 2023.

6

Quantifying the Economic Benefits of Precision Agriculture Technologies: A Mathematical Modeling Study

Deepak Kumar[1], Apexaben Rathod[2], Sachchida Nand Singh[3], Meena Y. R.[4], Rushil Chandra[5] and Ankur Gupta[6*]

[1]Department of Mathematics, Vivekananda Global University, Jaipur, Rajasthan, India
[2]Department of Agronomy, College of Agriculture, Parul University, Limda, Dist- Vadodara, Gujarat, India
[3]College of Agriculture Sciences, Teerthanker Mahaveer University, Moradabad, Uttar Pradesh, India
[4]Department of Civil Engineering, Faculty of Engineering and Technology, JAIN (Deemed-to-be University), Karnataka, India
[5]Symbiosis Law School, Nagpur Campus, Symbiosis International (Deemed University), Pune, India
[6]Department of Computer Science and Engineering, Vaish College of Engineering, Rohtak, Haryana, India

Abstract
Precision agriculture is a cutting-edge idea in farming management which utilizes digital tools to track and improve the output of farming goods. The interactivities between plants, animals, soil organisms, humans, and the ecology are the main emphasis of the agroecological approach. It seeks to improve the biomass, nitrogen, carbon, and water cycles as well as the biological activities that take place in the soil. In order to fertilize crops with the essential nourishments as part of precision agriculture, the study discusses the fundamental interdependence between the few elements controlling nutrient techniques in the soil. A modeling strategy

**Corresponding author*: ankurdujana@gmail.com
Deepak Kumar: ORCID: https://orcid.org/0009-0004-5076-0558
Sachchida Nand Singh: ORCID: https://orcid.org/0009-0009-9318-8244
Meena Y. R.: ORCID: https://orcid.org/0000-0002-4789-6938

Sabyasachi Pramanik, Niranjanamurthy M., Ankur Gupta and Ahmed J. Obaid (eds.) Mathematical Modeling in Agriculture, (103–114) © 2024 Scrivener Publishing LLC

is suggested that satisfies a number of conditions and standards. The quantities of the nourishments nitrogen, phosphorous, and potassium needed for the goal produced on every specific farm are determined using mathematical models to provide fertilizer recommendations. Giving consideration to all variables affecting nutrient fluxes in the soil, guidelines are provided for enhancing and upgrading models for calculating the demand for fertilizers.

***Keywords*:** Agroecological method, mathematical modeling, fertilizer needs, precision agriculture

6.1 Introduction

Definitions

Precision agriculture [1] is referred to as "a modern farming management concept using digital techniques to monitor and optimize [2] agricultural production processes" in the European Parliament's statement on "Precision agriculture and the future of farming in Europe." Here, optimization is the important concept. Precision agriculture involves evaluating the inside-farm soil alterations and modifying the fertilizer approach appropriately, as against uniformly distributing fertilizers across a whole farm. As a result, fertilizer consumption is optimized, shipping costs are saved, and the environmental effect is decreased. Precision agriculture and precision farming are often used interchangeably. Precision agriculture, sometimes known as PA, is the more commonly used phrase.

With a better knowledge of and control of spatial and temporal variability, precision farming is a framework to manage soils and plants that aims to decrease decision uncertainty. Consider it a systems approach to crop production with the objective of lowering choice uncertainty via improved knowledge and control of uncontrolled variance. To facilitate data from many origins and scales to support choices related to crop yield, knowledge from a variety of fields is commonly combined with information technology. The word "agroecology" [3] has three basic meanings: scientific field of study; social movement; and practice. Even if there are several definitions and techniques, the newer perspectives and facets added to agroecology as a technical field should aid in attempts to address the real issues of farming production due of more used frameworks. Considering and transdisciplinary research methodologies are both important. Natural ecosystems serve as an inspiration for agroecological methods. They concentrate on how humans, the environment, soil creatures, and plants interact. This implies that they improve the biomass, nutrient, carbon, and water cycles as well as the efficient use of natural resources. Due to improved plant and animal health and resilience, farmers may lower costs and external inputs (EIP-AGRI, 2020).

Important specifics

The term "precision agriculture" (PA) refers to a whole-farming management strategy which makes use of IT, satellite position information, remote sensing [4], and local data [5] collection. The objective of these automations [6] is to increase input returns while possibly minimizing ecological outcomes (EU-DG IP, 2014). Studies and research are required to further our understanding of PA's benefits and costs, particularly as it relates to the environment. Optimizing the usage of fertilizers, beginning with the three primary nutrients nitrogen, phosphorus and potassium, is a significant application of precision agriculture on arable land. These fertilizers are administered equally across fields at certain periods of the year in traditional farming. This results in inconsistent application across different locations.

The processing of a vast quantity of data—mainly agricultural data, gathered through various technical ways and of great significance to agriculturists and farming corporations—as well as the usage of decision algorithms, are major legal challenges connected with the methodical establishment of precision agriculture in Asia. By using a precision agricultural system (PAS), you may boost crop yield, crop profitability, and crop variability compared to the traditional uniform management used in the years before to PAS, and enhance soil and water quality. The authors draw the conclusion that more research will be done to assess a variety of PAS advantages, inclusive of gain, soil grade and water grade. For maximizing input returns when saving resources, precision agriculture analysis targets the creation of a decision support system for farming management.

One plant organism plus the environment it lives in make up the simplest plant ecosystem. On the basis of this idea, we define agroecosystems as human-managed plant ecosystems. Four categories of components—biological, climatic [7], soil, and biogenic—define these. In agroecosystems, the plants are interconnected with and constantly interacting with its non-living surroundings. An ontology or representation of farming depending on a framework perspective may be created using the agroecosystem idea, which is considered as an epistemological instrument. In order to analyze, evaluate, and anticipate solutions for farming design and management as well as land-usage strategy, it is deemed essential to develop sustainability indicators.

A wide or specific definition of agroecology is possible. Agroecology, broadly defined, often encompasses concepts related to an approach to agriculture that is more socially and ecologically conscious, putting equal emphasis on production and the ecological sustainability [8] of the productive technique. Agroecology, being specific, is the study of solely ecological processes that occur in agricultural fields, such as the interactions between predators and prey or crop and weed competition. It is not always easy to translate the findings of agroecological research from the plot, field, or farm level to the local, national, or worldwide extent that is significant for decision-takers. The findings are mostly misread or not utilized in the decision-taking techniques as a consequence of this gap.

There is a developing corpus of knowledge on sustainability transitions as a result of the escalating ecological challenges. According to [9], agroecology is being marketed as a strategy that may alleviate various food system challenges when solving climate change and advancing the sustainable development objectives. Compared to different techniques for sustainable development, agroecology is fundamentally dissimilar. It is established on bottom-up territorial techniques which help in having contextualized remedies to local maters. Even if agroecology is becoming more common and significant in Canada, still a tiny organization, it receives little assistance for growth. Particularly in compared to other nations (such as Brazil, Cuba, France, and Sweden) which actively incorporate agroecology into legislative frameworks, Canadian agriculture policy seems to be myopic.

According to the quantity of research papers and the proportion of times the notion was used, the writers of the various papers divided Europe's nations into three categories: a small group that made initial contributions that were entirely focused on science; a second group that tended to focus on science, followed by practice and movement to a smaller extent; and a third, a much more productive category where execution predominated, followed by science and movement to a lesser extent. A technology is expected to offer some sustainability advantages if it increases farmer production while posing minimal environmental risks. Agricultural systems that place a strong emphasis on these principles also often serve many functions within economies and landscapes.

Instead of focusing on each component independently, as is the case with the majority of traditional agricultural research, agroecological techniques examine the whole agroecosystem and the complex interactions that exist within it. All farmers value having high and consistent yields; nevertheless, agroecological approaches aim to optimize rather than increase productivity. In reality, computing success in terms of production optimization and ecological advantages could be challenging. The advantages of using an agroecological strategy are many. Crop programming is the creation of a collection of technical methods in enhancing managed ecological variables in order to achieve a specified high level of field crop output. Academician I.S.'s list of 10 guiding principles serves as the methodology for programming agricultural crops. In order to practice precision agriculture, it is necessary to fertilize crops with the three major nutrients N, P, and K. This article seeks to demonstrate a few fundamental dependencies between some of the elements influencing nourishment fluxes in the soil and those demands.

6.2 Method and Materials

The next ten years will see a revolution in farming due to the quick development of new technology. If digital [10] agriculture is actually going to provide answers to issues like how agriculture affects the environment, adoption of technology solutions becomes one of the main obstacles in manufacturing [11]. World food security is being harmed by factors such as climate change, decreased soil fertility, resource depletion (water, for example), the genetic fragility of industrial monocultures, and rising chemical agent usage. For a longer period of declining investing in farming research, current years have found a stronger dedication from countrywide and worldwide shareholders for research into farming and the techniques to make it more sustainable. This awareness is one of the reasons for this, among others.

It is necessary to cover and simulate the many processes involved in agricultural yield production while taking into consideration their unique traits and interdependencies. The linkages between the corresponding models and the hidden laws are accurate. Given its complexity, productive creation cannot be adequately modeled by a lone, complicated statistical model. The development of a system of models is necessary for the best management of the whole manufacturing process.

The agroecosystem, which is controllable, and the anthropogenic element, which is a control subsystem, make up the genuine production system for plants. The three essential components of the agroecosystem, as shown in Figure 6.1, are soil, sowing, and atmosphere. The material, energy, and information fluxes are the system's inputs, and the yield with its quantitative and qualitative properties is its output. The economy and technology are both part of the control subsystem, which is linked by a feedback contour.

Only timely access to precise and complete data on the plants, their growth, and the external ecology makes it feasible to control agricultural production operations. At three hierarchical levels, issues may be addressed

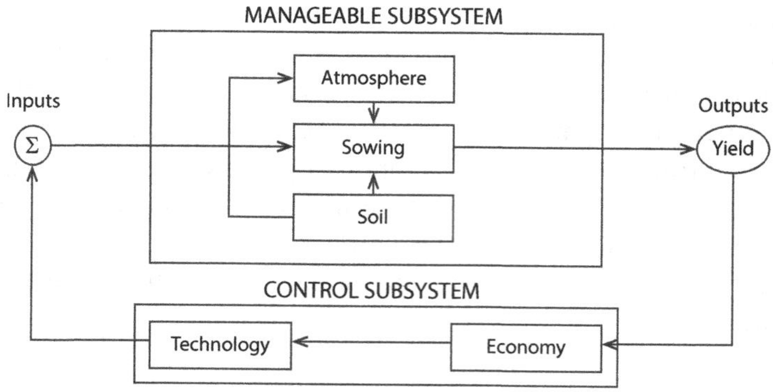

Figure 6.1 Actual system.

based on this knowledge utilizing mathematical models [12] and computer technologies to enable making appropriate, logical, and ideal judgments.

- Long-term planning and management of activities that will be relevant in five to ten years or more.
- Task management and planning on a technical and financial level for a single business year.
- Managing operations (organizational planning and management; farming tracking) to address issues pertaining to produce formation throughout the vegetative season.

The creation of a collected automated framework for managing the crops should be the urgent objective of agronomic research. The academician A's idea of "an electronic agronomist" would come to life with the development of a cybernetic yield management system. Yoffe's [13] transition into reality began in the late 1950s. Glogov (1984) identified two primary issues with farming as a science and a method of yield:

- The initial phase is to probabilistically propose the actual yield that may be expected for every particular field in a crop rotation system;
- The second step is creating technologies to achieve the desired yields, which also control yield creation during the growing season. The arrangement of production and management of the yield cycle are the two fundamental objectives in the managing cycle of material yield from the perspective of economic cybernetics [14].

- After the growing season and crop harvest, precise statistics on yield amounts may be gathered. However, planning and management duties in agriculture need an initial evaluation of these numbers.
- In order to predict yields throughout the growing season, there are two basic methods:

The subjective technique is based on either a broad forecast of yield amount or visual assessments of the stage of the sowing. It incorporates the process of expert evaluation.

- The measurements of several variables form the foundation of the objective approach. These techniques may be divided into three categories. Direct measurements of the sowings are included in the first category, whereas satellite or remote aerial photography is included in the second.
- The biological, climatic, soil and agrotechnical elements that have a substantial influence on the ultimate yield are taken into account by the third category of objective approaches, which use mathematical models. For the major agricultural crops, several similar approaches for yield forecasting have been developed. The bulk is empirical regression models that only include a few variables. They lack a consistent methodological foundation since they do not conduct agro-ecosystem analysis using the system approach.

Two key requirements must be met when building a mathematical model:

- The model must first be suitable, that is, it must match reality. The degree of this correlation might vary, but in all situations the framework should facilitate an approximation of the actual item, technique, or phenomena being represented.
- Secondly, the framework should be reasonably straightforward and solve the given issue with adequate accuracy and speed.

The successful fusion of these two requirements relies on mathematical modeling. The guiding concept should be the familiar "principle of simplicity" put forward by William of Ockham in the 16th century: "Entities are not to be multiplied beyond necessity" or "It is futile to do with more what can be done with fewer". All the aforementioned criteria are met by our method. The data from long-running farm fertilizer studies performed on the experimental farms of the N were utilized as the basis for this research. The data came from Poushkarov Institute of Soil Science, which is at the moment compiling a database, and the findings of fundamental studies conducted by a number of authors. The procedures adhere to the principles described above.

6.3 Conclusion and Results

According to [15], the linkages and interactions between the input (explanatory) and output (target) variables may form multivariate, non-linear agroecological phenomena. Additionally, whether or not scholars consider a particular patch of farmland to be a complex farming ecosystem such an agricultural region may very well be an ecological framework that is inherently complicated, where a number of different variables and circumstances often have a direct impact on food production. Thus, to analyze agroecological phenomena, analysis can require utilization of multivariate statistical approaches and non-linear mathematical models. These methods are far more difficult than traditional statistical techniques like analysis of variance and multiple linear regressions. A major flaw in most of the popular models is that the argument X on their right-hand side indicates the amount of mineral fertilizer that is applied without accounting for the amount of nourishments that are previously present in the soil.

According to the following hypothesis, the amount of a particular nutrient X is equal to the sum of its original level, X0, and the amount added by fertilizers F:

$$Y = Y0 + G$$

Some writers use the term "soil equivalent" to describe the amount of nutrients in the soil that are easily accessible to crops in a manner that is equal to the applied fertilizers.

The primary equation, which contains the statement "yield as an intrinsically non-linear function of macro element fertilization,"

$$X = f(Y) = [c(Y0 + F)](a(Y0 + F)b \exp].$$

This multiplicative equation, which combines a power function and an exponential function, is utilized to calculate the production in relation to the amount of the nourishment that needs to be supplied and the amount that is already present in the soil. As a result, it incorporates the homeostatic feedback mechanism. The N, P and K levels necessary for the goal production on every individual farm are provided by the fertilizer recommendation model. The balance sheet of obtainable nitrogen in the soil, N absorption by plants, and the coefficient of nitrogen utilization from fertilizers are used to calculate the amount of nitrogen fertilizer needed.

N rate calculation: $RN_{ij} = (a0 + a1.ZN + a2.Yi).cj$

Rate of nitrogen in kilograms per hectare (RN); field code number (i); soil type or soil group (j); and rate of nitrogen (ZN);

Yi, the goal yield; cj, the correction coefficient; and a0, a1, and a2 are the regression coefficients.

Example: Wheat planted on chernozem that has been leached

ZN=30, Yi=4,800 kg/ha, and cj=2.04.

RN is equal to (9.6743 - 0.0793 30 + 0.0122 4 800) 2.04 = 134.3 kg/ha.

The hypothesis that phosphorus and potassium optimum levels have superior agronomic effectiveness than yearly fertilization and have the benefit that it applies to soil fertility and all crops is supported by theoretical considerations and the findings of field fertilizer trials. P and K rates are intended to build up the soil's nutrients to their ideal levels.

Calculating the rate of P2O5:

$SP_{ij} = Aj.exp\ (Bj.Z\ 2)$ and $RP_{ij} = (SPij\ /\ T + U_{ij}).kj$, respectively, where:

Z2 is the quantity of accessible P determined by soil testing; SP_{ij} is the sorption capacity (amount needed to obtain the ideal concentration of 0.2 mg/kg P in 0.01 mol/l CaCl2); A,

Where: $RK_{ij} = U_{ij}\ /\ k1 + (Zj - Zi).k2\ /\ k3.T$

RK stands for rate of potassium in kilograms per hectare; U_{ij} stands for the amount of potassium absorbed by the crop; K1 stands for coefficient of K utilization; K2 stands for coefficient of calculation for the bulk weight of plow layer per hectare in tones; K3 stands for coefficient of the effect of the newly applied K fertilizer; Zj stands for optimal level of K required; Zi stands for available K determined by soil testing; T stands for time to build up a potassium level.

Example: Wheat cultivated on soil from a grey forest

T = 12 years, Y = 5,000 kg per hectare, U = 50, k1 = 0.80, k2 = 3, k3 = 0.60, Zj = 12, and Zi = 9.

RK = 11.5/0.80 + (12-9) X 3 /12 X 0.60 =15.625 X 10 =156.25 Kg/ha

Table 6.1 Determinants of nutrient movement in the soil.

Input		Output	
IN 1	Mineral fertilizers	OUT 1	Harvested product
IN 2	Manure	OUT 2	Crop residues
IN 3	Deposition	OUT 3	Leaching
IN 4	Biological N fixation	OUT 4	Gaseous losses
IN 5	Sedimentation	OUT 5	Erosion

RP = rate of phosphorus in kg/ha; T = time to build up a phosphorus level; Uij = quantity of phosphorus absorbed by crop; kj = coefficient of phosphorus aging. Bj = regression coefficients.

For instance: SP = 816; T = 12 years; Y = 4 800 kg/ha; U = wheat cultivated on leached chernozem

48; kag = 1.101.

RP = (816/12+48) 1.101 10 = 127.7 kg/ha

How to calculate the K2O rate

The mathematical models described here do not claim to account for all significant variables that affect how much of these components should be fertilized. Table 6.1 (Roy *et al.*, 2003) provides a more thorough explanation of the variables influencing the nutrient fluxes of substances in the soil and which must be taken into consideration when setting fertilizer rates of nitrogen, phosphorous, and potassium.

There is a clear need for further investigation in this area specifically, the effects of agricultural residues, gaseous losses, biological N fixing, and manure.

6.4 Conclusions

In order to further our understanding of PA and its cost-benefit analysis, more research is required particularly in regard to the effects on the environment. In comparison to the current traditional uniform management, precision agricultural system management will boost crop yield and crop profitability, reduce crop yield variability, and enhance soil and water grade. The connections and interactions between the factors that determine output and input might indicate agroecological phenomena that are both non-linear and multivariate. This totally solves the problems with modern decision support systems-based optimum fertilization. The frameworks for computing fertilizer suggestion described in this chapter may be applied extensively not only on the territory of Italy but also in different countries that are focused in maintaining and enhancing soil fertility as well as in achieving the planned yields of agricultural crops. The demand for the adoption of sustainable agriculture technologies is substantially met by this. To improve and develop new mathematical models that take into account a variety of parameters influencing nutrient fluxes in the soil, further study and experimental effort is required.

References

1. S. Pramanik, "An Effective Secured Privacy-Protecting Data Aggregation Method in IoT", in *Achieving Full Realization and Mitigating the Challenges of the Internet of Things*, Eds., M. O. Odhiambo and W.Mwashita, IGI Global, 2022, doi: 10.4018/978-1-7998-9312-7.ch008
2. S. Praveenkumar, V. Veeraiah, S. Pramanik, S. M. Basha, A. V. Lira Neto, V. H. C. De Albuquerque and A. Gupta, Prediction of Patients' Incurable Diseases Utilizing Deep Learning Approaches, *ICICC* 2023, Springer, doi.org/10.1007/978-981-99-3315-0_4
3. P. Dhoke, N. Purohit and P. P. Lokulwar, "Agroecology: Intelligent Implementation of Digital Farming Using Machine Learning & Deep Learning," *2023 World Conference on Communication & Computing (WCONF)*, RAIPUR, India, 2023, pp. 1-7, doi: 10.1109/WCONF58270.2023.10234967.
4. V. VidyaChellam, V. Veeraiah, A. Khanna, T. H. Sheikh, S. Pramanik and D. Dhabliya, A Machine Vision-based Approach for Tuberculosis Identification in Chest X-Rays Images of Patients, *ICICC*, 2023, Springer, doi.org/10.1007/978-981-99-3315-0_3
5. Reepu, S. Kumar, M. G. Chaudhary, K. G. Gupta, S. Pramanik and A. Gupta, Information Security and Privacy in IoT, in *Handbook of Research in Advancements in AI and IoT Convergence Technologies*, Eds., J. Zhao, V. V. Kumar, R. Natarajan and T. R. Mahesh, IGI Global, 2023.
6. D. Kaushik, M. Garg, Annu, A. Gupta and S. Pramanik, Application of Machine Learning and Deep Learning in Cyber Security: An Innovative Approach, in *Cybersecurity and Digital Forensics: Challenges and Future Trends*, M. Ghonge, S. Pramanik, R. Mangrulkar and D. N. Le, Eds., Wiley, 2022, doi: 10.1002/9781119795667.ch12
7. T. H. Ngọc, P. T. Khanh and S. Pramanik, Smart Agriculture using a Soil Monitoring System, in Advanced Technologies and AI-Equipped IoT Applications in High Tech Agriculture, Ed., A. Khang, IGI Global, 2023.
8. M. Sinha, E. Chacko, P. Makhija and S. Pramanik, Energy Efficient Smart Cities with Green IoT, in *Green Technological Innovation for Sustainable*

Smart Societies: Post Pandemic Era, C. Chakrabarty, Ed., Springer, 2021, DOI: 10.1007/978-3-030-73295-0_16\

9. G. S. Kuaban *et al.*, "Smart Sustainable Agrivoltaics Systems: The Future of Sustainable Agricultural Technology (Agri-tech) and Green Energy," *2023 3rd International Conference on Electrical, Computer, Communications and Mechatronics Engineering (ICECCME)*, Tenerife, Canary Islands, Spain, 2023, pp. 1-6, doi: 10.1109/ICECCME57830.2023.10253215.
10. B. K. Pandey, D. Pandey, V. K. Nassa, A. S. George, S. Pramanik and P. Dadheech, Applications for the Text Extraction Method of Complex Degraded Images, in *The Impact of Thrust Technologies on Image Processing*, Nova Publishers, 2023.
11. S. Choudhary, V. Narayan, M. Faiz and S. Pramanik, "Fuzzy Approach-Based Stable Energy-Efficient AODV Routing Protocol in Mobile Ad hoc Networks", in *Software Defined Networking for Ad Hoc Networks*, M. M. Ghonge, S. Pramanik and A. D. Potgantwar, Eds., Springer, 2022, https://doi.org/10.1007/978-3-030-91149-2_6
12. V. Veeraiah, V. Talukdar, Manikandan K., S. B. Talukdar, V. D. Solavande, S. Pramanik, A. Gupta, Machine Learning Frameworks in Carpooling, in *Handbook of Research on AI and Machine Learning Applications in Customer Support and Analytics*, Eds., Md S. Hossain, R. C. Ho and G. Trajkovski, IGI Global, 2023.
13. R. Kumar, S. Choudhary, J. Saha and S. Chakrabarti, "Theoretical Study on Vertically Coupled Stranski-Krastanov (SK) on Sub Monolayer (SML) InAs Quantum Dot Heterostructures Employing Ternary and Quaternary Barrier Materials," in *IEEE Transactions on Nanotechnology*, vol. 22, pp. 342-348, 2023, doi: 10.1109/TNANO.2023.3276210.
14. Y. Yang, B. Wen, X. Su, J. Huang and B. Liu, "Sliding Mode Fuzzy Control of Stochastic Nonlinear Systems Under Cyber-Attacks," in *IEEE Transactions on Cybernetics*, doi: 10.1109/TCYB.2023.3287029.
15. Y. Huang, S. S. Harilal, A. Bais and A. E. Hussein, "Progress Toward Machine Learning Methodologies for Laser-Induced Breakdown Spectroscopy with an Emphasis on Soil Analysis," in *IEEE Transactions on Plasma Science*, vol. 51, no. 7, pp. 1729-1749, July 2023, doi: 10.1109/TPS.2022.3231985.

7

Optimizing Resource Allocation in Agribusinesses: A Mathematical Modeling Approach Considering Legal Factors

Vishvendra Singh[1], Navghan Mahida[2], Anand Janardan Madane[3], Sudhakar Reddy[4], Parth Sharma[5] and Sabyasachi Pramanik[6*]

[1]*Department of Agriculture, Vivekananda Global University, Jaipur, Rajasthan, India*
[2]*Department of Animal Science, College of Agriculture, Parul University, Limda, Dist- Vadodara, Gujarat, India*
[3]*College of Agriculture Sciences, Teerthanker Mahaveer University, Moradabad, Uttar Pradesh, India*
[4]*Department of Physics, School of Sciences, JAIN (Deemed-to-be University), Bengaluru, Karnataka, India*
[5]*Symbiosis Law School, Nagpur Campus, Symbiosis International (Deemed University), Pune, India*
[6]*Department of Computer Science & Engineering, Haldia Institute of Technology, Haldia, West Bengal, India*

Abstract

Background information: Developing control plans for zoonotic illnesses is difficult due to the complicated dynamics of their dissemination as well as the requirement to deploy sophisticated multi-species monitoring and control strategies with constrained funding. As they may facilitate thorough quantitative description of disease transmission, mathematical models of disease transmission and particularly network models of disease spreading, show promise as tools for the formulation of managing policies.

**Corresponding author*: sabyalnt@gmail.com
Vishvendra Singh: ORCID: https://orcid.org/0009-0005-1722-3993
Anand Janardan Madane: ORCID: https://orcid.org/0009-0000-3905-6464
Sudhakar Reddy: ORCID: https://orcid.org/0000-0001-8207-3526

Sabyasachi Pramanik, Niranjanamurthy M., Ankur Gupta and Ahmed J. Obaid (eds.) Mathematical Modeling in Agriculture, (115–152) © 2024 Scrivener Publishing LLC

Methodology and key results: To analyze disease propagation and provide cost-effective monitoring and control, a layered dynamical network technique for zoonotic disease transmission and regulation is developed. Brucellosis transmission across wild animals, herds of cattle, and human subpopulations in a farming framework is used as a case study to build the model. A model is proposed that precisely measures disease calculation in associating with animal flocks of various species (such as cow flocks and kinds of wildlife for brucellosis). The model is then abstracted into a shape that enables both model recognition from data and complete targeted construction of several control capabilities. Then, based on current findings in the network control fraternity, approaches are established for these quantitative formulation of control regulations (which are aimed to the animal and human census), as well as for model recognition from snapshot and time-specific data.

Conclusions/Significance: It is shown that the modeling technique may provide quantitative insight into complete zoonotic disease control plans, allowing for the creation of policies that will lessen the impact of these illnesses. Numerous insights are gained, specifically for the brucellosis-transmission example, regarding the best allocation of resources among the points in the spread network (e.g., transhumance vs. sedentary flocks) and obtainable control potentials (e.g., vaccination, monitoring and killing, pasteurization of milk). Additionally, a tentative recognition of the brucellosis network framework is made utilizing past data, and the generated model's sturdiness is shown. Overall, our findings show that network modeling might be useful in creating control strategies for zoonotic illnesses.

Keywords: Mathematical modeling, agribusiness

Introduction

Zoonoses [1], contagious illnesses that may spread from animals to people, have a huge financial effect on both agricultural output and societal well-being. In the poor world, where zoonoses are particularly prevalent and have an impact, there is a dire need for low-cost but efficient solutions for their management and ultimate elimination. It can be difficult to control these zoonoses because doing so requires 1) an understanding of (and occasionally new development in) a particular zoonotic agent's surveillance, vaccination, and treatment capabilities in human beings and/or animals; 2) identification of modeling of the procedures and rates of spreading in every species and across species; 3) collaboration between the animal and public health sections, and 4) the capacity to develop the control frameworks required in the constraints established by the community's monetary and sociological conditions. By leveraging field knowledge that is specific to the area and illness, infectious disease experts have historically worked with governmental organizations to progressively build successful control and eradication plans.

It may be difficult to effectively describe the animal-human interface that promotes the transmission of zoonotic illnesses in this manner, which makes it difficult to allocate resources in the multi-species system.

Recent research has shown that mathematical modeling [2] may help professionals create control techniques, by making clear the roles different species played in the spread of the illness and enabling a priori comparison of the efficacy of alternative management efforts. However, work on utilizing mathematical models to explore zoonoses control is still not complete. The majority of the models in use is quite abstract and often depicts animal and/or human populations as one big, homogeneous entity. Additionally, rather than offering a full design for attaining optimum and reliable monitoring and control, the existing attempts often merely evaluate a small number of potential control options.

Over the last several years, network modeling of virus transmission has attracted a lot of attention. The network modeling pattern for harmful infections [3] constructs upon the traditional models for infection spreading in homogeneous populations; the network approach was specifically influenced by the identification [4] which spread patterns [5] are frequently structured and variable rather than homogeneous, and which manage capacities are selected. With an emphasis on knowing the task imposed by the network's topological form in its spread dynamics, quite a broad variety of studies was completed for network spread frameworks. The network spread models have furthermore been used as a framework for assessing workable targeted control strategies. Our team has recently researched the distribution of heterogeneous control resources in multi-group (network) models of viral dissemination. This research demonstrates how tailored controls may be created for high performance, as do comparable attempts on regulating human-engineered networks like traffic networks. They do this by using analytical techniques to identify certain network components (such as specific persons, subpopulations, or control capacities) which have an outsized influence on spread and then advise focusing control resources on these network elements. Even though these control-related insights appear relevant to zoonoses' mitigation, our findings so far were established utilizing general models for transmission and control, and possess only quantified spread in matters related to a single measure (the fundamental reproductive quantity), and did not include monitoring [6]. Instead of network-structured models, frameworks for zoonotic illnesses (consisting of brucellosis) was an easy compartmental framework with each compartment collecting homogeneous transmission may occur within a species as a whole or possibly for a specific age group. Models for zoonotic illnesses have only lately been created, and they represent the intricate spatial or community form of transmission. The community structure's involvement in spread has been characterized by these models, and viable control measures have been compared (mainly through simulation); but, the attempts do not allow for the orderly analytical creation of high-performance control mechanisms.

The research aim is for providing a thorough analysis of the modeling [7], monitoring [8], and cost-efficient handling of zoonoses by using and advancing a network-control theory mechanism to the prevention of viral propagation. In a particular application, the authors investigate modeling and scheme of brucellosis monitoring and management in a typical farming environment in a region with limited resources, like sub-Saharan Africa. In order to achieve it, a network framework for brucellosis transmission in animal flocks and to human census is grown. This model records the technique of transmission of the zoonotic bacteria, permits for the representation of realistic monitoring and control methods and their expenditures, and calculates the effectiveness of the control technique in relation to the infection's monetary and social consequence. Following a generic formulation of the model, we describe methods for estimating critical network-framework parameters from sparse implemented data, including both straightforward heuristic methods and innovative network-estimation tools from the control sciences. The form of monitoring and control measures is explored using the parameterized models with the purpose of proposing appropriate targeted control/monitoring techniques and enhancements to current techniques. In order to make the created strategies somewhat resilient to errors and constraints in the models and accessible data, a specific emphasis of the structure is to acquire straightforward patterns into high-performance techniques which do not depend on model specifications.

In numerous respects, this study extends previous research—including our own prior work—on network modeling of virus propagation. It first makes clear the use of a layered network structure in the graphic representation of spread across many species and breeds. Second, it thoroughly designs the spread control capabilities by carefully modeling a species of zoonoses monitoring and control capacities and using a network control theory technique. This network control theory technique is useful as it allows for the methodical creation of various control and monitoring potentials in a multi-faceted network to satisfy various performance requirements or enhance a performance measurement. Since extremely few and diverse control resources should be utilized in several zoonotic-infection control situations, we feel that this extensive design skill is crucial for the zoonotic disease reduction issues which are examined here.

Methods

In order to clearly illustrate the ways disease-based features may be gathered utilizing the modeling system and to facilitate the enhancement of numerical control rules for this specific underappreciated zoonotic infection, we find it to be most instructive to establish the suggested layered-network modeling system for zoonotic diseases utilizing an application of brucellosis transmission. Let's start by quickly going through the techniques for simulating the spread of brucellosis and creating control measures. Due to variations in the farming system, flock and pasture dimensions, availability, and monetary resources, transmission dynamics and control capacities/expenditures in various places having higher brucellosis incidence (for example, in the Philippines) differ dramatically from herd to herd.

In order to guide systematic monitoring and the development of focused control tactics, a framework for transmission at the flock level (which also takes into account a number of unique human subpopulations) appears promising. The flock and subpopulation-level contact-network framework we provide is a linear multi-group framework depending on population-dynamics notions. We improve the previous models of this kind in two ways: 1) we explicitly record genuine multi-faceted control capacities and expenditures as network formation alterations and feedback controls, and 2) we discriminate geographic spread characteristics across numerous species. Depending on this observation, the authors apply and expand a family of currently built techniques for formation and controller design in complex dynamical networks to create the best allocation rules for heterogeneous resources to reduce the spreading of zoonotic diseases in a way that takes advantage of community (network) form. Outcomes which are adjusted to recall variations in farming methods and those that best balance the costs of disease in the human and animal census are of particular note. In addition, using snapshot and time-course data, we use system identification methods to estimate the spread model's parameters. These techniques are specifically utilized to parameterize the brucellosis spread model utilizing time-course information from the Jackson bison flock and snapshot information on bovine brucellosis in Africa. Following a short summary of the brucellosis as an example, the authors show created nominal network framework for the spread of brucellosis across various animal flocks and human subpopulations. The modeling of control costs and efforts is then shown. Finally, strategies for model identification and control design are briefly reviewed.

A Framework for the Transmission and Command of Brucellosis: A Case Study Overview

Various species of the zoonotic bacteria known as brucellosis are known to infect people, as well as animals (such as cattle, small ruminants, pigs, camels, and bison, wildlife (such as bison, elk, and caribou), and humans. Abortion, decreased fertility, decreased milk production, and abscess development are common symptoms of the chronic illness in animals that affects the reproductive system. These animals can transmit brucellosis to humans, who experience various discontinuous illness, utmost exhaustion, and uneasiness over the course of weeks or some months. In some cases, the disease progresses to a chronic condition with the potential for relapse and a variety of problems (like joint/bone concerns, gastro-intestinal concerns and abortion). If illness is common, it can provide a large social and economic effect because of the decreased livestock farming output as well as the loss of production and human life. B. abortus is the major prevalent distress of Brucella in cattle and other species. B. abortus is mostly spread by touch with aborted fetuses or, in human beings, consumption of raw cattle commodities (like unpasteurized milk). While a larger portion of the people in underdeveloped nations may get infected by eating of raw goods, pastoralists are often exposed to both methods of transmission.

Brucellosis Nominal Transmission Modeling

The created model analyzes the prevalence of brucellosis (quantity of infections) in various animal species' separate herds as well as the commonness in human subpopulations. In particular, brucellosis disease is modeled in N sorts of animals, labeled 1,...,N, which may comprise N species, or perhaps N breeds if transmission features are varied. Let's suppose that Mi herds or groups, denoted by the numbers 1,...,Mi, are existent for species i. We think that taking into account a lone disease class for every flock is adequate in the first controller design that is investigated here since the illness is often chronic in cattle and mostly damages the reproductive system. In particular, the quantity of infected individuals in flock j of species i at a period t is monitored through time and is indicated as xij(t) for i1,...,N, j1,...,Mi. Having the justification that, across the time horizon of interest, economic and resource-limit drivers normally maintain flock numbers reasonably consistent, the cumulative quantity of individuals in the flock is considered fixed; the flock dimension is represented by Nij. Moreover, the quantity of brucellosis-infected humans is monitored in P human subpopulations with the labels w1,...,P that vary in their interactions with animals.

Thus, the authors first create a statistical prediction framework for the dynamics of the disease quantities xij(t) and zw(t), without the use of designable controls, in the nominal situation. In the mathematical epidemiology literature, deterministic differential-equation frameworks (sometimes referred to as multi-group frameworks) are often used to simulate disease-spreading dynamics at the size of flocks or subpopulations, and the authors apply and improve this modeling prototype here. Shown the aforementioned processes of transmission, it is obvious that both transmission within a flock and transmission across flocks (of a single or more species) whose elements mix are possibilities. For this reason, we simulate both forms of transmission in this article. (We point out that inter-flock transmission can be more frequent for flocks engaged in transhumance or those which are housed in close quarters with other herds in a production system.) The occurrence of sexual contact and/or contact with infected birth material (and thus the disease rate) must slightly scale with the pairwise interactivities between infected and non-infected flock masses. In the flock j of species i, the infection rate due to inter-herd interactions is modeled as proportional to the product of the infected population and the non-infected population, i.e., xij(Nijxij). Even in places where brucellosis incidence is high, the infected population often makes up just 25% of the entire population. As a result, it is often accurate to assume that the non-infected population is about equivalent to the entire flock population Nij. In these situations, the rate of infection at time t because of transmission in the flock may be modeled by the linear function aijNijxij(t), where 0*f*aij*f*1 is a breed-based and flock-based scaling constant which captures possibility of spreading by interaction (and shows, for example, span of the infection's existence in an aborted fetus, the occurrence of the bacteria in excreta, or the dimension of the farm for grazing. These linear approximations in transmission are well-inspired for usage in controller design and have been utilized often in network models of spread. Here, the nonlinear population-dynamics frameworks were utilized to validate designs and identify model parameters while the linear approximation was employed for monitoring and control design.

A transmission from different flocks of similar species is then included in the model. Think about the transfer of a species i from a flock k to a flock j, where j=k. The range of combination between the two flocks (or, much categorically, the interactivity of the individuals from herd k with those from flock j by combining the flocks or through different ways, like the procurement of an animal) affects the transmission rate for this pair, which is determined by the decontaminated population of flock j and the contaminated population of flock k. The interactivity scaling parameter gij, ik1 denotes similar commingling as must occur within a lone flock, and it is used in the model to quantify the level of interaction and 1 is usually significantly smaller than gij,ik. The inter-flock infection rate is modeled as aijgik,ijNijxik(t), using the same population assumptions that result in a linear model for within-herd transmission (from k to j). Because of transmission from different flocks of the similar breed, herd j's overall infection rate is therefore fwzw(t) is rated. The following differential equations combine these rates to represent the quantity of affected people in each of the human-being subpopulations:

$$y'_{ij} = (\beta_i M_{ij} - \alpha_i) y_{ij}(t) + \sum_{k \neq j} \beta_i h_{ik,ij} M_{ij} y_{rk}(t) + \sum_{r \neq i} \sum_{k} \delta_{ri} h_{rk,ij} M_{ij} y_{rk}(t) + v_{ij}(t) \tag{7.1}$$

Similar to intra-species transmission, the rate of infection from herd k of species r to herd j of species i is predicted, but taking consideration for different infectivities. In particular, brigrk,ijNijxrk(t) is the linear model for this infection rate. Therefore, r=i k brigrk,ij Nijxrk(t) represents the overall infection rate brought on by transmission from different species or breeds.

The model moreover accounts for variations in infected numbers in a flock brought on by 1) natural mortality or, less often, remission, and 2) the introduction of contaminated animals from outside the modeled framework (for example, via the buying of the animals by a farmer). It is widely known that the rate of decline in infective caused by death or remission is related to the quantity of infective in the flock. In particular, in flock j of species i, cixij in the framework, where ci is a species-dependent death/remission rate, give the rate at which infectious decline via natural mortality. Second, for each flock j of breed i, the infected rate brought on by the injection of diseased.

The aforementioned framework is quite abstract in that only a single infection state for each species is considered, and the stochastics [9] of transmission are completely disregarded. Although we have focused on the linear approximation under the premise that the infected percentages are minimal, the model dynamics are generally nonlinear (and the authors may apply the nonlinear approximations as required). This rudimentary representation of the dynamics is captivating because it 1) enables systematic controller design and the evolution of basic understandings of resource allocation (whereas a thorough simulation model may be utilized to test the design), and 2) highlights the task of network form in spreading and spread control. The portrayal of brucellosis transmission from cattle to the humans before control is a second key component of the nominal model. Similar to the animal model, differential-equation models of transmission are produced by representing the infection rate at each period in order to depict the infection in the human population. A linear population - dynamics framework is once again employed to explain the spread in order to approximately record these infectivity properties in a manner which allows control design.

The infected rate for the human subpopulation i at time t is given by the linear combination of the infectious numbers in different herds representing a subpopulation of humans, and as people do not spread the germs, the self-loops for these vertices are always negatively weighted. Figure 7.1 depicts a spread graph as an example. We remark that precise spread graph specification for relevant cases is crucial since the spread graph determines the contact network (i.e., the transmission rates) across modeled populations. Let's be clear that there are often three ways to express the spread graph: 1) the djk,i captures the rate of infection brought on by the herd k of the species j. Along with two human subpopulations, the bacteria are stored in animals.

A conventional system-ID method can be utilized to conclude the spread graph, or, for large-scale networks, current features of distinct network topologies (for example, "small-world topologies," where every subunit is within some steps of each other in the graph though relatively

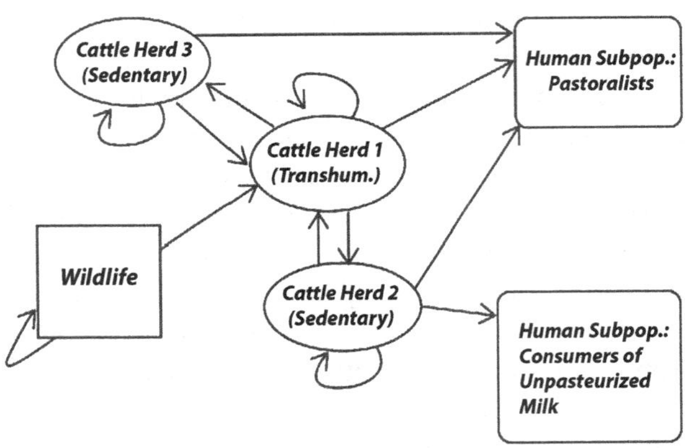

Figure 7.1 Spread graph example. Brucellosis spread chart.

sparse connectivity) may be utilized. Control-strategy planners can utilize intimate knowledge of a farming framework to find flock connection and suggest transmission rates. The authors would justify and explain the description of the spread graph which we have utilized for this instance in the illustrated case that is shown in the results.

Modeling Disease Costs and Control Capabilities

Next, the monitoring and control capacities 1) within the animal population and 2) for transmission from animals to people are explicitly represented in the expanded nominal multi-group model for brucellosis transmission. Controlling the spread of disease in the animal population can be done in a number of ways, such as vaccination, monitoring for diseased animals, culling them, limiting herd commingling (by restricting trade or movement of livestock), and improving farm sanitation. In general, these different control options can be seen as having an impact on the dynamics of the model in two ways: 1) they change the rate at which a flock/subpopulation spreads brucellosis to a different flock, or they alter the model parameters (equivalently, the edge weights in the nominal graph); and 2) they force the removal of infected individuals at specific times, i.e., they alter the state variables xij and zw.

Vaccination: There are brucellosis vaccines available for small ruminants and cattle. By administering a vaccination to a herd, the disease-proneness of the herd's members is reduced. Abstractly, our model incorporates vaccination capabilities as follows: vaccination is understood as scaling every nominal transmission rates to a specific flock of a species by a constant vij between 0 and 1. This means that Equation 1 now reads vijNijgij,rk for the transmission coefficient from the flock k of the species r to the flock of interest. With vij1 denoting no vaccination and smaller vij denoting more efficient vaccination schemes, the constant vij, which measures the efficacy of the vaccination approach, may generally take on one of many discrete values. Given that brucellosis symptoms are ambiguous, surveillance is not a simple task. In general, to test for brucellosis having higher fidelity, either the organism must be identified using fluid or tissue specimens for culture, or diseased animals must be identified using serological tests; these tests often compromise between sensitivity and specificity and may be expensive. Typically, depopulation is utilized as a form of control via surveillance. Implementing the control in the framework alters the self-loop weights in the network-graph's viewpoint.

Limiting commingling between herds and improving cleanliness are two more strategies for reducing the spread of brucellosis among cattle. These additional control measures, such as vaccination and monitoring/killing controls, may readily be described as changing the nominal model and, therefore, the nominal interaction graph in different manners. The transmission rates between herds will be reduced specifically by minimizing commingling across herds (or, alternatively, by lowering the edge weights between various vertices in the interaction graph), which will help to decrease the interaction weights gij,rk. Depending on how commingling is restricted, the network's interaction weights may be constrained or just a few transhumance herds may be impacted. Equation 1's infection rates within herds will decrease as a result of better housing and sanitation for cattle, which may be depicted by scaling these infection rates preventing the spread of disease from cattle to people. Human brucellosis is a serious, often fatal illness that must be prevented at all costs. There is no human brucellosis vaccination. Instead, three main strategies are used to prevent the spread of brucellosis from livestock to humans: pasteurization of milk products, particular meat preparing, surveillance and control programs, and a decrease in transmission to those who handle livestock products by better sanitation and instruction in safe operation. The main goal of these control measures is to lower the rates of transmission from some (or maybe all) of the animal herds to one or more human subpopulations. In other words, the controls act as weights between 0 and 1 to scale the nominal rates of transmission djk,wxjk(t). For instance, the transmission coefficients from all milk-production flocks to census which have historically consumed unpasteurized products will scale with proper pasteurization of milk, specifically altering the coefficients from djk,w to hdjk,w where h denotes the efficiency of the pasteurization techniques. It is important to keep in mind that, despite their apparent simplicity, the techniques for avoiding transmission to humans may need large investments in underdeveloped nations (requiring, for example, the creation of a cold chain from source to delivery).

Creating a Cost Model and Confronting the Challenge of Control Design

In general, creating successful control methods requires striking the right balance between expenses associated with disease prevalence and control costs. Alternately, control design might be seen as a limited-resource approach to reducing infection generality or related expenditures. Models for the cost of infection as well as the cost of implementing the different control measures are therefore required in order to appropriately develop control strategies. Every diseased person in every flock and human subgroup at each specific period is considered as suffering an expenditure, and these expenditures are added together to generate the total expense of infection. Equation 1 specifically models the rate of infective removal from flock j of breed i as the nominal rate scaled by a constant sij greater than 1, i.e. as sijcixij. In this case, sij1 stands for the scenario in which there is no monitoring or culling (removing patients at the nominal rate owing to death and remission). While arbitrarily big sij corresponds to fully effective and instantaneous monitoring and culling, greater sij reflects surveillance/culling program that is becoming more successful. It is important to note that each herd and human subpopulation may have a different incremental cost of infection. For example, an infection of milking cows or sheep may result in a higher cost. The most important factor is often the overall infection cost over time, Cinf T C(t)dt.

The cost of control is considered additionally to the expenses associated with brucellosis spread. The expenditure is believed to contain three components for every potential control activity, such as monitoring utilizing serological testing succeeded by killing: 1) a worldwide overhead expenditure for the framework required to conduct the control plan, 2) a per-herd (or per-human) expenditure, and 3) a per-activity cost, a cost per affected member of the herd. This cost represents the extra expense of facilitating controls to the flock (that can vary depending on the dimension and kind of the flock). For a testing and killing approach, there are fixed costs for setting up the testing laboratory infrastructure, a cost for checking every flock, and expenditure to the agriculturist for every animal that is culled. The expenses associated with each adopted control action are added together to get the overall cost of control. Ccont stands for the cumulative expenditure of control over a certain length of time. The whole cost of a control approach, designated as the full cost Cfull, is calculated as the cumulative infected and control expenditures over a certain span of time, or Cinf zCcont. For a (limited or infinite) time horizon of interest, an efficient control technique is one in which the whole expenditure is reduced or at least adequately minimal. Achieving a low-cost or high-performing control approach is the aim of control design. On the contrary, controls moreover need to be created to lower the expenditure of infection, subject to a cap on the expenditure of control (defined by the resources obtainable for this particular infection's resources). In both formulations, the authors emphasize which in addition to (and often as a precondition for) cost reduction, reaching a fundamental reproductive number R0 smaller than 1 is necessary, and this additional limitation is routinely observed in our design. The following examples show how the spread and control framework's dynamics and related expenditure model work.

Analysis, Design, and Parameterization Techniques

By employing brucellosis as a case study, we have previously described a layered network framework for zoonotic infection transmission which shows promise as a tool for the methodical draft of multifaceted control capacities. To be more precise, the formulation enables the use of certain novel network control theory approaches for the systematic blueprint of constrained control resources to get high performance (lower spreading expenditure) or, alternatively, to decrease overall resource and spread cost. The method is especially encouraging since it has the potential to be used in low-data scenarios, which are typical in the transmission of zoonotic diseases. This is because the approach's straightforward model structure allows for the parameterization of the model using sparse historical data. The network-control-theory techniques for higher-performance drafting and framework parameterization are discussed in this section. To enable a broad audience, the authors omit technical explanations and algorithmic attributes and instead direct the audience to particular outcomes in the engineering literature for these. Our goal is to offer the techniques in adequate feature which the audience comprehends the ideas and fundamental techniques fundamental technique.

The authors emphasize that the modeling technique allows for straightforward simulation and inquiry of the spreading dynamics, given a particular control regulation, before outlining the approaches for control design and parameterization. We specifically point out that both the linear and nonlinear differential equation frameworks may be numerically determined utilizing common derivative-approximation techniques, and may be easily implemented using, for example, the MATLAB program. The linear framework moreover makes it simple to analyze dynamics, including computing steady-state and transient dynamics in closed form and computing dynamics' properties like the fundamental reproduction number. The approach is directly dependent on the traditional analysis of linear systems or linear differential equations.

Overview of the Control and Surveillance Design

Recently, a significant effort has been made by control-systems engineers to design networks' surveillance and control capabilities. The "science of networks" which natural scientists have developed is closely related to this field of controls engineering study. Here, as an example more broadly of zoonosis-control framework, the authors seek extension and implementation of the novel network control approaches to handle the brucellosis mitigation challenge mentioned before. Although for simpler models than the one taken into consideration here, network controller design approaches have previously been used to spread-control issues in prior work. The techniques utilized to create controls for the zoonotic-disease frameworks are the same as they were in earlier research. Due to the broad variety of control capabilities taken into account and the notable variations in the spread characteristics of the human and animal populations, the expenditure computation and a group of accessible control capabilities are more complex than in prior studies. The method of network control theory utilized here merges the tools of optimization having ideas from graph theory and a structural knowledge of linear systems. Following is a brief rundown of the process.

1. Firstly, the design issue is approached using traditional optimization techniques. In other words, we take into account the challenge of minimizing the whole cost Cfull while taking into account the design factors (such as vij, sij) within the restrictions placed on these parameters. The traditional Lagrange-multiplier approach may be used to tackle this minimization issue, which after identifying the best design requires resolving a set of (nonlinear) equations. It is important to note that the cost function is mostly represented in matters related to implicitly specified functions, mostly the eigen values of the state matrix of the linear system (such as the fundamental reproductive amount). Thus, computing the partial derivatives of Cfull related to the design parameters necessitates the use of implicit differentiation techniques, particularly eigen value-sensitive equations. After obtaining the Lagrangian formulation, we proceed in two directions: numerical solution and structural analysis of the optimum.
2. A number of common recursive solvers, like in the MATLAB software package (or those that may be readily created by

hand), can be used to numerically solve the Lagrangian. These techniques may be shown to provide a globally optimum design with some wide criteria on the network topology and expenditure specification (that ensures convexity of the optimization issue).

3. Often, rather than obtaining a numerical design, it is considerably extra informative to identify structural traits of higher-performance or optimum designs. It is generally true for the zoonotic disease minimization issues that we are considering because socio-political and geographic issues can make it challenging to implement a precise policy and to parameterize a model, making an optimal design more of a recommendation than a workable procedure. Furthermore, by comprehending the traits of high-performance drafting, the authors may explicitly state the contribution of the network's graph structure to control. For purposes similar to them, characterizing the system/graph formation of higher-performance architectures has been a main focus of our current study. By obtaining relationships between the optimal design parameters and the state matrix of the linear dynamics from the Lagrangian, it is possible to achieve this structural characterization of high-performance design. In other words, the dynamics portrays a form on the optimal resource allocation. Considering that the network graph specifies the kinetics of the illness propagation, we may thus instantly link the ideal design parameters given the network graph's structure (such as the vertices' degrees, etc.). We may also connect the dynamics that result from using the best controller to the structure of the nominal graph. By doing this, we are able to create higher-performance management systems which are resistant to application and designing flaws using relatively basic graphical insights. The results section will provide a presentation of the results of this design process.

Network Model Identification and Validation for Zoonoses

The design technique proposes that higher-performance spreading reduction is greatly chosen; specific flocks or subpopulations need unequal control resources as they irregularly influence spread (or spreading expenditure), either due to their local dynamics or their network communications, as we will go into more detail in the results section. While getting basic insights into these targeted resource allocations has been the main focus of this article, precise modeling of zoonotic-agent spreading may enable the implementation of increasingly sophisticated spread control techniques by allowing for better precise performance of the spread consequence of the network elements (flocks or subpopulations). Additional study is required on extracting the framework's parameters from data in order to generate reliable models for spread control.

The authors have examined the framework-recognition job for the brucellosis spreading implementation from two perspectives: firstly, from the perspective of identifying a framework for a lone flock that can be used to construct a network model; and, secondly, from the perspective of recognizing the full network dynamics or significant statistics therefrom. In specifically, a lone-flock model was parameterized utilizing temporal data from the Jackson bison herd, time-snapshot flock-size, and seroprevalence data from multiple African nations. Data [10] utilized for parameterization specifically consists of 1) mean prevalence vs. mean dimension in smaller, mid-sized, and larger flocks for numerous areas in Africa, and 2) time-course occurrence and herd-size data during a 10-year span for the Jackson bison flock. This parameterization attempt makes use of very common system identification heuristic methods. We observe that these individual-herd models allow the authors to register various above-mentioned findings on spread-control approach since they also indicate the spread-consequence of a flock (albeit not its precise topological form). Finally, after sufficient information on the spread of brucellosis within an agricultural network has been gathered, new approaches for network identification may be implemented with reference to the complete parameterization of a network framework. This comprehensive network recognition is reserved for further development.

Results

The outcomes of the modeling and control technique described here are best shown in the circumstance of a particular case of brucellosis spreading, in our opinion. An instance which is typical at a smaller scale of mostly farming societies having both transhumance and sedentary herding (like in Kenya or Zimbabwe) is used to demonstrate model dynamics and control design. Moreover, a non-intensive farming framework uses a single breed of cattle, with 12 transhumance flocks averaging 75 animals each and 10 sedentary herd's average. Each model has 40 cattle. Assuming that farming methods (including cow density on grazing fields) are comparable amongst the herds, same within-flock transmission rates are designed:

As shown below, the transmission rates were derived by model-recognition. Additionally, in the specified case, every transhumance flock is supposed to mix with seven other flocks for a portion of the year (particularly, three months), including five different transhumance herds and two sitting flocks. There are two sections of the human population that are susceptible to infection: 1) pastoralists, who are infected by cattle at a high rate; and 2) non-farming clients of raw milk products, who are infected by cattle at a lower rate but still considerable incidence. It is anticipated that each herd will have an equivalent effect on the various human population groupings.

Indicative Model

Figures 7.2 and 7.3 shows the nominal spread graph that depicts the contact network before control. Figures 7.2 and 7.3 depict implementations of the transmission of the disease across the cow herds under two distinct beginning scenarios, one with a lone infectious agent in a transhumance flock and the different one with a lone infectious agent in a seated flock. As predicted, the virus spreads widely under both beginning circumstances since the nominal model's fundamental reproductive number, R0, in the animal population is larger than 1. When the virus starts in a transhumance herd, it spreads significantly more quickly. It's interesting to note that the illness ultimately spreads to transhumance herds, even when it starts in a sedentary herd. Humans see in many instances that humans are the ones that started the infection in the biggest herds that are sedentary and transhumance. We took this action because bigger herds show a quicker infection's initial development, which causes the illness to spread more quickly (and more easily to see) across the network. The authors emphasize that while the development will be slower if the diseases were started in little flocks, the qualitative response characteristics will remain the same. The framework moreover demonstrates that during the first phases of disease, the disease rate in the human subpopulation engaged in animal husbandry increases quickly. It's worth making a few points concerning the nominal model.

The graph is produced to impose a degree distribution, but in this constraint is chosen at random. The specific transmission rates utilized in the instance are achieved via recognition of the brucellosis spread framework.

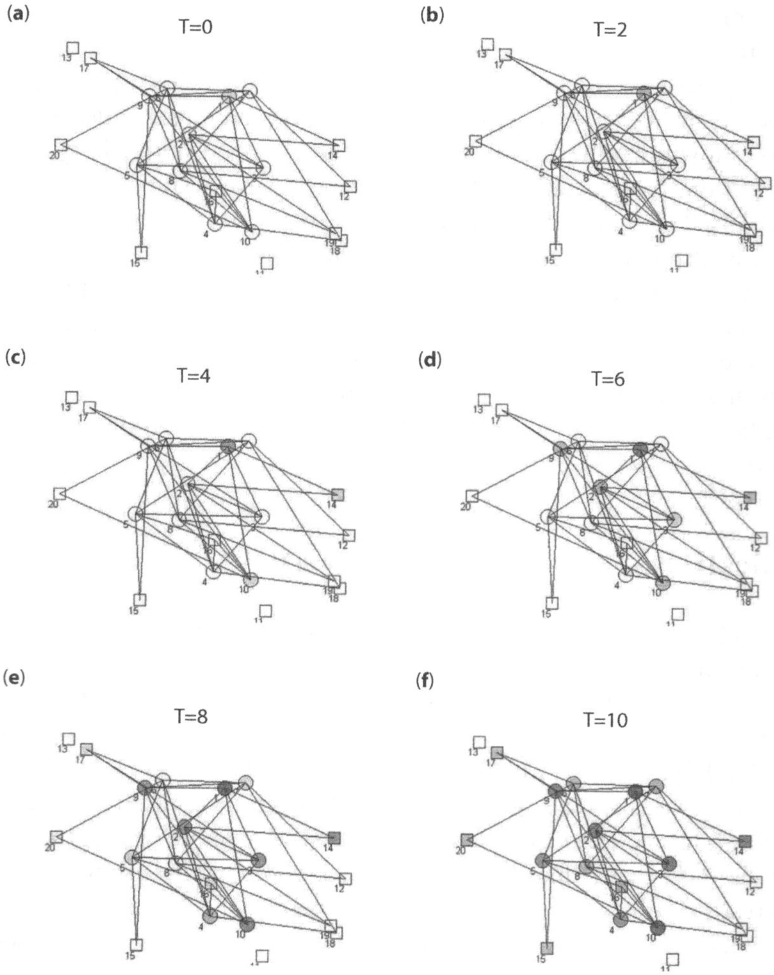

Figure 7.2 Nominal spread model simulation based on an initial infection in a transhumance herd. In a non-intensive agricultural system, brucellosis in cow herds is mimicked under the fictitious condition that no restrictions are used. A network of ten transhumance herds (shown as circles) and ten stationary herds (represented as squares) is examined in this tiny example. The herd sizes change, but the intra-herd transmission conditions remain the same. Transhumance herds mix with sedentary and other transhumance herds, as shown by the spread graph superimposed over the dynamics. Although different models might be used instead, in this example, the probability and frequency of spread across herds are stated based on a distance measure between the herds. The biggest transhumance herd (Herd 1) has a modest number of sick animals before the simulation starts. The dynamics of the spread over time are shown (months are represented on the time axis), and the intensity of the red color for each herd indicates how infected it is. As would be predicted, the virus spreads swiftly since the fundamental reproductive number (R0) for the spread is larger than 1.

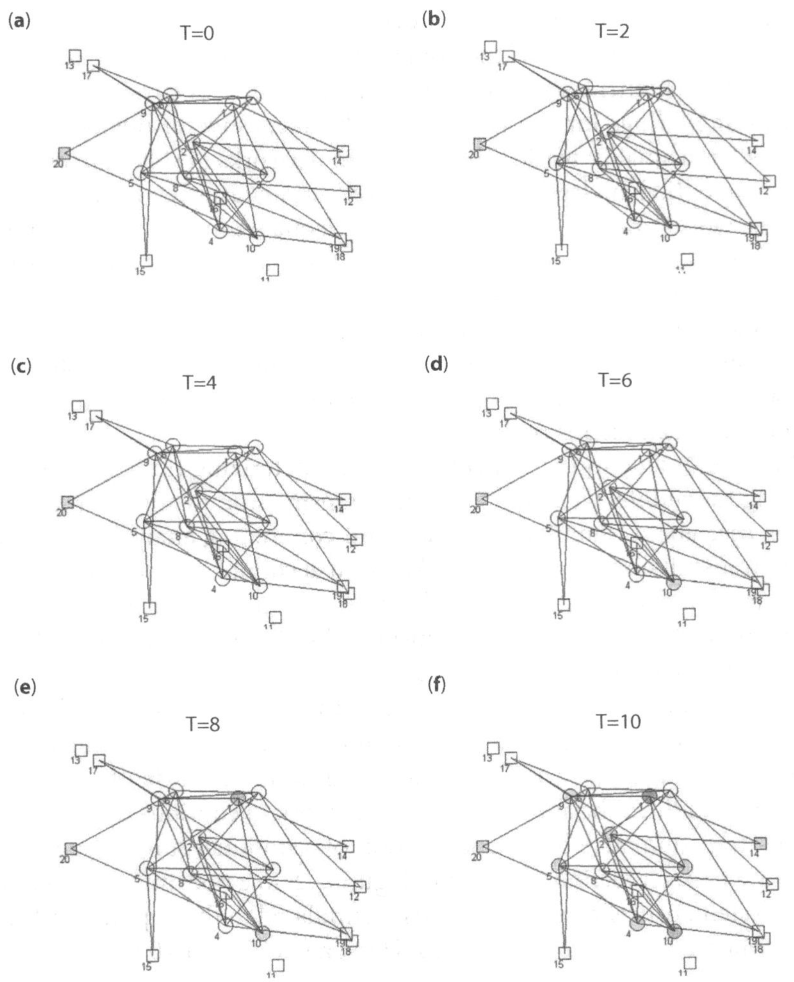

Figure 7.3 Modeling the notional spread of an initial illness in a sedentary herd using simulation. For the 20-herd example, the basic model for brucellosis propagation is again simulated without control, but this time the illness starts in a small, sedentary herd (herd 20). The illness once again spreads widely, but much more slowly this time. It's interesting to note that, despite the infection having started in a small, sedentary herd, brucellosis spreads more widely in vast migratory herds than in small, sedentary ones. This model feature is consistent with field assessments of the incidence of brucellosis in pastoralist groups, such as those in Ethiopia's Tigray area.

In the 21-flock instance that was previously selected, the authors observed that the graph topology of the contact network is one where 50% of the vertices have higher degree, and the remaining vertices have lower degree. Thus, our findings for the instance can be suggestive of outcomes if the topology is a smaller world. These random networks [11] having enforced lower- and higher-degree vertices have certain aspects with the usual smaller-world graphs [12] if they are of adequate dimension. We emphasize, though, that the analysis techniques (and subsequent control design and recognition techniques) may be used despite of the graph topology [13], and that the particular topology the authors have selected is influenced more by our knowledge of the farming operations controlling the spread of brucellosis than by the classic model classes taken into account in the complex-systems review.

Remark 2: Due to the difficulties in parameterizing transmission rates relating to wildlife, we have eliminated transmission to and from animals in the case above. Simulating models and conducting analyses are simple when transmission to animals is taken into account. As an example, we've done implementations in which the rate of transmission between every flock and the wildlife reserve is considered to be the same. As predicted, the frequency of illness rises as the rate of transmission to and from the wildlife reservoir rises. The authors hypothesize that in reality, the rates of transmission between every flock and the reservoir of wildlife are best described as random variables, showing variation in the depth of interaction with wildlife. The authors allow future work to continue development.

In a transhumance herd, a simulation of a notional spread framework from a starting infection is shown in Figure 7.2. In a non-intensive agricultural system, brucellosis in cow herds is simulated in the hypothetical situation when controls are not utilized. Here, a network of 10 transhumance flocks (represented by spheres) and 10 stationary flocks (represented by squares) is taken into consideration. The herd sizes differ, but the intra-herd transmission circumstances are generally equivalent. Transhumance herds mix with sedentary and non-transhumance flocks, as seen by the spread graph (which is superimposed over the dynamics). Here, a distance measurement between the herds is utilized to specify the likelihood and frequency of spread across herds, while other models may be employed instead. The biggest transhumance herd (Herd 1) has a limited number of sick animals at the start of the simulation. The dynamics of the spreading through time are depicted (months are represented on the time axis), and the severity of the red color for each flock represents the degree of infection in that herd. Since the virus spreads rapidly and the fundamental reproductive number R0 is bigger than 1, this is to be anticipated.

Figure 7.3 shows the simulation of a hypothetical spread model starting with an infection in a herd that is sedentary. For the 20-herd example, the basic framework for brucellosis transmission is once again implemented without a control, but the illness starts in a smaller, seated flock (flock 20). The sickness returns, but this time it spreads much more slowly. Even though the virus first appeared in a small sedentary herd, it is interesting to note that brucellosis spreads more quickly in big migratory herds than in smaller or sedentary ones. This model feature is consistent with field assessments of the frequency of brucellosis in pastoralist groups, such as those in Ethiopia's Tigray area.

Control Strategy Modeling

The 20-herd example compares a number of immunization tactics, namely those that target transhumance flock vs. those which split resources between transhumance and seated flocks. In this illustrated example, the cost required for each vaccination of a herd is considered to be the same. Specifically, flocks are portrayed seeing vaccinated at a particular frequency. (In practice, the relative expenditure of immunizing sedentary and transhumance herds can occasionally vary considerably. In some situations, it may be extremely difficult to immunize transhumance herds, whereas in others, medical professionals can take facility of their movability by setting up vaccination facilities at a position frequently visited by these flocks. For convenience of demonstration, we assume equivalent costs, although the design may be accomplished with various cost structures as well.) Comparing three vaccination regimens with equivalent total resource costs, including one which vaccinates both seated and transhumance flocks as well as one which vaccinates exclusively sedentary herds. The fundamental reproductive ratio R0 that may be computed from the state matrix spectrum of the linear model dynamics is compared for the three immunization procedures. The outcomes are shown in Figure 7.4. If resources are few, vaccination of transhumance flocks, but not uniform vaccination or immunization of stationary herds, may lower the fundamental reproductive amount under 1. When additional resources are available, uniform vaccination finally outperforms transhumance-herd vaccination alone and is preferred (and lowers R0 to 1). However, in the transhumance herds, chronic illnesses are still often seen.

Improved surveillance methods that enable quick and inexpensive decentralized surveillance/culling are predicted to considerably enhance brucellosis control. The framework enables assessment of the advantages of

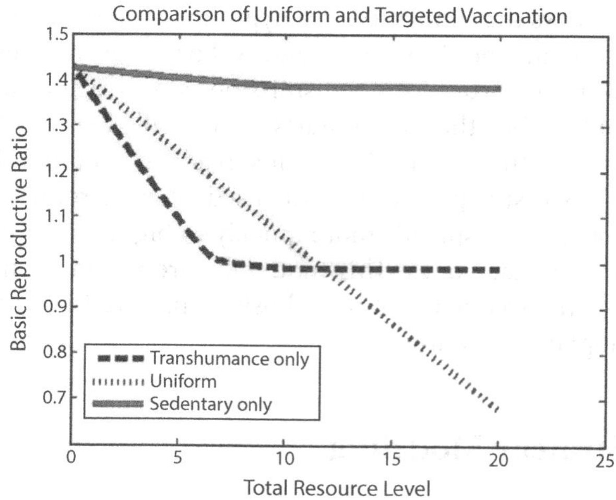

Figure 7.4 Comparing immunization programs quantitatively. A vaccination program that covers all herds is contrasted with one that just covers sedentary herds and transhumance herds. The immunization program aimed at the transhumance herds works better at low resource levels than the uniform program, while the uniform program gains efficacy at increasing resource levels. Both are more effective than sedentary-herd immunization.

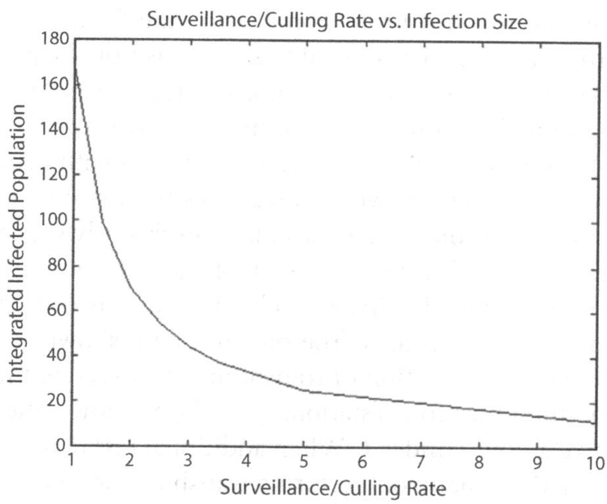

Figure 7.5 Examination of a policy for monitoring and control. Enhancing brucellosis surveillance protocols to enable quick/low-cost distributed monitoring and killing is a crucial regulation objective. The framework provides calculation of infection expenditures as a function of the monitoring and killing rate, and consequently illustrates the expenditure advantage of upgrading monitoring/killing strategies.

quicker surveillance/culling. For the 20-herd scenario, a consistent monitoring/killing regulation is taken into account. It is established how an infection cost (particularly, the population of diseased animals integrated over time) depends on the pace of monitoring and culling. Figure 7.5's dependence illustrates the advancement in spread mitigation brought on by quicker surveillance/culling.

Figure 7.4 shows a quantitative analysis of vaccination regulations. A consistent vaccination program is contrasted with ones that solely target sedentary cattle and transhumance herds, respectively. The immunization program aimed to transhumance herds works well than the uniform program at low resource levels, whereas the uniform program performs better at greater resource levels. All are superior to vaccination of sedentary herds.

A practitioner might assess if the additive expense to develop newer monitoring/killing strategies is worthwhile given the decrease in spread expenditure brought on by these advancements using characterization.

Here, for the 20-herd network, we adopt a homogeneous monitoring/killing technique and compute the aggregated infected magnitude (the amount of affected animals added together during the course of the illness). In R0 to be less than 1 in this illustration, a killing rate of a minimum of 0.43 (43% of the diseased population annually) is required; nonetheless, culling rates beyond this threshold have a considerable impact on the extent of the overall infection.

Optimized Approaches

According to the techniques described above, the modeling approach we have proposed enables systematic optimization [14] of control resources. As an example, we have discovered that, depending on the quantity of available resources, the best vaccine resource allocation minimizes the primary reproductive amount in the illustrated case. Figure 7.6 shows the ideal primary reproductive amount and the percentage of resources provided to the transhumance flocks. Figure 7.7 also depicts how the vaccination strategy has affected the prevalence of infection among pastoralists. The effectiveness of an ideal monitoring and killing approach (relating to the obtained primary reproductive amount R0) is also shown in Figure 7.8 in contrast to the optimal vaccination techniques (Figure 7.6). Last but not least, the 20-herd network takes into account the concurrent framework of monitoring/vaccination capacities and pasteurization to avoid transfer to humans. This straightforward example examines the best design (in terms

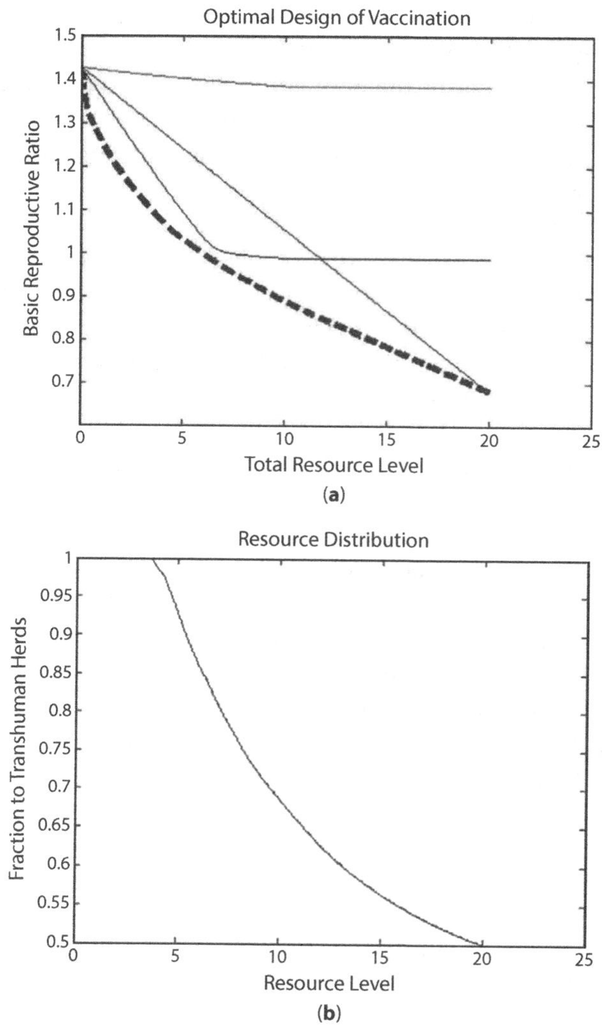

Figure 7.6 Creating the best possible immunization program. The top figure displays the fundamental reproductive number for the 20-herd example when the best immunization schedule is followed. The performance of the only-transhumance, only-seated, and constant immunization rules—all of which were produced in Figure 7.4—is shown by the three bright, solid lines here. A bold, dashed line indicates how well the optimum resource allocation performed. Additionally, the optimal proportion of resources allotted to sedentary herds is shown.

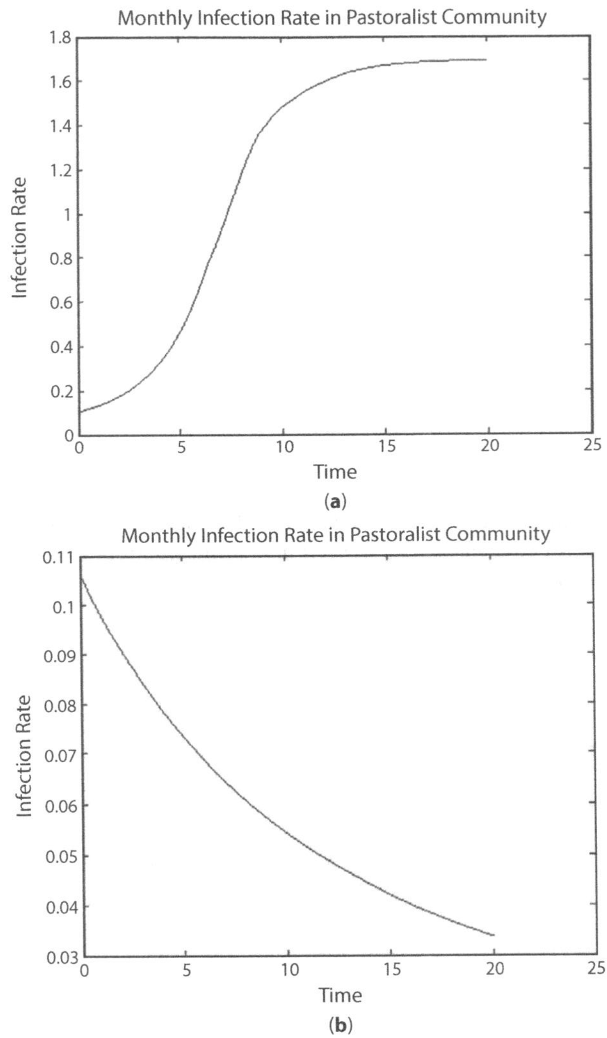

Figure 7.7 The spread framework determines the rate of human disease. In an epidemic of brucellosis when the initial level of infection in the cattle population was low, the rate of disease in the pastoralist human subpopulation is depicted. The scenario where no control is utilized is shown in the top figure, and the scenario where the best animal vaccination strategy is followed is shown in the bottom figure. It is essential to note that the infection rate was simulated using the nonlinear spread model, since the linear model rapidly loses accuracy in the absence of a control.

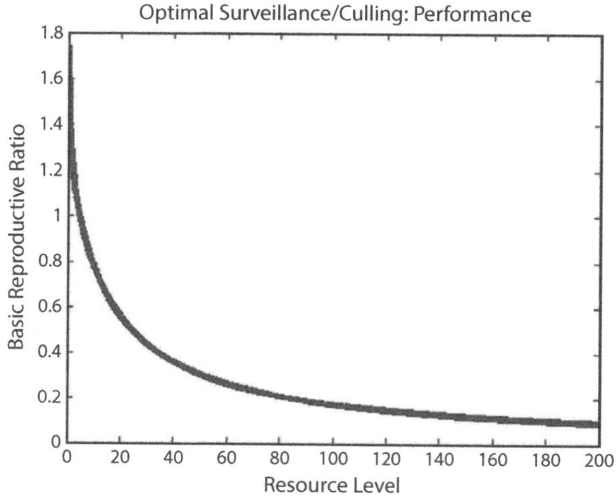

Figure 7.8 Fundamental reproduction ratio for ideal layout. For the 20-herd example, the fundamental reproductive amount for the best monitoring/control strategy is depicted as a function of resource level.

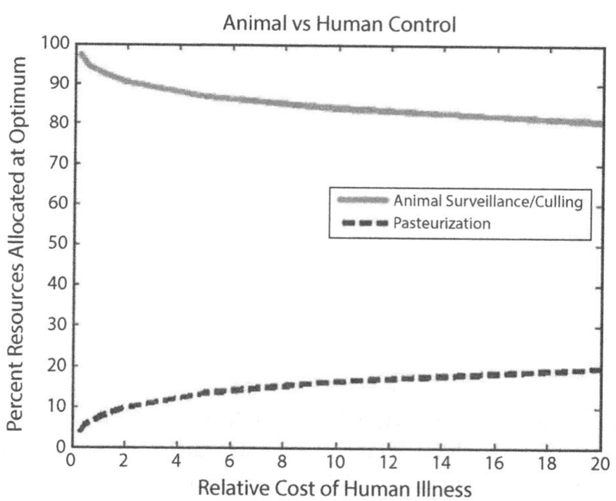

Figure 7.9 Animal-level and animal-to-human controls are compared. Using the two-herd scenario, the distribution of resources between animal-level control strategies and animal-to-human control rules is investigated. In particular, the best way to allocate resources between animal monitoring/control and pasteurization is established, given a certain proportional cost (per person) of human infection vs. animal infected. The relative cost is displayed against the distribution of resources. More funding for pasteurization may be advantageous if the costs of treating human sickness are greater, but the majority of funds should still go toward animal control.

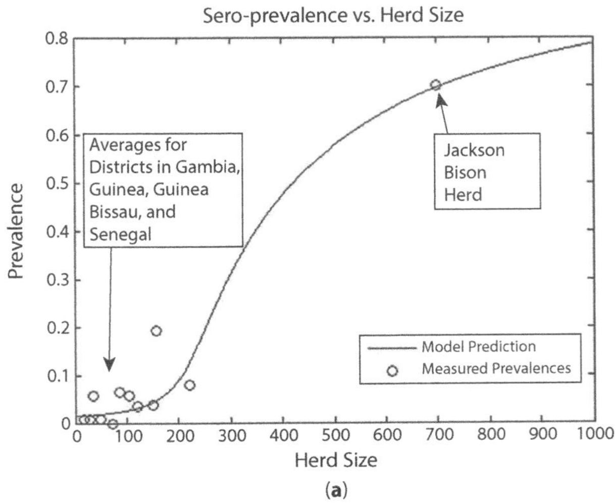

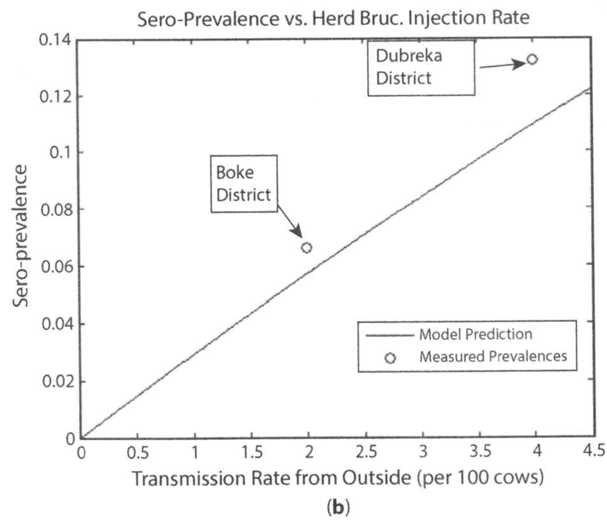

Figure 7.10 Model recognition using data from snapshots. A nonlinear SIR framework for brucellosis transmission in a flock was built utilizing a heuristic approach, using seroprevalence and snapshot herd-size data from the Jackson Bison Flock and many West African nations. The model's predictive power for seroprevalence in relation to herd size is shown at the top. Additionally, using a description of the common agricultural techniques, one may approximate the number of inter-flock contact and, thus, the comparative rate of outer-flock disease for two non-intensive agricultural regions in Guinea with comparable flock numbers. After accounting for this fluctuation, the framework is depicted to facilitate a more accurate picture of prevalence (bottom).

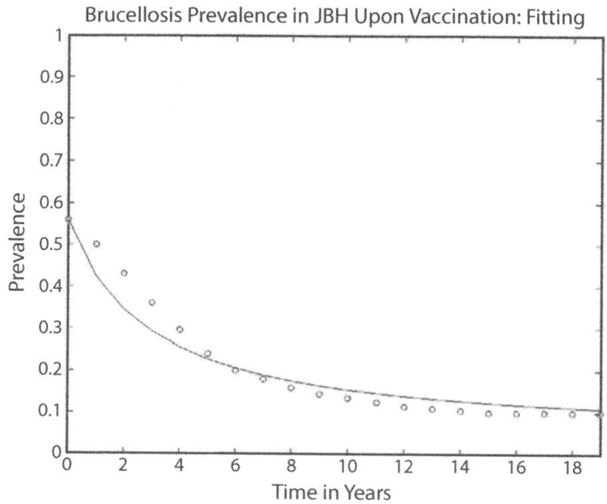

Figure 7.11 Identification of the framework using time-course data. After a vaccination program is started, a SIR framework for brucellosis transmission is found using time-course data from the Jackson bison flock. Although this straightforward framework is not as precise as the multi-state framework, it is enough for the extensive policy-design initiatives carried out in this study.

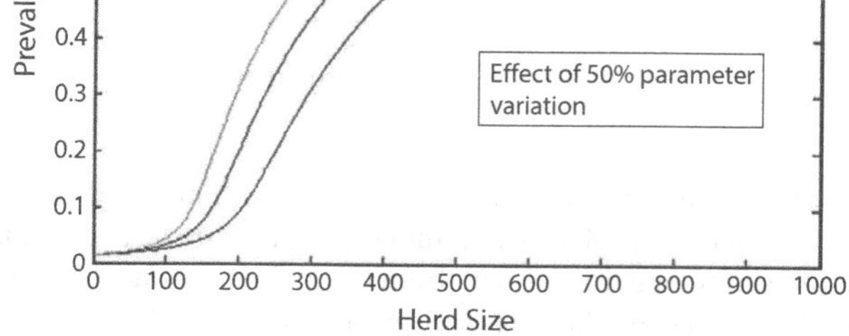

Figure 7.12 Hardness. The resilience of the model to parameter modifications is important since there is relatively little and unclear data available for model parameterization. We have examined the model's capacity to forecast the prevalence of brucellosis in a single herd when there is up to a 50% inaccuracy in each recognized model parameter as a first step in this approach. The graphic above demonstrates that even with this heterogeneity, the model is still able to accurately estimate the frequency of brucellosis in a particular herd.

of cumulative infected dimension estimation) of various human vs. animal disease expenditures. The expenditures of immunizing each herd and pasteurizing the milk from every flock are supposed to be equivalent. The outcomes are depicted in Figure 7.9.

Parameterization

The methods for determining the model of bovine brucellosis using snapshots and time-courses mentioned above were used. Figures 7.10 and Figure 7.11 show the findings of the identification. In the illustrated case, the discovered parameters have been utilized to determine the notional intra-herd and inter-herd transmission rates. Analyses of a surveillance and control regime are shown in Figure 7.5. An important policy objective is to enhance brucellosis surveillance methods to enable quick and affordable dispersed monitoring and culling. The model makes it possible to calculate infection expenditures as a function of monitoring and killing rates, which shows the financial benefits of developing surveillance and culling methods.

When the best vaccination strategy is utilized, the fundamental reproductive number for the 20-herd example is displayed in the top plot. The performance of the only-transhumance, only-seated, and consistent immunization rules as described in Figure 7.4 is shown here by the three bright, solid lines. Using a bold, dashed line, the performance of the ideal resource allocation is emphasized. Additionally, the optimal resource allocation for inactive cattle is presented.

Discussion

We have found it useful to provide an example to demonstrate the outcomes of our process. However, it is important to note that the technique has wider applications: the network modeling design developed may be utilized to quantitatively capture zoonotic infection transmission in a specific farming framework or community, and as a result, may be utilized for control and monitoring policy design for this transmission. The tractability of the model not only enables specific designs for instances but also comprehensive insights into brucellosis management and monitoring framework (and various zoonotic infections), as well as intellectual understanding into framework parameterization from past data. Let's talk about a few bigger conclusions drawn from the models.

Figure 7.7 shows the spread model's estimate of the human infection rate. In a brucellosis epidemic having an initial lower level of disease in the cattle census, the rate of infection in the pastoralist human subpopulation is demonstrated. The top picture depicts the scenario in which no control is utilized, while the bottom figure depicts the scenario in which the best animal vaccination strategy is used. It is foremost to observe that the infection rate was simulated using the nonlinear spread model since the linear model soon loses accuracy in the absence of controls. Since the mathematical intricacies are not essential to our growth, we highlight that several of these ideas permit formal explanations utilizing the model optimization technique. We also provide a quick summary of the text after this debate.

Wide-Ranging Perspectives on High-Performance Control

The brucellosis controller design challenge, as well as zoonotic-disease control more broadly, may be studied thoroughly using network control approaches. The many facets of the control design work that may be handled using these techniques are listed below, along with some of the most important conclusions drawn from the design. These perceptions are intended to guide policy choice.

1) Based on the structure of the spread graph, it is possible to identify the optimal [15] or higher-performance [16] subdivisions of control resources among the animal flocks. High-performance designs are preferable when the costs of resource allocation and the costs of infection brought on by herd prevalence are comparatively uniform.

Figure 7.8 shows the fundamental reproductive ratio for ideal design. For the 20-herd example, the fundamental reproductive amount for the best monitoring/control strategy is presented as a function of resource level tend, to the degree permitted by the restrictions on resources and management, to balance the spread-consequences of the flocks. As a result, until the spread effects are equalized, finite resources (maybe for immunization, monitoring, killing, or various objectives) must be directed to flocks having higher spread consequence (summation of outgoing edge weights from that flock's vertex on the graph). If more resources are available, it is best if they are split evenly among the herds. The design gets somewhat more complex when limitations hinder the distribution of resources (or more resources) to certain herds. In this case, some herds must be given additional resources in order to reduce the spread-consequence of their limited connections. Figure 7.7's ideal vaccine strategy for the illustrated scenario shows these resource allocation features. As anticipated, until sufficient resources are available to ensure equal spread effect, the obtainable resources are allocated to the transhumance flocks that in this scenario have a larger spread consequence. A steady, transhumance-exclusive, or seated- exclusive resource allotment scheme performs noticeably worse than the optimum allocation strategy. If an aggregated disease dimension metric is employed rather, findings are similarly quite comparable.

2) The design technique enables concurrent creation of numerous methods and contrast of various spread-control tactics in the animal census (e.g., vaccination vs. monitoring/killing). For the sake of illustration, we compared an ideal vaccination approach (Figure 7.6) with an optimal monitoring and killing strategy (Figure 7.8). A detailed understanding of control and cost characteristics, such as vaccination effectiveness and relative-cost information, is necessary for a rigorous comparison of numerous methods (for example, the two techniques in the instance). But even a cursory inspection of Figures 7.6 and 7.8 reveals considerable distinctions between the two approaches. In example, one discovers that a monitoring and culling approach may lower the primary reproductive amount casually close to 0, i.e., it may swiftly eradicate the illness, but a vaccine strategy can only do so up to a certain threshold. Therefore, monitoring and culling will need to be implemented, maybe at a greater expense, if a sufficiently quick decrease in spread is required (for example, in a situation where the zoonotic pathogen is very harmful to humans and moreover a methodical transmitter). This balance between immunization and cost will change as cost and efficacy metrics become available.

Comparing animal-level and animal-to-human controls is shown in Figure 7.9. In the two-herd scenario, the distribution of resources between animal-level control strategies and animal-to-human control rules is addressed. The ideal resource allocation between minimal surveillance/control and pasteurization is established, specifically under the assumption of a certain proportional expenditure of human disease compared to animal disease (per person). The relative cost is shown against this resource allocation. Additional pasteurization funding is advantageous if the costs of human sickness are greater, but animal control should still get the majority of funding and explicit surveillance/control tactics may be used.

As a function of resource and infection costs, the design process may determine the tradeoff between providing resources to human population classes as opposed to the animal census. The necessity of comparing strategies that allocate control resources to reduce transmission among animals with those which impede transmission to people is underscored by the enormous variation in brucellosis (and various zoonotic illnesses) transmission in the human and animal census. These comparisons are made possible by the recently established network design methodologies, which in turn enable the proper allocation of resources for animal-level management and control of animal-to-human transmission. Let's use our concurrent framework of monitoring/vaccination potentialities and pasteurization to avert transmission for the 20-flock instance network as a demonstration. The majority of the resources in this example are used to prevent infection at the animal level, protecting both human and animal census. Still, as is to be awaited, the percentage of resources used for pasteurization rises as it is assumed that the cost of human illness is higher than the expenditure of disease in cattle. It is not unexpected that controlling the animal population has such a positive impact since doing so efficiently lowers illness prevalence in both animal and human populations. The framework technique permits for the parallel creation of human and animal level plans and the presentation of these frameworks across a span of undisclosed parameters, albeit the details of the subdivision would differ depending on the characteristics of the framework challenge.

4) The fundamental resource-allocation concepts may be improved to provide proactive or reactive zoonotic disease mitigation methods. Techniques which dynamically allot resources depending on these patterns may be devised when real-time calculation of disease computation (or expenditures) trends is obtainable. For a first attempt in this regard, the reader is directed to [17]. The article [18] in particular offers a methodical strategy for dynamic resource allocation depending on usage of present model identification using snapshot data, Figure 7.10. Utilizing snapshot flock-dimension and seroprevalence data from numerous African nations and from the Jackson Bison Flock, a nonlinear SIR framework for brucellosis transmission in a flock was created utilizing a heuristic approach. The model's capacity to forecast seroprevalence in relation to flock dimension is depicted (above). Additionally, given two non-intensive agricultural areas in Africa with comparable flock numbers, it is possible to approximately estimate the quantity of inter-flock contact and, thus, the comparative rate of outside-flock disease from a description of the common agricultural techniques. Once this variance is taken into account, the framework is demonstrated to facilitate a good approximation of prevalence (bottom).

Talking About Parameterzing Models

Using the findings shown in Figures 7.10, 7.11, and 7.12 as a backdrop, let's share some insights on the model identification methodology:

1. Execution of the ratio between the per-person infected rate and the remission rate (via cure or death and substitution) is made possible by the snapshot data, for just one herd. Figure 7.10 shows how the generated model represents a mean herd dimension vs. mean prevalence curve for bovine brucellosis for areas in Africa and for the Jackson Bison flock with reasonable accuracy.
2. The time-course data also enables inference of the complete per-individual infected rate, allowing for the identification of a complete single-herd model when combined with the snapshot data-dependent survey. The time-course data for the Jackson bison flock are shown in Figure 7.11 to illustrate how the model might represent them. We should point out that in the 20-herd example, we employed the parameterized model.
3. The parameterization of the single-herd model was achieved estimating an equivalent rate of interactivity/disease from outside the flock since we lacked quantitative information on the rates of disease in every flock from outside the flock. The parameterization result is obviously inaccurate because of this shoddy assumption. With regard to the data from African flocks, the authors may get a few approximate understandings of which flocks are much vulnerable to infection from outside the herd since they contact with each other more often (e.g., share similar feed lots). We also show in Figure 7.10 how additional data on outside infected rates might result in a much more precise parameterization of the framework.
4. Provided the much sparse and extremely changeable data that are obtainable for parameterization, the framework's resistance to parameter changes is crucial. As a preliminary phase in this way, the authors looked at the model's propensity to forecast lone-flock brucellosis commonness when each detected model parameter was up to 50% off. Figure 7.11 demonstrates that despite this variation, the

model is still reliable for forecasting single-herd brucellosis prevalence.

5. The unpredictability of the data used to parameterize the model emphasizes the significance of taking stochastics into account while simulating the spread of illness. Both inherent transmission variability and unpredictability/changeability in framework parameters can have a considerable influence on the dynamics of the model and, as a result, alter control strategy. The authors are sure that the key perception gained via the framework technique will hold true even in the face of uncertainty since the design methods we have followed exhibit strong resiliency to uncertainty. Still, the authors see strengthening the framework to account for dynamics and parameter uncertainties, as well as applying these models for more accurate survey and design, as an essential succeeding step.

Conclusion

In order to track the transmission of zoonotic agents at a flock/subpopulation level, a network modeling tool has been developed, presented, compared, and used to develop management techniques to halt the spread of zoonoses. The modeling and management of brucellosis spread in animal and human census served as the case study that allowed for the detailed development of the presented technique. The goal has been to identify the model's parameters using historical data.

The modeling and controller design work must be seen as a first phase toward developing thorough zoonoses control rules from statistical techniques. The rules proposed by our technique need to be evaluated in implemented flocks and it's important to carefully consider the social and political repercussions of control policies and spread mitigation when defining costs for a zoonosis of interest. However, we think that this basic investigation has produced some encouraging results.

1. The transmission of a zoonotic infection across subgroups of a single or various species may be shown in straightforward, visual terms using network modeling. The spread dynamics may be simulated and subjected to a straightforward quantitative examination using the layered network model we

created. The architecture for layered network modeling naturally accommodates a range of control mechanisms.

2. The network modeling strategy which the authors have proposed here enables methodical and significant thorough resource planning for population management in both the animal and human census. In particular, newly created techniques from the literature on network control theory may be used to construct heterogeneous control capacities for the linear network model under study in order to reduce a resource and spread expenditure. The authors emphasize that there is further considerable work to be completed in order to confirm that the design's abstract regulation can really be implemented and provide performance that is consistent with what the model predicts.
3. Network modeling techniques provide enough information to compare and construct diverse control measures, such as those that rely on culling and surveillance versus those that rely on vaccination, and those which take into account the topology of connections between the flocks. It is possible to acquire both descriptive and calculable forecasts and designs.
4. Despite the fact that optimum controller design needs a concise spread dynamics and cost model, our framework technique may nevertheless supply valuable understanding in the absence of such models. First, the technique offers straightforward conclusions about ideal designs that can be put into practice without the need for a detailed model, such as the idea that greater resources should be allocated to herds that are big and well knit. Second, the designs produced by the technique exhibit some level of resistance to changes in the model parameters.
5. We have presented some early findings about the parameterization of the network framework using data and have moreover noted few difficulties in properly resolving the parameterization issue. Model parameterization remains a key difficulty when applying techniques like those suggested here due to the finite data on the transmission of numerous zoonotic illnesses. However, our early attempts at parameterization are promising, as the model exhibits some degree of resistance to parameterization errors.

6. It is obvious that stochastics in transmission have a very large influence on the spread dynamics when transmission across tiny heterogeneous groups is taken into account (as in our model). In example, inherent variations in interpersonal interactions that promote spread as well as environmental variables that alter transmission patterns may have a significant influence on the spread dynamics. Unsurprisingly, zoonotic disease prevalence like brucellosis exhibit high flock-to-flock variability, which may be described by taking stochastics in transmission into account. Because it is extremely challenging to parameterize these stochastic frameworks having limited data and as our primary goal in this first design effort is to gain very basic insights into regulation design, even the deterministic multi-group framework can be adequate, the authors have chosen to ignore stochastics in transmission. We see the creation of stochastic models for transmission as a superb scientific undertaking of paramount significance.

References

1. M. S. Hossain, U. S. Salsabil, M. M. M. Syeed, M. M. Rahman, K. Fatema and M. F. Uddin, "SmartPoultry: Early Detection of Poultry Disease from Smartphone Captured Fecal Image," *2023 20th International Joint Conference on Computer Science and Software Engineering (JCSSE)*, Phitsanulok, Thailand, 2023, pp. 345-350, doi: 10.1109/JCSSE58229.2023.10202054.
2. Ashisha G. R., Anitha Mary X., T. George, Martin Sagayam K, Unai Fernandez-Gamiz, Hatıra Günerhan, M. N. Uddin and S. Pramanik, Analysis of Diabetes Disease Using Machine Learning Techniques: A Review, *Journal of Information Technology Management*, 2023.
3. S. Pramanik and S. Bandyopadhyay, Identifying Disease and Diagnosis in Females Using Machine Learning, in *Encyclopedia of Data Science and Machine Learning*, Ed. John Wang, IGI Global, 2023, doi: 10.4018/978-1-7998-9220-5.ch187.
4. V. Jain, M. Rastogi, J. V. N. Ramesh, A. Chauhan, P. Agarwal, S. Pramanik, and A. Gupta, FinTech and Artificial Intelligence in Relationship Banking and Computer Technology, in *AI, IoT, and Blockchain Breakthroughs in E-Governance*, Eds., K. Saini, A. Mummoorthy, R. Chandrika and N. S. Gowri Ganesh, IGI Global, 2023.
5. S. Praveenkumar, V. Veeraiah, S. Pramanik, S. M. Basha, A. V. Lira Neto, V. H. C. De Albuquerque and A. Gupta, Prediction of Patients' Incurable

Diseases Utilizing Deep Learning Approaches, *ICICC*, 2023, Springer, doi.org/10.1007/978-981-99-3315-0_4

6. V. VidyaChellam, V. Veeraiah, A. Khanna, T. H. Sheikh, S. Pramanik and D. Dhabliya, A Machine Vision-based Approach for Tuberculosis Identification in Chest X-Rays Images of Patients, *ICICC*, 2023, Springer, doi.org/10.1007/978-981-99-3315-0_3
7. A. D. Andrushia, T. M. Neebha, A. T. Patricia, K. M. Sagayam, S. Pramanik, Capsule Network based Disease Classification for VitisVinifera Leaves, Neural Computing and Applications, doi: 10.1007/s00521-023-09058-y, 2023
8. A. Roy and S. Pramanik, A Review of the Hydrogen Fuel Path to Emission Reduction in the Surface Transport Industry, *International Journal of Hydrogen Energy*, https://doi.org/10.1016/j.ijhydene.2023.07.010, 2023
9. R. Jayasingh, J. Kumar R. J. S., D. B. Telagathoti, K. M. Sagayam and S. Pramanik, "Speckle noise removal by SORAMA segmentation in Digital Image Processing to facilitate precise robotic surgery", *International Journal of Reliable and Quality E-Healthcare*, vol. 11, issue 1, doi: 10.4018/IJRQEH.295083, 2022
10. D. Samanta, S. Dutta, M. G. Galety and S. Pramanik, A Novel Approach for Web Mining Taxonomy for High-Performance Computing, *4th International Conference of Computer Science and Renewable Energies (ICCSRE'2021), 2021*, DOI:10.1051/e3sconf/202129701073.
11. R. Bansal, A. J. Obaid, A. Gupta, R. Singh, and S. Pramanik, "Impact of Big Data on Digital Transformation in 5G Era", *2nd International Conference on Physics and Applied Sciences (ICPAS 2021)*, doi:10.1088/1742-6596/1963/1/012170, 2021.
12. B. Bansal, V. Nisha Jenipher, R. Jain, R. Dilip, M. Kumbhkar, S. Pramanik, S. Roy and A. Gupta, "Big Data Architecture for Network Security", in *Cyber Security and Network Security*, Eds., Wiley, 2022, https://doi.org/10.1002/9781119812555.ch11
13. R. Anand, J. Singh, D. Pandey, B. K. Pandey, V. K. Nassa and S. Pramanik, "Modern Technique for Interactive Communication in LEACH-Based Ad Hoc Wireless Sensor Network", in *Software Defined Networking for Ad Hoc Networks*, M. M. Ghonge, S. Pramanik and A. D. Potgantwar, Eds., Springer, 2022, https://doi.org/10.1007/978-3-030-91149-2_3
14. S. Choudhary, V. Narayan, M. Faiz and S. Pramanik, "Fuzzy Approach-Based Stable Energy-Efficient AODV Routing Protocol in Mobile Ad hoc Networks", in *Software Defined Networking for Ad Hoc Networks*, M. M. Ghonge, S. Pramanik and A. D. Potgantwar, Eds., Springer, 2022, https://doi.org/10.1007/978-3-030-91149-2_6
15. K. Dushyant, G. Muskan, A. Gupta and S. Pramanik, "Utilizing Machine Learning and Deep Learning in Cyber Security: An Innovative Approach", in *Cyber Security and Digital Forensics*, M. M. Ghonge, S. Pramanik, R. Mangrulkar, D. N. Le, Eds., Wiley, 2022, https://doi.org/10.1002/9781119795667.ch12

16. B. K. Pandey, D. Pandey, S. Wairya, G. Agarwal, P. Dadeech, S. R. Dogiwal and S. Pramanik, "Application of Integrated Steganography and Image Compressing Techniques for Confidential Information Transmission", in *Cyber Security and Network Security*, https://doi.org/10.1002/9781119812555.ch8, Wiley, 2022.
17. J. Li, D. Wang, Z. Yang and M. Liu, "HEGANLDA: A Computational Model for Predicting Potential Lncrna-Disease Associations Based on Multiple Heterogeneous Networks," in *IEEE/ACM Transactions on Computational Biology and Bioinformatics*, vol. 20, no. 1, pp. 388-398, 1 Jan.-Feb. 2023, doi: 10.1109/TCBB.2021.3136886.
18. M. M. I. Khan and G. Nencioni, "Resource Allocation in Networking and Computing Systems: A Security and Dependability Perspective," in *IEEE Access*, vol. 11, pp. 89433-89454, 2023, doi: 10.1109/ACCESS.2023.3306534.

8

Modeling the Dynamics of Agricultural Cooperatives and Legal Implications for Farmer Organizations

Shiv Shankar Shankar[1], Prashantkumar Zala[2], Ashutosh Awasthi[3], Ezhilarasan G.[4], Sukhvinder Singh Dari[5] and Soma Bag[6*]

[1]Department of Mathematics, Vivekananda Global University, Jaipur, Rajasthan, India
[2]Department of Agricultural Extension, College of Agriculture, Parul University, Limda, Dist- Vadodara, Gujarat, India
[3]College of Agriculture Sciences, Teerthanker Mahaveer University, Moradabad, Uttar Pradesh, India
[4]Department of Electrical and Electronics Engineering, Faculty of Engineering and Technology, JAIN (Deemed-to-be University), Karnataka, India
[5]Symbiosis Law School, Nagpur Campus, Symbiosis International (Deemed University), Pune, India
[6]Asadtala Nivedita Kanya Vidyamath, Nandigram, West Bengal, India

Abstract
This abstract summarizes the findings of mathematical modeling that examined the impact of structural flaws on the operational dependability of a network of rice-growing systems on farms over an extended period of time. The foremost rice system was constructed in the Kuban in 1929, and it has been 90 years since then. Systems for growing rice that have been in use in the Krasnodar Territory for more than 30 years are now being employed. As a result, it is essential to support the expenditures of preservation and maintain scientifically. To do this, a numerical experiment was conducted on a variety of structural flaws in the on-farm network of rice systems. Depending on farm research, statistical modeling was done for the

**Corresponding author*: bagsoma2016@gmail.com
Shiv Shankar Shankar: ORCID: https://orcid.org/0000-0002-5487-0912
Ashutosh Awasthi: ORCID: https://orcid.org/0000-0002-5731-9018
Ezhilarasan G.: ORCID: https://orcid.org/0000-0002-5335-2347

Sabyasachi Pramanik, Niranjanamurthy M., Ankur Gupta and Ahmed J. Obaid (eds.) Mathematical Modeling in Agriculture, (153–162) © 2024 Scrivener Publishing LLC

two major typical GPR sounding flaws—a transverse fracture that runs through the drainage structure and the destruction of the well's sidewalls. The SCAD Office calculation complex was used for all computations. The mathematical modeling revealed that the operational dependability margin of the drainage pipeline with a well reduces as the structure is used over time. To determine the suitability of the solid-state framework of the stress-strain state at greatest water pressure with actual farm testing under different boundary conditions, mathematical modeling of a sewerage system having a well without faults and damages was developed. The functioning dependability of the construction of the on-farm network of rice systems is identified and defined using the data from the technical study of operational reliability. Implementing engineering, technological, and land reclamation methods for sustainable and efficient water conservation, rational usage, and growth of the whole water management complex is necessary to ensure the operational dependability of rice systems.

Keywords: Agriculture, legal methods, mathematical model

8.1 Introduction

The Kuban River's watershed is now featured by a higher degree of integrated economic [1] growth and the extent of irreversible water resource extraction in Lower Kuban. The repayment of water resources is not offered at the current level of irreversible extraction, particularly during the dry season. Additionally, there is adequate non-compliance with hygienic and ecological emissions [2]. The first rice system was constructed in the Kuban in 1929, and it has been 90 years since then. The rice sector is now experiencing a remarkable resurgence in the Krasnodar Territory; for instance, in 2015, more than 170,000 hectares were designated for rice production. As a result, the lower parts of the Kuban River have almost no free water resources reservoir [3] in Krasnodar, for future growth.

All of these outstanding issues show that solutions for enhancing the practical dependability of rice systems are required for their continued usage in the southern Russian water sector. Old rice systems are still in use in the Krasnodar Territory and the Republic of Adygea, so it is required to analytically support the expenditures of upkeep and regulate. To do this, numerical statistical modeling of the consequences of different, most familiar flaws in building the rice systems on their continued functional dependability was required. Using the Lower Kuban as an example, it is important to conduct the technical explanation of the strategy for carrying out different farm surveys of everlasting rice systems using harmless assessing instruments. The execution of engineering, technical and farm recovery methods for sustainable [4] and fruitful water conservation, logical usage, and growth of the water management system as an aggregate is the application of the working dependability of the forms of rice systems.

8.2 Resources and Techniques

Using an OKO-2 GPR, field assessments of the rice systems [5] were conducted. Since the drainage structure is visible and readily available for examination and measurements, it was inspected in the greatest depth.

However, the area of the building that was submerged while it was in use sustained significant damage; the pipeline was completely wrecked from the junction to the water consumption well. A transverse fracture developed as a result of the disintegration of concrete rings. The reality that the concrete in such location is in the major unfavorable circumstances and causes an aggregation of faults and solid disintegration is the reason why it was destroyed. Metal hooks used to clear clogged drainage structures that are closed cause damage to the pipeline's inner surface and butt joints.

Depending on farm research, statistical modeling was done for the two such typical flaws found following GPR sounding [6]: the destruction of the well's sidewalls and a transverse fracture running through the drainage system. The SCAD Office Complex [7] was used for all computations; solid-state [8] frameworks of the stress-strain condition of the drainage form were constructed.

The goal was to determine the suitability of the solid-state framework of the stress-strain [9] state at greatest water pressure having operating field tests for various boundary conditions in the formulation of mathematical modeling [10] of the model of a drainage structure [11] having a well sans defects and damages (Figure 8.1).

These mathematical modeling techniques highlight the suitability of the stress-strain model status of the well-connected drainage system. Figure 8.2 shows the figure of the final displacements from a 3-D [12] study of the stiffness of the drainage pipeline [13] having the well that accounts for contact proximity.

It was discovered that there is a loss of stability of the bearing [14] parts right in the epicenter of the gap of a drainage pipe rupture during the statistical modeling of a drainage pipeline having a well and stability deprivation when laid on a soil base. The structure is subject to perpendicular horizontal loads and increasing vertical stresses over its length. It is a deprivation of stability on all sides of the horizontal and vertical axes as a result of the crack's length and opening lengthening.

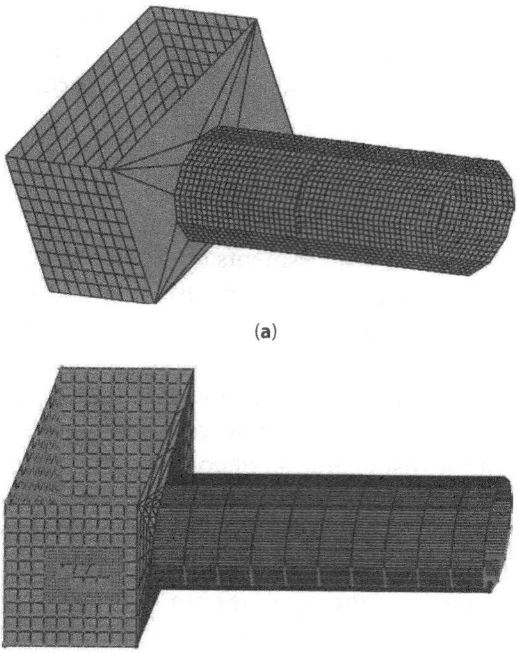

(a)

(b)

Figure 8.1 The calculation plan for the solid-state drainage pipeline having a well: (a) having a transverse fracture; (b) with the sidewalls of the well being destroyed.

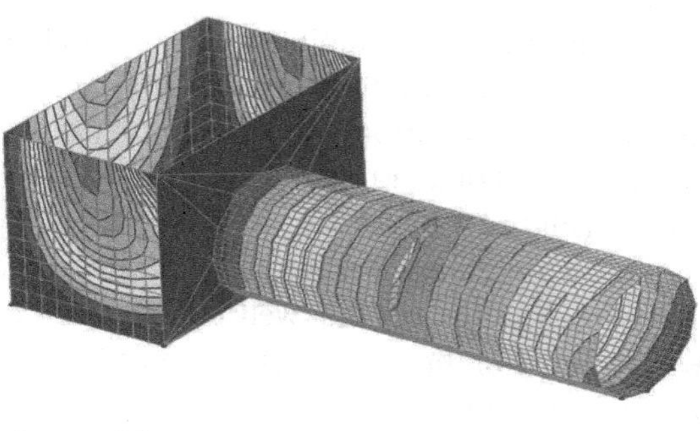

Figure 8.2 A simulation of the impact of a transverse fracture in a drainage design as a function of cumulative displacements.

Voids [15] were seen beneath the ring and to its left at the connection between the rings, but the butt joints' sealing was compromised, allowing water to escape into the pipeline's base's soil. Tunnels and holes have developed on the exterior of the bottom portion of the building as a result of the oxidizing [16] reaction that happens when damp soil comes into touch with concrete. The concrete along the examined structure of the water-lifting structure has a porous structure and a distinctive hue, according to the research.

In contrast to buildings that have been in use for less than 20 years, it was observed that the concrete exterior has a deep chisel mark. The starting and end rings of the construction have a significant number of fractures, exposing the reinforcement there. Cracks that harm the well's side faces as well as those that run parallel to the lower side of the wall are present. The research showed that horizontal stresses are perpendicular to the frame's axis and that vertical stresses [17] grow down the pipeline's axis toward the well. The model, which was thought of as a thin-walled shell, was subjected to calculations of the stress-strain state while taking into consideration the circumstances surrounding its support. It was feasible to prove that the development of a crack is harmful for the functioning of the structure by determining the critical stresses of the gap of a transverse crack [18] that crossed the drainage form and the limiting sizes of its gap.

The crack opening length must be more than one-third of the pipeline diameter to reach the intensive danger criterion. The earth foundation will erode as a result of a minor crack forming, endangering the building (Figure 8.3).

While horizontal displacements are minimal, vertical displacements [19] grow vertically down the hole. The degradation of concrete and bare reinforcement has a major impact on the drainage pipeline with a well's operability. There were serious fissures [20] at the connection where the rings of the first and second sections joined; these cracks had an aperture of up to 15-20 mm and indicated concrete delamination, which exposed the reinforcement and caused the concrete body of the building to collapse. The vertical faces of the well's sidewalls also had concrete fissures in them. The piece of the rings in the second compartment has been entirely

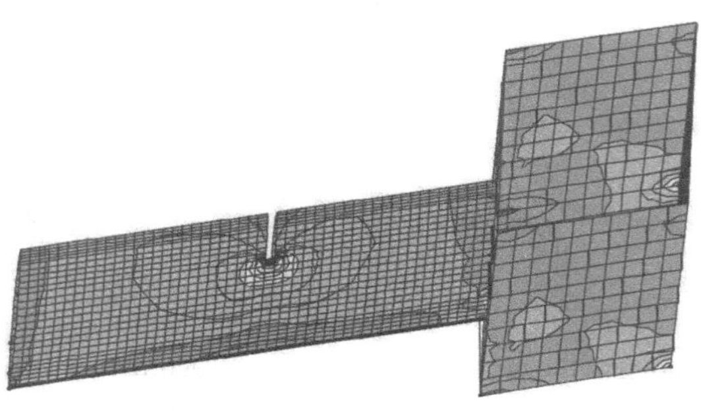

Figure 8.3 The transverse crack's critical stresses as it crosses the drainage structure.

destroyed by inclined, almost horizontal fissures. The top edge of the hole was 23° misaligned from the vertical axis, the fracture opening ranged from 0.5 to 1.8 mm, and the crack margins were horizontally displaced by 0.7 to 1.1 mm.

It was discovered that there is a depreciation of stability of the supporting components happening in the center of the drainage pipe that functions similar to a bending beam, during statistical [21] modeling of the effect of flaws on the side of the well of the drainage formation. Figure 8.4 displays the representation graphically.

Displacements are negligible and have no impact on the overall functionality of the structure while the destruction of the side parts of the well causes flaws. The concrete of the well's sides and bottom is completely saturated with water, and the parts that were subjected to the examination have fissures that range in width from 0.4 to 1.2 mm. There were two cracks on the south side (1.4 and 1.2 mm wide) and three cracks on the north side (0.3, 0.5, and 0.9 mm width) of the second compartment. The third compartment of the structure has seven fractures, none of which locations could be explained by external forces. One or two fractures were discovered in the midst of the compartments that came after.

A scheme of equal greatest horizontal stresses with a drainage pipe having a well is shown in Figure 8.5; the stresses are at their maximum allowable levels. The link between the well and the drain pipe might be hazardous as the load grows.

Figure 8.6 displays the highest equivalent horizontal strains across the drainage pipe when the well is present. The stresses have maximum allowable values, as shown in Figure 8.5. The connection between the well and the drainage pipe might be hazardous as the load grows.

Figure 8.4 A simulation of the impact of a fault in the structure of the demolition of a drainage form well's side, expressed as the total displacements.

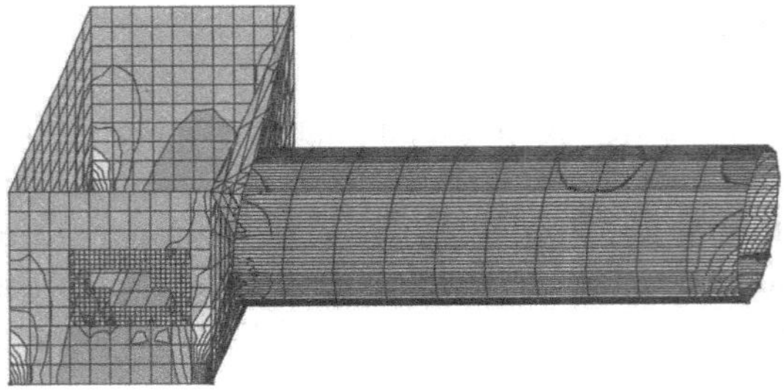

Figure 8.5 How to simulate the impact of a flaw that results in the destruction of a drainage structure's side using equal maximum horizontal stresses along the structure.

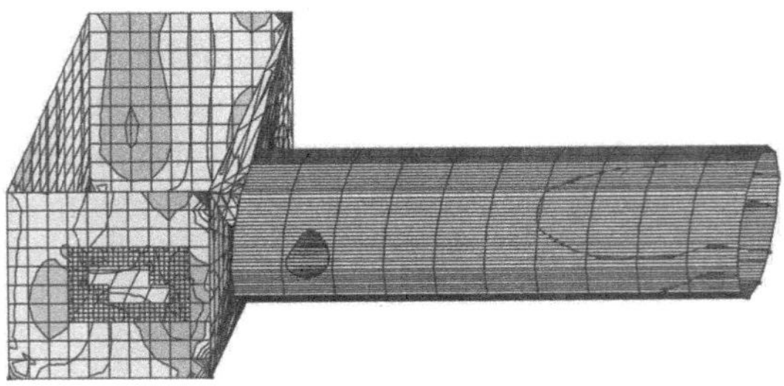

Figure 8.6 How to simulate the impact of a fault that results in the destruction of a drainage structure's side using equal maximum horizontal stresses throughout the structure.

Each fracture on the eastern side correlates to a gap on the western portion, which is situated just across from the first, since they all entirely cut through the part of the building. This demonstrates unequivocally the existence of a fracture in the below concrete connecting the opposing fissures in the well's sides. The protective layers' thickness is not always met. B 50 concrete was used in the building of the concrete. The identification of zones of defect formation and damages on the assisting components of the drainage pipeline [22] having a well—that can hold similar category of faults—implanted the building of a sufficient calculated framework for different operational parameters. For the functioning of the structure, an intense danger threshold has been defined, beginning from a diameter of 76 mm for the creation of gaps and deconsolidation of strengthened concrete. It has been shown that using statistical modeling to evaluate the capacity of rice system forms may enhance the impacts of operational dependability on control quality.

8.3 Conclusion

The mathematical modeling revealed that the operational dependability margin of the drainage pipeline with a well reduces as the structure is used over time. Mathematical modeling was done to portray numerous faults that were found in the GPR sounding and defined by a transverse crack running through the drainage system and the destruction of the well's sidewalls. These modeling techniques highlight how accurate the stress-strain state model of a drainage pipeline with a well. The computations for the spillway's decrease of stability and drawdown produced findings in the formation of surplus von-Mises and absolute movements, both horizontally and vertically. The rigidity of a drainage pipeline having a well was examined in three dimensions while accounting for contact support. The research showed that horizontal stresses are perpendicular to the structure's axis and that vertical stresses grow down the pipeline's axis toward the well. The operational reliability of rice system structures was identified and defined using the data from the technical examination of operational reliability.

References

1. J. Bartak and A. Jastrzębska, "Detecting Regimes of Economic Growth with Fuzzy Concept-Based Models," in *IEEE Access*, vol. 11, pp. 11664-11674, 2023, doi: 10.1109/ACCESS.2023.3241850.
2. A. Roy and S. Pramanik, A Review of the Hydrogen Fuel Path to Emission Reduction in the Surface Transport Industry, *International Journal of Hydrogen Energy*, https://doi.org/10.1016/j.ijhydene.2023.07.010, 2023.
3. N. Rathnayake, U. Rathnayake, I. Chathuranika, T. L. Dang and Y. Hoshino, "Projected Water Levels and Identified Future Floods: A Comparative Analysis for Mahaweli River, Sri Lanka," in *IEEE Access*, vol. 11, pp. 8920-8937, 2023, doi: 10.1109/ACCESS.2023.3238717.
4. M. Sinha, E. Chacko, P. Makhija and S. Pramanik, Energy Efficient Smart Cities with Green IoT, in *Green Technological Innovation for Sustainable Smart Societies: Post Pandemic Era*, C. Chakrabarty, Eds., Springer, 2021, doi: 10.1007/978-3-030-73295-0_16
5. P. Pal, R. P. Sharma, S. Tripathi, C. Kumar and D. Ramesh, "NSGA-III Based Heterogeneous Transmission Range Selection for Node Deployment in IEEE 802.15.4 Infrastructure for Sugarcane and Rice Crop Monitoring in a Humid Sub-Tropical Region," in *IEEE Transactions on Wireless Communications*, vol. 22, no. 6, pp. 3643-3656, June 2023, doi: 10.1109/TWC.2022.3220146.

6. E. Heggy *et al.*, "Probing Shallow Aquifers in Hyperarid Dune Fields Using VHF Sounding Radar," in *IEEE Transactions on Geoscience and Remote Sensing*, vol. 61, pp. 1-22, 2023, Art no. 4505822, doi: 10.1109/TGRS.2023.3306286.
7. S. Jeyapandi, A. E. Muthu, N. Vallileka, R. Niranjana, K. Saravanan and I. Sakthidevi, "Building Resilience in Shopping: Secure Shopping System for Pandemics Like Covid-19," *2023 4th International Conference on Smart Electronics and Communication (ICOSEC)*, Trichy, India, 2023, pp. 692-700, doi: 10.1109/ICOSEC58147.2023.10275894.
8. Dhamodaran, S. Ahamad, J. V. N. Ramesh, S. Sathappan, A. Namdev, R. R. Kanse, and S. Pramanik, Fire Detection System Utilizing an Aggregate Technique in UAV and Cloud Computing, *Handbook of Research on Thrust Technologies' Effect on Image Processing*, IGI Global, 2023.
9. S. Xue, J. Kwon, Y. Guo, T. Garg, M. D. Sumption and E. W. Collings, "Compressive Stress-Strain Behavior of REBCO Coated Conductors and Cables," in *IEEE Transactions on Applied Superconductivity*, vol. 33, no. 5, pp. 1-6, Aug. 2023, Art no. 4800706, doi: 10.1109/TASC.2023.3247370.
10. S. Dhamodaran, S. Ahamad, J. V. N. Ramesh, G. Muthugurunathan, K. Manikandan, S. Pramanik, and D. Pandey, in *Handbook of Research on Thrust Technologies' Effect on Image Processing*, IGI Global, 2023.
11. J. L. Svensen, C. Sun, G. Cembrano and V. Puig, "Model Predictive Control of Urban Drainage Systems Considering Uncertainty," in *IEEE Transactions on Control Systems Technology*, vol. 31, no. 6, pp. 2968-2975, Nov. 2023, doi: 10.1109/TCST.2023.3286648.
12. V. Talukdar, A. Z. Kurniullah, P. Keshwani, H. Khan, S. Pramanik, A. Gupta, and D. Pandey, *Load Balancing Techniques in Cloud Computing, Emerging Trends in Cloud Computing Analytics, Scalability, and Service Models*, IGI Global, 2023.
13. X. Zhou *et al.*, "Underground Pipeline Mapping from Multipositional Data: Data Acquisition Platform and Pipeline Mapping Model," in *IEEE Transactions on Geoscience and Remote Sensing*, vol. 61, pp. 1-13, 2023, Art no. 4504113, doi: 10.1109/TGRS.2023.3294518.
14. J. Chen, W. Hu, D. Cao, Z. Zhang, Z. Chen and F. Blaabjerg, "A Meta-Learning Method for Electric Machine Bearing Fault Diagnosis Under Varying Working Conditions with Limited Data," in *IEEE Transactions on Industrial Informatics*, vol. 19, no. 3, pp. 2552-2564, March 2023, doi: 10.1109/TII.2022.3165027.
15. F. Steiner, M. Hirman and P. Rous, "Analysis of the Effect of Soldered Surface Roughness on Void Formation Using Computed Tomography," *2023 46th International Spring Seminar on Electronics Technology (ISSE)*, Timisoara, Romania, 2023, pp. 1-5, doi: 10.1109/ISSE57496.2023.10168450.
16. Y. Wang, Z. Xia and M. Wong, "Dependence of the Transfer Characteristics of Metal-Oxide Thin-Film Transistors on the Temperature of an Extended

Oxidizing Heat-Treatment," in *IEEE Electron Device Letters*, vol. 44, no. 3, pp. 452-455, March 2023, doi: 10.1109/LED.2023.3240566.
17. K. Puthran, C. Barth, A. Ballarino, A. Devred and T. Arndt, "Onset of Mechanical Degradation due to Transverse Compressive Stress in Nb3Sn Rutherford-Type Cables," in *IEEE Transactions on Applied Superconductivity*, vol. 33, no. 5, pp. 1-6, Aug. 2023, Art no. 8400406, doi: 10.1109/TASC.2023.3241568.
18. L. Fan, S. Li, Y. Li, B. Li, D. Cao and F. -Y. Wang, "Pavement Cracks Coupled with Shadows: A New Shadow-Crack Dataset and a Shadow-Removal-Oriented Crack Detection Approach," in *IEEE/CAA Journal of Automatica Sinica*, vol. 10, no. 7, pp. 1593-1607, July 2023, doi: 10.1109/JAS.2023.123447.
19. S. Li *et al.*, "Long-Distance and Continuous Horizontal Displacement Sensor of Deep Excavation Based on Distributed Optical Fiber Sensing," in *IEEE Sensors Journal*, vol. 23, no. 20, pp. 24608-24616, 15 Oct.15, 2023, doi: 10.1109/JSEN.2023.3312947.
20. S. Kavitha, K. R. Baskaran and B. Dhanapriya, "Explainable AI for Detecting Fissures on Concrete Surfaces Using Transfer Learning," 2023 *International Conference on Inventive Computation Technologies (ICICT)*, Lalitpur, Nepal, 2023, pp. 376-384, doi: 10.1109/ICICT57646.2023.10134145.
21. P. K. Mall, S. Pramanik, S. Srivastava, M. Faiz, S. Sriramulu and M. N. Kumar, FuzzyNet-Based Modelling Smart Traffic System in Smart Cities Using Deep Learning Models, in *Data-Driven Mathematical Modeling in Smart Cities*, IGI Global, 2023, doi: 10.4018/978-1-6684-6408-3.ch005
22. D. Mondal, A. Ratnaparkhi, A. Deshpande, V. Deshpande, A. P. Kshirsagar and S. Pramanik, Applications, Modern Trends and Challenges of Multiscale Modelling in Smart Cities, in *Data-Driven Mathematical Modeling in Smart Cities*, IGI Global, 2023, doi: 10.4018/978-1-6684-6408-3.ch001

9

Optimizing Agroforestry Systems for Sustainable Agriculture: A Mathematical Modeling Approach

Beemkumar Nagappan[1], Aakriti Chauhan[2], Chandni Mori[3], Praveen Kumar Singh[4], Shilpa Sharma[5] and Sabyasachi Pramanik[6]*

[1]Department of Mechanical Engineering, Faculty of Engineering and Technology, JAIN (Deemed-to-be University), Karnataka, India
[2]Department of Agri-Business Management, Vivekananda Global University, Jaipur, Rajasthan, India
[3]Department of Horticulture, College of Agriculture, Parul University, Limda, Dist- Vadodara, Gujarat, India
[4]College of Agriculture Sciences, Teerthanker Mahaveer University, Moradabad, Uttar Pradesh, India
[5]Symbiosis Law School, Nagpur Campus, Symbiosis International (Deemed University), Pune, India
[6]Department of Computer Science & Engineering, Haldia Institute of Technology, Haldia, West Bengal, India

Abstract
The current study discusses the statistical and mathematical tools currently available for hazard characterization. In addition to other prospective approaches for threat evaluation of chemicals in food and nutrition, the modern statistical modeling and significant methodologies utilized for supervisory decision-taking in Asia are discussed. For the creation of fresh and enhanced quantitative approaches, a framework is created. A gap analysis identifies the methods' weak points and potential areas for development. Future research is needed, and this critical appraisal defines those requirements. Our research leads to the conclusion that statistical

**Corresponding author*: sabyalnt@gmail.com
Beemkumar Nagappan: ORCID: https://orcid.org/0000-0003-3868-0382
Aakriti Chauhan: ORCID: https://orcid.org/0009-0002-0935-2347
Praveen Kumar Singh: ORCID: https://orcid.org/0009-0001-4832-0025

Sabyasachi Pramanik, Niranjanamurthy M., Ankur Gupta and Ahmed J. Obaid (eds.) Mathematical Modeling in Agriculture, (163–188) © 2024 Scrivener Publishing LLC

modeling of the dose-response connection will enhance the threat evaluation procedure. The utilization of a needed amount of dosage classes to create a variety of various response levels is certainly necessary for an acceptable mathematical modeling portrayal of the dose-response connection. If an adequate design is utilized, this need not always result in an expansion in the overall quantity of animals used in the research. When available, data specific to a chemical, such as information on its mode of action or toxicokinetics, should be utilized to characterize dose-response relationships. It is decided that a variety of techniques are required to ensure that the technique chosen is the most perfect for the data provided and for the risk feature problem. A lone way of hazard presentation would not be adequate for every type of risk evaluation. Future dosage adjustments response characterization has to be more explicit about how much ambiguity and variability there is in the output.

Keywords: Benchmark, probabilistic techniques, toxicokinetic models, hazard characterization, risk assessment

9.1 Introduction

Low molecular weight substances, macronutrients, entire foods, innovative foods, nutritional supplements, micronutrients, and food processing might all theoretically be harmful to human health. Therefore, risk investigators, public health administrators, and the food business have all continued to be interested in human food safety. The necessity for toxicological and planned studies fit for risk evaluation was highlighted by the existence of minute quantities of chemicals in human food which are acknowledged to be dangerous at large levels. The primary goal has been the process of hazard recognition, hazard portrayal, and exposure evaluation, which results in the determination of a safe level for human exposure.

In order to explain modern art of mathematical modeling and significant approaches which may be utilized for supervisory decision-taking in Europe, the approaches explained in the succeeding sections was put together. The current procedures of the US Food and Drug Administration, the US Environmental Protection Agency, the Joint FAO/WHO Expert Committee on Food Additives [1], etc., are contrasted. Risk investigators, public healthcare officers, and the food industry should benefit from this and the industry in producing objective quantitative risk characterization and in using suitable testing, estimate, and assessment methods.

The working party of the Food Safety in Europe (FOSIE)—Concentrated steps on Risk evaluation of Chemicals in Foodstuff and Diet—was tasked with conducting a rigorous and in-depth evaluation of the significant threat characterization approaches utilized in food safety evaluation. It was accomplished by reviewing the state-of-the-art mathematical [2] and statistical [3] approaches now in use for hazard feature, and by outlining prospective future growth in the usage of quantitative [4] techniques. The operating class's goal was to provide a model for the creation of newer and enhanced significant techniques for evaluating the threat of compounds in foodstuff [5] and nutrition. A gap analysis revealed areas where the current processes may be improved, refined, and made more effective. The need for further study was defined in light of this evaluation. It's important to note that although the quantitative approaches presented below generally focus on chemicals found in foodstuff, like food pollutants or naturally present contamination, they are, in theory, moreover relevant to different classes, like nutrients and complete foods. Major components of mathematical modeling for determining foodstuff safeness were the focus of the work of the mathematical modeling topic group. Models for exposure and risk that used information from human hygienic [6] research were not discussed. The following frameworks were created and evolved for toxicity data which may employ implemented human data; however, they are not thought to be frameworks that are useful in the domain of hygienic. In this issue, [7] will discuss epidemiological risk modeling based on conventional analytical epidemiology techniques, while [8] will discuss exposure modeling. Modeling and significant survey should be dependent on data whose quality was guaranteed. We briefly summarize the current situation of significant risk evaluation methodologies for foodstuff safeness in the US before the outcomes of this coordinated study are presented.

European Union (EU) techniques

The idea of tolerable daily consumptions in foodstuff supplements was created by scientists working for the Joint FAO/WHO Expert Committee on Food Additives (JECFA) in the last century. It was an evolution of the FDA's job from the 1970s. The method makes the assumption that a greatest secure dosage, or a dose without major negative health consequences, may be discovered via research on experimental animals or people, and that the "safe" consumption of dose for the human census is then derived by dividing this dose by the relevant safety factor(s). At this time, the EU uses this conventional method to evaluate substances in food. The test procedures used follow the rules for testing chemicals established by the Agency for Economic Utility and Development, which the EU approved as part of its regulations on contaminant materials and compositions. In 1989, a guideline for industry provided a summary of the minimal criteria for data for the evaluation of foodstuff ingredients. Otherwise, there are currently no well-accepted quantitative techniques. The EU Scientific Organization on Foodstuff was founded in 1980 to provide the organization with advice on various issues pertaining to the safety of people's wellness and protection emerging from the intake of foodstuff, particularly with regard to the constituent of foodstuff, techniques that have the potential to change foodstuff, the usage of foodstuff compositions and different processing facilities, and the existence of impurities. When new food components, such constituents, are discovered and when toxins are found, evaluations are conducted. When new information is brought to the Committee and is anticipated to alter the previous assessment, the evaluation may be reexamined. The EU Scientific Organization on Contamination, Eco-contamination, and the Ecology publishes studies on the hazards that industry contaminations and various exposure elements pose to people's health and the environment.

The methodology used for defining carcinogens in Europe varies depending on whether an official body is analyzing environmental issues or only food-related ones. How thoroughly the technique of action of the poisonous response is comprehended affects the EPA's strategy. The use of mathematical modeling for low-dose extrapolation, assuming linearity, is necessary if the mechanism of activity is not properly well characterized or if the dosage response is depicted to be linear in characteristic. If it is known that the mechanism of action has a dosage response suggestive of a threshold, a margin of security feature method is applied, related to that utilized for effects other than cancer. The organizations advise reducing carcinogens in the diet "As Low as Reasonably Achievable (ALARA)" or to "Irreducible Levels" because they believe that quantitative lower-dose threat estimation has an appropriate scientific foundation.

Since 1975, many organizations have assessed food additions and pollutants and issued their assessments. For several years, the US, the Committee of Europe, and different worldwide committees have considered how safe flavorings are. The Committee of Europe has created records of flavoring compounds that are either utilized as flavoring chemicals in and of themselves or as sources from which flavorings are made (first edition in 1970). Depending on the Committee of Europe, the Scientific Committee on Food (SCF) has a brief record of the most "active principles with limit values in final food." This was incorporated in the Directive on Flavors for Foodstuffs, and their immediate usage as flavoring agents is not included in this list. Structure-activity relationships (SAR) must be taken into consideration when evaluating chemically defined flavorings when their chemical formation is comparable to those of substances with well-known harmful and organic effects. The "Threshold of regulation" idea was used by the JECFA (44th conference) in 1995 to provide an evaluation method depending on chemical organization, expected metabolism, and exposure from FDA examinations of gathering migrants. It was dependent on the job of [9], who analyzed data from the Registry of Harmful Consequences of Harmful Elements, Assessment of Food Additives (PAFA) database, and the [10] carcinogen potency database. These databases contained information on elements with sub-chronic and persistent toxic [11] expanded on this technique. In instances when they both took into account the same components; the SCF has concurred with the JECFA assessments of flavors.

Since 1976, organizations have conducted toxic assessments of monomers and various beginning materials to be utilized in the production of plastic products that come into proximity with food in the context of packaging goods. Additionally, the relocation of heavy metals from ceramics and the byproducts of cellophane's decomposition was taken into account and directed by directives. In 1990, a manual for presenting data for review appeared. When assessing chemicals and materials that come into contact with food, the SCF does not use the "Threshold of Regulation" premise of the US FDA. The SCF has issued papers on the requirements for submitting material for assessment prior to marketing as well as on considerations to be considered when evaluating the well-being of new foods. Toxicological data is analyzed conventionally in line with the acceptable daily intake (ADI) idea in a decision tree [12] that also takes nutritional considerations into account.

Description of this chapter

The authors go through eight techniques for foodstuff safety/threat evaluation which vary in terms of the data they need, how difficult they are, how well they can be used in various circumstances, and the kind and caliber of risk assessments [13] they provide. Each approach is evaluated using a consistent framework. The statistical models and significant techniques pertinent to the particular technique are first introduced. Following this, the data needed and consideration of subpopulations are taken into account. The method's advantages, disadvantages, and shortcomings are then explored. Insofar as this information was made available, the application of the procedures is explained, and then, the adoption by supervisory authorities is documented. At the conclusion of each section, gaps and research requirements are identified.

Section 9.2 discusses the usage of structural-active data that may be used when the chemical lacks sufficient toxicity data. The assessment of the so-called Threshold of Toxicological Concern (TTC) is a practical method to define a reference dosage for regulation if the amount of exposure to people is exceedingly low. Without chemical-specific toxicity data, a safety evaluation may be conducted if human-being consumption of a food element is less than the TTC. The threshold techniques in Section 9.3 are provided with the presumption that there is a level of exposure below which the food chemical does not create a physiologically significant effect. The use of quality uncertainty conditions to guess an ADI or its analogous is described, along with an explanation of the usage of no-observed-adverse-effect-levels (NOAELs) or comparable exposure amounts as points of departure. Even though we now have a poor understanding of the biological techniques associated with the creation of the harmful consequence, this strategy, which employs default uncertainty factors, may be used.

Section 9.4 addresses categorical regression. Despite its bounded usage up to date, this technique enables the accumulation of all detrimental consequences that have been detected into a single regression study. In addition to producing an estimation of a point of departure for the use of uncertainty factors, it moreover offers details on toxicity at greater dosages. Chemical-specific adjustment factors (CSAFs) [14] are a development of the UF technique used with threshold approaches. This approach, which is detailed in Section 9.5, is seen to be a significant improvement on past default techniques since it directly applies toxic and biological comprehension specified to the molecule of concern. In the sections that follow, while finding a probabilistic distribution of an ADI or conducting lower-dose calculation from a standard dose, the improvement that follows from swapping out the standard UFs for CSAFs is acknowledged. Section 9.6 discusses common mathematical approaches for estimating low-dose danger. These may be used if it is thought that there is always a physiologically significant danger at every level of exposure. In the past, it was possible to compare different compounds utilizing easy statistical frameworks like the one-hit or linear multistage framework, but the real numerical results were much more influenced by the framework than by the data. The authors demonstrate how the shortcomings of the strategy call for the growth of procedures that make greater use of biological principles and all available dose-response data.

The next three Sections (9.7-9.9) provide additional approaches and frameworks which utilize every available data more thoroughly, but which are only helpful if there is enough data. Section 9.7 discusses the benchmark dosage, also known as the BMDL, which is a mathematical lesser confidence limit of a dose which causes a pre-established alteration in response to environment. This technique may provide low-dose extrapolation a model-defined starting point. A genuine basic organic no-adverse-effect level is estimated using probabilistic risk assessment, which is covered in Section 9.8, using probability distributions of the uncertainty components. Risk characterization may be significantly enhanced if data on toxicokinetic [15] and toxicodynamic processes that are mechanistically based are available. Using summary tables, the authors offer the findings of a common assessment and contrast of the methodologies shown in Sections 9.2 through 9.10. We analyze the methods to determine their key traits, data needs and acceptability by governing authorities. We talk about accuracy, consistency, and predictive power. The thorough formulation of the research requirements identified previously in each department is found in Section 9.11. Finally, Section 9.12 formulates and draws the review's findings.

9.2 Relationships Between Structure and Activity (SAR) and the Level of Toxicological Involvement

If there is a pressing requirement for an evaluation, such as post-accident, or if the level of human display is exceedingly lower, structure-activity relationships (SAR) [16] can be employed in the absence of appropriate toxic data on the chemical. Regardless of the chemical's toxicological characteristics, there may be zero danger in the field of food safety if the display is much less. The method was mostly used to evaluate flavoring agents and packaging migrants.

The foundation of structure-activity connection analysis is the idea that a compound's biological activities are connected to its construction and the functional categories that are situated in it. It is a potent technique, particularly when utilized to forecast a chemical's activity based on information for a group of compounds with similar structures or for a particular and distinct target.

The pharmaceutical industry has made considerable use of SAR to forecast receptor binding, prospective therapeutic effects, and kinetics. SAR has been used most often for risk assessment when identifying hazards, such as for evaluating possible genotoxicity and in the field of ecotoxicology. Depending on the examination of the capabilities of a broad variety of actual substances, SAR may be used to forecast the potential toxicity of a novel chemical. SAR, depending on information on the oral LD50 estimates for more than 2,500 compounds, was able to accurately predict the toxicity for 95% of the compounds, while this method performed less well when used to predict ocular irritation. The least reported untoward effect level from chronic bio-assays was used in a study of a smaller database (234 different compounds) to establish a similar prediction [17]. There are several computer-dependent expert systems that can apply SAR to various data sets. Modeling dosage response does not use these techniques. In order to provide data without the need of *in vivo* animal testing, an aggregation of SAR and PBTK modeling was investigated.

There are two meanings associated with the word "threshold." It is believed that there is a "threshold" in the dose-response curve for most types of toxicity because the dosage or exposure level must be reached before an unfavorable reaction would occur (the validity of this is examined in Section 9.3). It is also a pragmatic "threshold for concern" or a "threshold for regulation," which refers to an exposure that is so low that it is not toxicologically significant.

Models and procedures

Given the shortage of chemical-dependent toxic data, the establishment of safety on the support of structure-based thoughts is a pragmatic technique used to prevent the needless use of test subjects and human resources for the investigation of insignificant dangers. The method is based on the conclusion that there is an appropriate safety margin and that the human exposure is quite lower than the hazardous levels of structural correlation. The safety margin would need to account for both potential differences in potency relative to structural counterparts as well as the typical characteristics of deduction from animal toxic data (see below). Regarding the migration of food-containment materials, the FDA in the United States recommended using a nutritional concentration of 0.5 ppb as a regulatory threshold in 1993. Exposures from assembling in the food lower this level will not be of concern. This figure was selected because it was 200 times lower and 2,000 times lower than the non-pesticide dosages that induce toxicity in animals, respectively.

In accordance with the Federal Register (1993), "substances whose chemical structures give rise to a reasonable expectation that they may be carcinogens will be excluded from review under the proposed regulation." However, the threshold for rules was contrasted with potent data on recognized toxins since it was plausible that an untested substance may be a carcinogen. In most instances, this exposure level would be linked with a continuance cancer [18] threat of less than two in a million if the migrant were an unrecognized carcinogen, according on survey of the dosage response for 430 known toxins (see later). As a result, data on both threshold and non-threshold factors supported the value of 0.5 ppb.

The JECFA evaluation system for groups of flavoring ingredients is an improvement on the FDA methodology. The system uses a number of exposure thresholds depending on the chemical's structural characteristics. The Cramer decision tree splits chemicals into three categories based on the chemical composition and the presence of functional groups.

Class I: Materials have straightforward chemical constituents, quick metabolic processes, and little oral toxicity.

Class II: Have lesser benign constituent characteristics than those of compounds in Class I, but they do not imply toxicity. Reactive functional groups may be present in substances of this type.

Class III: Substances with structural characteristics that make it impossible to make a strong first assumption of safeness or which even strongly imply notable toxicity.

The NOAEL distribution in all compounds studied in chronic or subchronic investigations was used to determine the thresholds for agents in structural classes I, II, and III. The 5th percentile of the NOAEL distribution was obtained after an analysis of the distributions of NOAEL estimates for compounds in every class. The 5th percentile NOAEL estimate from the class data was divided by the customary "safety" factor of 150 to determine the threshold for each class (see Section 9.5 below). 137, 28 and 448 compounds, respectively, provided the 5th percentile NOAEL estimates for structural classes I, II, and III, which were 3.0.

Human vulnerability limits of 1900, 550, and 95 mg per human being per day, estimating 60 kg body weight, are obtained by multiplying 0.92 and 0.16 mg per kg body weight per day. The applicability of the class I–III thresholds for substances which are neurotoxins, advancing toxins, or immunotoxins has been questioned. [19] examined the NOAELs for substances demonstrating these various kinds in a recent study. Additionally, the 20 chemicals in the [20] investigation with NOAELs lower than the class III level were assessed for their effects.

The JECFA's flavor rating system incorporates an extra daily human exposure limit of 1.6 mg. It would not apply to substances that are metabolized to safe products nor have any information on their toxicity from structural analogs or the substance itself. This threshold was calculated by looking at dose response data for recognized carcinogens and extrapolating it to intakes that are thought to have a one in a million risk using a very conservative model. The risk distribution of 10–6 threat estimations and the likelihood that an untested material will turn out to be a toxic were used to determine the threshold. JECFA ruled that the technique and

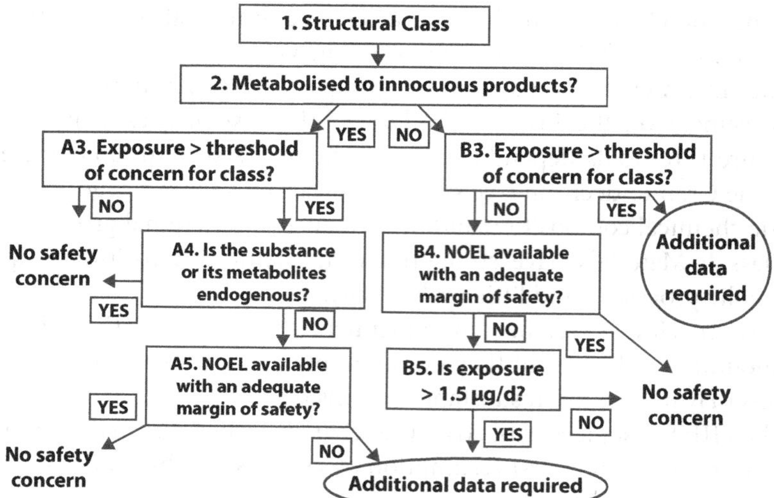

Figure 9.1 Decision-making framework utilized by the JECFA to assess flavoring chemicals (derived from the JECFA's 51st Report; WHO, 2000).

the criterion will not be used to substances having structural warnings for genotoxicity notwithstanding the cautious approach employed to establish the threshold of 1.6 mg per day. Depending on structural and metabolic factors, exposure, and a series of queries, the prospect for harmfulness of flavoring chemicals is assessed (Figure 9.1). The JECFA investigated 47 main alcohols, carboxylic acids and esters as flavoring agents. Most of these compounds were thought to be metabolized to harmless byproducts. As the predicted intakes were lower than the relevant levels for the structural class, 41 of the chemicals were deemed not to pose a safety risk. The remaining six drugs were assessed depending on a NOAEL in a two-year animal feeding trial for five endogenous compounds and one additional substance.

Subpopulations

Except for the threshold and non-threshold techniques utilized to calculate the thresholds of scrutiny for various structural classes, subpopulations are not specifically taken into consideration. The usage of these structure-dependent criteria is structural and exposure dependent. To use the Cramer decision tree [21] to identify the structural class, the chemical structure must be known. Exposure estimations are often dependent on models and forecasts since thresholds are typically applied to chemicals for which there aren't enough data points available.

Possibilities, constraints, and shortcomings

The merit of this approach is that it makes it easy to determine if the existing exposures pose a safety risk. The key flaws are the differences between the thresholds that were derived from the databases [22] and the assumptions made for the chemical under assessment's structure-activity connection. Another challenge is that the evaluation is based on exposure approximates from the safety review, and a change in exposure will need a reevaluation.

Applicability

Although in reality JECFA uses it for chemicals lacking structural alarms for genotoxicity and having minimal exposures or for constituents which create typical endogenous metabolites, the approach (Fig. 1) is theoretically applicable to all substances.

Acceptance by authorities and regulatory bodies

The SCF (1999) in the EU recognized the JECFA approach in general, and EU law (EC, 2000) now includes its usage for the assessment of chemically described flavoring compounds. The SCF has, however, temporarily disallowed the use of the TTC idea. Despite being obtained using the very cautious linear low-dose extrapolation model nor JECFA or the SCF have received the 1.6 mg per day threshold for potential genotoxic carcinogens. This is most likely because purposely introducing any quantity of a genotoxic substance to the human diet is undesirable. In the absence of good dose-response data on the substance, the value of the 1.5 mg per day threshold for the evaluation of natural genotoxicants in the food may offer a reasonable threshold for risk management.

Research gaps and needs

When determining the structural class of the flavoring chemical and thresholds, the Cramer decision tree is a crucial component of the JECFA decision technique. Given current toxicological and metabolic evidence, this older paper should be revalidated.

9.3 Threshold Approaches

It is examined that there exists a level of exposure lower than which no physiologically major result may be created for certain dangers, such as non-genotoxic effects. In these circumstances, it is standard tendency in the US to assess the amount of exposure without noticeably harmful results. Still, it is not necessary to utilize an absolute threshold in a scientific manner to use the following techniques. An acceptable threshold can be seen as a question of biology or regulation instead.

Models and procedures

The approach used is the implementation of safety elements or uncertainty elements that permit for inter-specimens contrasts and for human-being variability, to a dose which is thought to be a replacement for the threshold. Lower-dose calculation of the dose-response connection is never utilized in threshold results. The outcome is regarded as human exposure, which wouldn't have any physiologically significant negative healthcare results. An instance of a methodical risk evaluation is safety assurance, where the degree of threat is thought to be "biologically insignificant" as a substitute to a precise methodical calculation.

A rough threshold for toxicity in the animal research is defined utilizing the dose-response relationship as the foundation for safety assurance. The NOAEL is often employed as a stand-in for the threshold (WHO, 1999b). Although not every known outcome is negative and will serve as the foundation for a significant risk evaluation, the phrase no-noticed-resultant level (NOEL) is frequently employed. As a result, the NOAEL has become increasingly prevalent in recent years, even if the NOAEL and NOEL are exchangeable in the context of JECFA assessments since the phrase "adverse" is contained in the acuteness of NOEL. The NOAEL is a threshold of exposure where handled animals do not notably deviate from untreated control animals in biologically relevant assessments of the unfavorable results recognized at greater doses. The NOAEL is a dosage that has no discernible activity; therefore in actuality it is thought to be near to the threshold in animals. Despite not being employed quantitatively, the dose-response connection above the NOAEL aids in determining whether or not the effects seen at various dosage levels are possibly to be therapy-based. Comparing test groups to controls, a significant effect is produced. As a result, this estimate obviously exceeds the threshold. To account for this in risk assessment [23], extra uncertainty factors are often included (see below). Like the NOAEL, the lowest-observed-adverse-effect-levels (LOAEL) is an implemented finding that depends on the study's model. For databases which don't permit for the estimation of a NOAEL, the standard dosage will provide a much scientifically plausible approach to risk assessment.

Subpopulations

The factor of 10 for human variability, which takes into account "at risk" populations including children and the elderly, may be thought of as a way to consider for differences within the human census. The survey of human variability by [24] showed that the 11-fold component was a sufficient default, but which circumstances can be imagined in which the compound may exhibit metabolic properties which will significantly enhance human variability.

Due to their relative immaturity in comparison to adults, it has been suggested that babies and children may be a sensitive subpopulation. This incompleteness can primarily result in how well they can get rid of the substance (toxicokinetics) or how their intended organ reacts to it (toxicodynamics). There isn't much in the longtime response data on people which have looked at this, for obvious reasons. The usage of an extra 11-fold feature for babies and kids (as needed by the Foodstuff Quality Safety Act in the United States) does not seem to be valid depending on thought of the customary 100-fold feature utilized to account for specimen's differences and human volatility, according to a survey of literature. However, if there were no studies on reproductive and developmental toxicity, if the present testing procedures were deemed insufficient, if the results generated in neonates and young were mainly irreparable, or if a "severity" factor was launched for risk control, then an additional factor might be appropriate. It is crucial that the justification for adding an additional component for mainly delicate subgroups be founded on established scientific ideas and knowledge.

The choice of suitable uncertainty factors may be impacted by the database's suitability (WHO, 1994). In order to account for database flaws, like the lack of a NOAEL or a chronic survey in animals, extra uncertainty features can be utilized for contaminants for which a risk management can need to be conducted on a non-ideal database.

Possibilities, constraints, and shortcomings

This method's key advantage is that it is straightforward and practical, and it can be easily used to a variety of datasets. However, each of the numerical values used to calculate an ADI, TDI, or RfD, which is the substitute for the threshold (often the NOAEL), has limits, as does the uncertainty or safety feature used to calculate the substitute for the threshold. The biggest issue having using the NOAEL is that it might be lower, at, or over the threshold resulting on the research design's antecedents.

Group size: The more people in the group, the more sensitive it will be. The current recommended flock dimensions in testing recommendations provide the optimum agreement between practicability and sensitivity.

Sensitivity of the reverse consequence measurement: the NOAEL will be lower if the detection technique is more sensitive. Studies that are presented for regulatory reasons should adhere to GLP or be documented in major details to give quality assurance to those doing the risk evaluation since studies conducted using subpar or insufficient methodologies would provide a higher NOAEL.

Dosage spacing: The choice of the dosage levels administered to the animals and the strength of the result at the LOAEL are factors that affect the NOAEL in actual datasets. The NOAEL will likely underestimate the threshold when the doses are widely separated in proportion to the slope of the dose-response curve and the dosage above the NOAEL only has a marginal impact. The NOAEL may be at or over the threshold if the doses are spaced much nearly together.

When the highest dosage under study is the sole dose exhibiting a negative effect, there is a question if this is a genuine reaction or a chance "false positive" finding. In these cases, a conservative evaluation will be made by considering the dosage below the peak dose as a NOAEL. The NOAEL technique offers no information concerning risk in excess of the ADI. If the straightforward usage of the NOAEL were to be substituted, wherever feasible, by a method which used much of the dose-response data and offered a much reliable metric, like the BMD, the risk assessment process would have a stronger scientific foundation. As the benchmark dosage may be placed at a phase of response which is regarded as beneficial, the usage of a standard dose renders the location or presence of a threshold meaningless.

The figure of 100 was chosen because it is equivalent to an 11-fold feature in human variability and an 11-fold feature for interspecies differences. The effectiveness of this 110-fold figure as a universal default that is used to very diverse chemicals which can exhibit a variety of toxicokinetic or toxicodynamic characteristics was assessed in various recent studies. These assessments serve as post hoc evaluations of the applicability of the "uncertainty factors" chosen in the 1960s, prior to more current developments in toxicology and risk evaluation. The evaluations have traced a number of circumstances in which they may be either insufficient (for instance, if the molecule's metabolism exhibits a genetic polymorphism) or surplus (for example, when the compound just barely absorbs from the stomach). As a result, replacing the basic default factor with an amount which was more notably tied to the database undergoing consideration, like the usage of a chemical-based improvement feature, would provide the risk assessment procedure a more scientific foundation.

Applicability

The standard method used by every agencies doing risk evaluation for compounds thought to display threshold results is the application of 110-fold and 10-fold uncertainty features to the NOAEL. The method is easily adaptable, and documented principles and procedures for its use have been created (WHO, 1987). To determine the amount of human exposure that will be free from noticeable side effects, the NOAEL stated on a body weight consideration (like mg/kg body weight/day) is split by an uncertainty component (or safety component). There is a similar underlying methodology, even if regulatory authorities use different nomenclature (NOEL vs. NOAEL; acceptable everyday consumption (ADI) vs. tolerated daily intake (TDI) vs. reference dose.

Research gaps and needs

For determining the sufficiency and limits of inter and intra-species default uncertainty features, it is important to continue studying the data on toxicokinetics, toxicodynamics, and in main response in animals and persons. A probabilistic implementation of uncertainty factors and/or the creation of categorical default factors connected to pathways and processes (see Section 9.5) may both enhance the usage of default uncertainty factors. An important risk management problem is the proportion of a census which will be enclosed by the usage of a default feature. A census distribution framework and the standard deviation, sometimes known as the geometric standard deviation, may be used to make estimates. A significant evaluation of the appropriateness of the uncertainty features on a case-by-case intent will be possible thanks to the growth of this technique utilizing chemical-based data or category defaults.

9.4 Reciprocal Analysis

Models and procedures

Categorical regression is a technique which was suggested for significant dose-response survey for non-cancer toxicity data. A statistical approach which may be utilized to calculate the possible healthcare threat from chemical vulnerability is categorical regression. The approach may calculate the odds which a certain group of severity can happen at a particular dosage level using regression of ordered categories of toxic severity. Though one cannot see the result firsthand, still one might calculate the likelihood that it would happen.

According to [25], categorical or regression incorporates statistical regression on the implemented dosages linked to several categories of overall toxicity severity. These severity levels are often associated with a qualitative classification [26] system that distinguishes between various degrees of outcome. It is possible to anticipate effect severity by combining dosages and their related effects from several researches using this kind of severity scale. By doing this, it is possible to consider all harmful effects that have been classified rather than just the critical effect. This approach combines descriptive, continuous, and dichotomous data into a unified severity grading system. Additionally, in certain circumstances, this mathematical technique may use exposure length and concentration (dosage) as independent variables which affect response as toxicity information from different research may be integrated and utilized in this method, even when the exposure times are different. The regression's findings may then be utilized to calculate an exposure level based on health for risk assessment. In its use, a similar estimate is similar to a reference dosage. This dosage may be assessed for any specific exposure length of interest if the framework is fitted to the findings from (sufficient) surveys having various exposures.

Figure 9.1 shows an illustration of how categorical regression may be used to predict toxic data from both human and animal datasets. These categories of regression severity were modified using information for methoxychlor taken from [27]. As a function of both equal human dosages in mg/day and exposure period as a proportion of lifetime, the data from research conducted on humans and animals are combined for the sake of simplicity. One dosage group from one research is shown by each symbol on the figure. It should be noted that a categorical regression may incorporate data from several target organs. The graph's solid line illustrates how the categorical regression model makes use of information from seemingly unrelated sources to determine the best fit curve for a NOAEL across the course of exposure, here, a human lifetime. The diagram moreover shows how toxicity may be predicted by combining dosage and duration.

The data needed to use the categorical regression framework are collected by classifying the severity of effects linked to different dosages (or exposures). The number of adversity categories is limited since the majority of toxic surveys only employ some dosage levels. Though the sorts of results in the test census should be comparable, basically, with only some studies with results throughout the dose-response curve would be enough to build a categorical regression. Naturally, the more data that are provided, the better the curve's form will be, and the more precisely the model's associated confidence intervals may be determined.

A case study was created by [28] utilizing the carbamate pesticide aldicarb. Categorical regression may be utilized to evaluate threats above the RfD, as the case study illustrated. The results from experimental animals corroborated the findings from investigations on humans. The two data

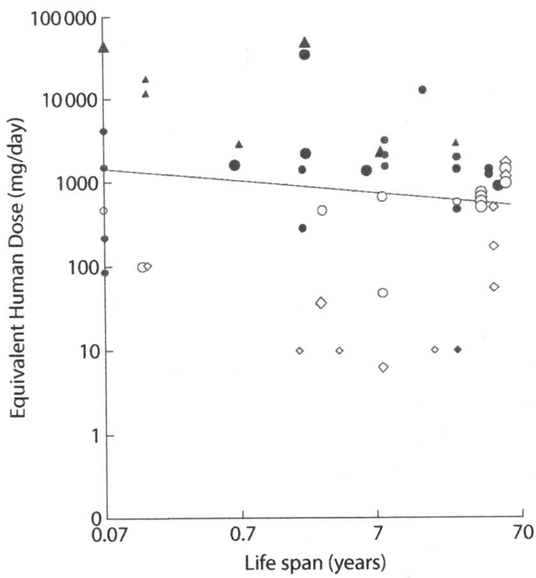

Figure 9.2 Data on the oral toxicity of methoxychlor for humans and animals were modeled using categorical regression. Outcome levels are depicted for various targets. To roughly determine the comparable human dosage, body surface area has been used to translate animal doses. Dose spans are distributed by the relevant specimens' life duration to produce an amount that when multiplied by 72 years (the estimated mean human life span) results in the interrelated location on the x-axis. Thus, a lifetime exposure survey in any specimens will be represented by 72 years. Survey functionality is indicated by the dimension of the related symbols.

sets from two distinct human exposure trials which assessed blood cholinesterase stoppage underwent final analysis. If blood cholinesterase control is not taken into account, these findings showed that a larger probable risk calculation of untoward results is 0.009% at a dosage that is ten times larger than the RfD. The greatest probable risk estimate for adverse effects, where blood cholinesterase control of 25% or greater is regarded as an unfavorable outcome, is 0.1% at a dosage ten times the recommended dose (RfD). Although it was highlighted that the use of this categorical regression was framework-based for evaluating threat above the RfD, a statistical technique of this kind can be useful in addition to widely used risk assessment techniques. The ability to define the exposure-response relationship for a chemical and assist the threat assessor in choosing the best data set and framework is another interesting use for categorical regression.

Subpopulations

The categorical regression technique has the benefit of allowing data from much research, even those involving several species, to be pooled. Though the framework would enable threat estimate, it doesn't certainly solve the problem of sensitive subpopulations. If data are available from such subpopulations, categorical regression might be employed to handle sensitive subpopulations.

Needs for data

A homogeneous categorical regression model makes the assumption that, except from the variables (dose and span), all experimental groups received the same treatment. The diverse character of the data may be compensated for by the modeling software when the parameters are different since stratification will mean over the random survey outcomes. In a categorical regression study, both continuous and quantitative data may be included, but their derivation and assignment to severity categories must take different factors into account. Quantitative [29] data need a toxicologist's classification choice based on the endpoint's characteristics. While keeping in mind of continuous data, the response magnitude of every severity category must be established, and when it is exceeded, the continuous data is classified into the associated extreme class. In the published toxicological literature, categorical effect assessments are often encountered, like reports of histopathological outcomes or hematological or enzyme function abnormalities. If it is possible to give them an extreme class and analyze them in conjunction with different data which was suitably sorted by severity categorization, additional data which doesn't automatically suit this definition can moreover be utilized.

Possibilities, constraints, and shortcomings

The method's ability to include every applicable surveys and toxicological judgment into the computation of a NOAEL or other originating point for threat evaluation is one of its strengths. Additionally, it enables the quantification of the level of uncertainty around with estimation. The statistical [30] power of the study may be dramatically increased by the risk assessor by merging several research of little individual utility. The risk assessor may use this approach to incorporate different research; forecast exposures associated to different degrees of effect severity, and utilize many independent variables to define a particular response. By interpolating or extrapolating the fitted model, categorical regression enables the user to forecast the outcome class in the issue of any given dosage. Estimating the outcome for a bracketed time period or aggregation for which there are no data is an illustration of this. The risk assessor may use the model to forecast exposures at associated responses for various intensities and make decisions to safeguard a population from ever-worsening effects.

Additionally, categorical regression may provide details regarding possible toxicity at doses higher than the ADI, TDI, or RfD used in risk assessments. The NOAEL technique has limitations since it only considers "safe" or "acceptable" consumption levels. Information is required to assess the urgency of a problem in cases when exposures may exceed these thresholds (for example, when drinking water is polluted, when public utilities should stop supplying it?). As they are point estimations created to be lower than the population threshold for toxicity, ADI or RfD provide little insight. One of the benefits of categorical regression is that it may be utilized to predict likely threat higher than the RfD or RfC and offers information on rising toxicity with increasing dose rate. There are currently a number of articles on this technique using interesting substances.

A significant drawback of employing a categorical regression framework is that significant toxicological judgment is required to assess the extremity of the outcome and to consistently categorize the outcomes from various researches. It is particularly true when merging many studies. This restricts the method's applicability and may bring subjectivity into the procedure, particularly when used to disparate endpoints. The classification of experimental doses and their associated effects into severity levels is not standardized and requires some kind of criterion harmonization. When the regression involves a variety of negative effects, a serious issue arises.

While aggregating outcomes of reproductive toxicity having neurological or renal outcomes, for example, a rating of the various effects would be very subjective and include non-scientific assessments. A goodness-of-fit methodology may be deceptive, for instance, by not confirming correctness in the extrapolation and perhaps lacking the capacity to distinguish between or reject models. Additionally, confidence bands might obscure important hypotheses, particularly when integrating data from different species, time periods, or both.

Acceptance by authorities and regulatory bodies

Before implementing usage recommendations, there has to be an acceptable time span to develop the framework and identify any difficulties, as with any newer methods to threat evaluation. This mathematical technique has a lot of potential, but before regulatory bodies feel at ease using it, the difficulties (as outlined above) must be adequately addressed.

Research gaps and needs

Although promising, this mathematical model has only been used sparingly up to now. As a result, there are an amount of concerns which require being resolved, by creating use rules or by carrying out more researches to verify the framework.

(i) Will the approach reach vulnerable census?
(ii) Which rules must be followed in determining extremity and categorizing extremity?
(iii) Which rules must be followed while merging researches? For not including research? For approving or disapproving certain endpoints?
(iv) Which rules should be followed when substituting this model for other recognized risk assessment techniques?
(v) Should a policy or directive be put in place to evaluate and take the results of the categorical regression framework?
(vi) What is the fitted model's slope to be explained? Is it noise, study variation, or something else entirely?

Provided biological heterogeneity, different end-point criteria, and perhaps insufficient power to create the model, a categorical regression may not be viable with fewer investigations.

9.5 Chemical-Specific Adjustments

Models and techniques

The mathematical model used to estimate the danger of a food chemical depends on the scientific evidence available on its toxicity and biological processes. This replaces the rather unsophisticated default techniques utilizing the 100-fold factor as a product of two 10-fold factors (see Section 9.3) with more scientifically plausible, quantitative, dose-response assessments using chemical-specific data for compounds with rich data.

In spite of dose-response survey, using animal toxicology data is better prevalent; a model was established to include significant data detailing the organic techniques which assign to a response. In the framework, every interspecies and inter-individual 11-fold feature is split to account for contrasts in toxicokinetics (that finds chemical delivery to the chosen site) and toxicodynamics (that finds target site reaction to chemical) (Figure 9.3). Each subfactor represents one uncertainty source. The International Project on Chemical Safety assisted the splitting of the 11-fold features with the human variability factor given equal weighting for toxicokinetics and

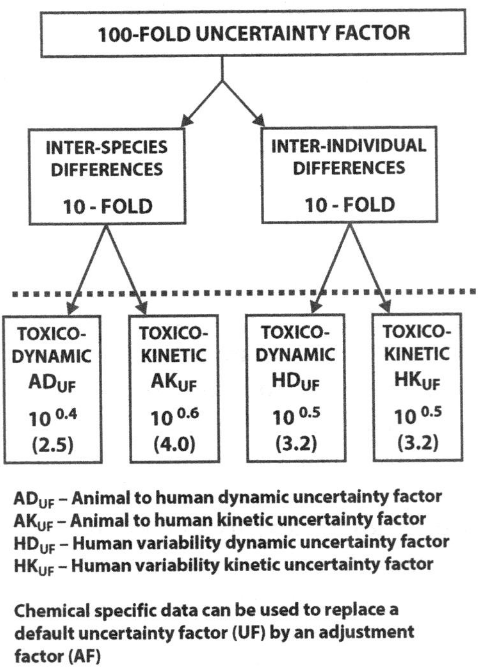

Figure 9.3 The 101-fold uncertainty feature divided into toxicokinetics and toxicodynamics.

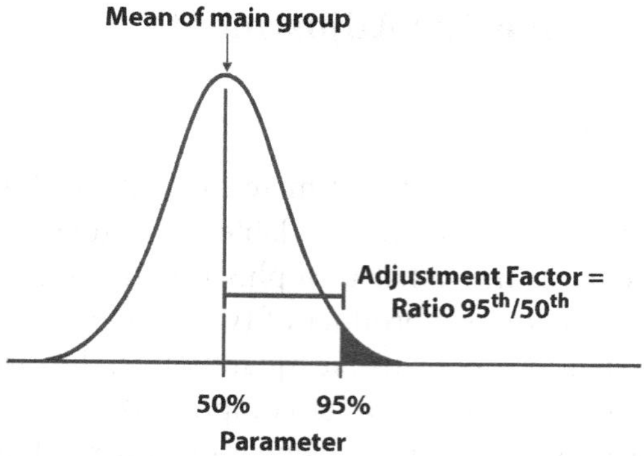

Figure 9.4 Chemical-specific CSAF creation. The CSAF is calculated by dividing the average by the parameter estimate for a predefined census percentage, like the 95th percentile.

toxicodynamics (Figure 9.3) that is consistent with Renwick and Lazarus' 1998 analysis. Chemical-specific data may substitute any of the four default subfactors (Figure 9.3).

Human variability in a toxicokinetic or toxicodynamic parameter is best seen as a distribution, determining the population covered by the CSAF. A CSAF for human variability is the ratio between the population mean or median and the parameter value corresponding to a pre-defined and acceptable percentage of the population covered, such as the 95th, 97.5th, or 98%.

Conclusion

1. Hazard characterization is not suited for all risk assessments, food categories, or databases. To match the endpoint discovered and the breadth and grade of dose-response data, a variety of methods are needed. So lower molecular weight compounds, micro-nutrients, dietary supplements, macronutrients, entire foodstuff, innovative foodstuff, and foodstuff processing can necessitate particular significant procedures.
2. As intended, almost all toxicity studies may identify hazards and provide the NOAEL, the dosage predicted to cause no harm. Current testing guidelines may not yield reliable statistical dose-response data.
3. Mathematical modeling of the dose-response connection involves many doses with various response levels.

4. Statistical dose-response modeling will enhance threat evaluation. To improve modeling, procedures should be adjusted to increase treatment groups. An adequate design can minimize the amount of animals investigated when the quantity of treatment (dose) flocks is increased.
5. Dose-response characterization should include chemical-based data on the method or approach of activity and/or toxicokinetics, such as PBTK model data. Toxicokinetic data may convert external doses to target-organ doses. A PBTK model may be extended into an organically-dependent dose-response framework using mode of action data.
6. By generating chemical-specific adjustment factors, the default (11-fold) uncertainty features may be modified. This information may help clean the distribution of extrapolation factors utilized in statistical assessments.
7. The dose-response feature must reflect output uncertainty and variability. Probabilistic analysis may characterize uncertainty and supply an output, like ADI, that corresponds to the 5th or 10th percentile of the uncertainty distribution of the output. Probabilistic analysis helps framework how conservatism affects output.
8. Simple default methods like the NOAEL and uncertainty features or linear lower-dose extrapolation are frequently implemented, but they have limitations.

References

1. X. Zhuo, F. Yan, M. Zhao and Y. Guo, "Design and Application of Food Additive Terahertz Fingerprint Library," *2023 42nd Chinese Control Conference (CCC)*, Tianjin, China, 2023, pp. 1303-1309, doi: 10.23919/CCC58697.2023.10240303.
2. D. Pradhan, P. K. Sahu, N. S. Goje, H. Myo Tun, M. M. Ghonge, R. Rajeswari and S. Pramanik, "Security, Privacy, Risk, and Safety toward 5G Green Network (5G-GN)", in *Cyber Security and Network Security*, Wiley, 2022, doi: 10.1002/9781119812555.ch9
3. R. Anand, J. Singh, D. Pandey, B. K. Pandey, V. K. Nassa and S. Pramanik, "Modern Technique for Interactive Communication in LEACH-Based Ad Hoc Wireless Sensor Network", in *Software Defined Networking for Ad Hoc Networks*, M. M. Ghonge, S. Pramanik and A. D. Potgantwar, Eds., Springer, 2022, https://doi.org/10.1007/978-3-030-91149-2_3

4. S. Choudhary, V. Narayan, M. Faiz and S. Pramanik, "Fuzzy Approach-Based Stable Energy-Efficient AODV Routing Protocol in Mobile Ad hoc Networks", in *Software Defined Networking for Ad Hoc Networks*, M. M. Ghonge, S. Pramanik and A. D. Potgantwar, Eds., Springer, 2022, https://doi.org/10.1007/978-3-030-91149-2_6
5. P. T. Khanh, T. H. Ngọc, and S. Pramanik, Future of Smart Agriculture Techniques and Applications, in *Advanced Technologies and AI-Equipped IoT Applications in High Tech Agriculture*, Ed., A. Khang, IGI Global, 2023.
6. S. Shrimali and C. Teuscher, "A Novel Deep Learning-, Camera-, and Sensor-Based System for Enforcing Hand Hygiene Compliance in Healthcare Facilities," in *IEEE Sensors Journal*, vol. 23, no. 12, pp. 13659-13670, 15 June 15, 2023, doi: 10.1109/JSEN.2023.3271297.
7. Reepu, S. Kumar, M. G. Chaudhary, K. G. Gupta, S. Pramanik and A. Gupta, Information Security and Privacy in IoT, in *Handbook of Research in Advancements in AI and IoT Convergence Technologies*, Eds., J. Zhao, V. V. Kumar, R. Natarajan and T. R. Mahesh, IGI Global, 2023.
8. V. Veeraiah, V. Talukdar, Manikandan K., S. B. Talukdar, V. D. Solavande, S. Pramanik, A. Gupta, Machine Learning Frameworks in Carpooling, in *Handbook of Research on AI and Machine Learning Applications in Customer Support and Analytics*, Eds., Md S. Hossain, R. C. Ho and G. Trajkovski, IGI Global, 2023.
9. S. Kayalvizhi, N. Akash Silas, S. Pothirajan and T. Krishnakumar, "Product Constituents and FDA regulations for Consumer Safety Using OCR," *2022 OPJU International Technology Conference on Emerging Technologies for Sustainable Development (OTCON)*, Raigarh, Chhattisgarh, India, 2023, pp. 1-7, doi: 10.1109/OTCON56053.2023.10113917.
10. M. Alouzi, F. Al-Kamali, C. D'Amours and F. Chan, "Direct Conversion of Hybrid Precoding and Combining from Full Array Architecture to Subarray Architecture for mmWave MIMO Systems," in *IEEE Access*, vol. 11, pp. 35457-35468, 2023, doi: 10.1109/ACCESS.2023.3264838.
11. A. K and N. Bharadvaja, "Molecular docking studies of flavonoid's pattern of inhibition against Wnt/β-catenin signaling," *2023 International Conference on Computational Intelligence and Sustainable Engineering Solutions (CISES)*, Greater Noida, India, 2023, pp. 575-579, doi: 10.1109/CISES58720.2023.10183501.
12. Y. Yan *et al.*, "Integration of Canopy Water Removal and Spectral Triangle Index for Improved Estimations of Leaf Nitrogen and Grain Protein Concentrations in Winter Wheat," in *IEEE Transactions on Geoscience and Remote Sensing*, vol. 61, pp. 1-18, 2023, Art no. 4404118, doi: 10.1109/TGRS.2023.3277456.
13. V. Jain, M. Rastogi, J. V. N. Ramesh, A. Chauhan, P. Agarwal, S. Pramanik, and A. Gupta, FinTech and Artificial Intelligence in Relationship Banking and Computer Technology, in AI, IoT, and Blockchain Breakthroughs in

E-Governance, Eds., K. Saini, A. Mummoorthy, R. Chandrika and N. S. Gowri Ganesh, IGI Global, 2023.

14. S. Praveenkumar, V. Veeraiah, S. Pramanik, S. M. Basha, A. V. Lira Neto, V. H. C. De Albuquerque and A. Gupta, Prediction of Patients' Incurable Diseases Utilizing Deep Learning Approaches, *ICICC*, 2023, Springer, doi.org/10.1007/978-981-99-3315-0_4
15. V. K. Gupta, A. Gupta, M. K. Singh, A. Gupta and A. Tomar, "Toxicity Detection Methodology of Adverse Outcome Pathways Using Physicochemical Properties and Machine Learning Approaches," *2023 4th International Conference for Emerging Technology (INCET)*, Belgaum, India, 2023, pp. 1-7, doi: 10.1109/INCET57972.2023.10170360.
16. T. R. Noviandy, A. Maulana, G. M. Idroes, I. Irvanizam, M. Subianto and R. Idroes, "QSAR-Based Stacked Ensemble Classifier for Hepatitis C NS5B Inhibitor Prediction," *2023 2nd International Conference on Computer System, Information Technology, and Electrical Engineering (COSITE)*, Banda Aceh, Indonesia, 2023, pp. 220-225, doi: 10.1109/COSITE60233.2023.10250039.
17. V. VidyaChellam, V. Veeraiah, A. Khanna, T. H. Sheikh, S. Pramanik and D. Dhabliya, A Machine Vision-based Approach for Tuberculosis Identification in Chest X-Rays Images of Patients, *ICICC*, 2023, Springer, doi.org/10.1007/978-981-99-3315-0_3
18. A. Gupta, A. Verma and S. Pramanik, Security Aspects in Advanced Image Processing Techniques for COVID-19, in *An Interdisciplinary Approach to Modern Network Security*, S. Pramanik, A. Sharma, S. Bhatia and D. N. Le, Eds., CRC Press, 2022.
19. V. F. Annese, V. Galli, G. Coco and M. Caironi, "Eat, Test, Digest: Towards Diagnostic Food for Next-Generation Gastrointestinal Tract Monitoring," *2023 9th International Workshop on Advances in Sensors and Interfaces (IWASI)*, Monopoli (Bari), Italy, 2023, pp. 236-240, doi: 10.1109/IWASI58316.2023.10164549.
20. Ramchandran, V., and Gernand, J. M. (August 23, 2022). "Evaluation of Risk and Uncertainty for Model-Predicted NOAELs of Engineered Nanomaterials Based on Dose-Response-Recovery Clusters." ASME. *ASME J. Risk Uncertainty Part B*. March 2023; 9(1): 011205. https://doi.org/10.1115/1.4055157
21. S. Pramanik, Intelligent Farming Utilizing a Soil Tracking Device, in *Convergence of Cloud Computing, AI and Agricultural Science*, Eds., A. K. Sharma, N. Chanderwal and R. Khan, IGI Global, 2023.
22. P. T. Khanh, T. H. Ngọc, and S. Pramanik, Future of Smart Agriculture Techniques and Applications, in *Advanced Technologies and AI-Equipped IoT Applications in High Tech Agriculture*, Ed., A. Khang, IGI Global, 2023.
23 S. Pramanik, "An Effective Secured Privacy-Protecting Data Aggregation Method in IoT", in *Achieving Full Realization and Mitigating the Challenges of the Internet of Things*, Eds., M. O. Odhiambo and W. Mwashita, IGI Global, 2022, doi: 10.4018/978-1-7998-9312-7.ch008

24. N. Kanwal, F. Pérez-Bueno, A. Schmidt, K. Engan and R. Molina, "The Devil Is in the Details: Whole Slide Image Acquisition and Processing for Artifacts Detection, Color Variation, and Data Augmentation: A Review," in *IEEE Access*, vol. 10, pp. 58821-58844, 2022, doi: 10.1109/ACCESS.2022.3176091.
25. S. Pramanik and S. Bandyopadhyay, Identifying Disease and Diagnosis in Females Using Machine Learning, in *Encyclopedia of Data Science and Machine Learning*, Ed. John Wang, IGI Global, 2023, DOI: 10.4018/978-1-7998-9220-5.ch187.
26. S. Pramanik and S. Bandyopadhyay, Analysis of Big Data, in *Encyclopedia of Data Science and Machine Learning*, Ed. John Wang, IGI Global, 2023, doi: 10.4018/978-1-7998-9220-5.ch006
27. P. Leepheng, D. Phromyothin, C. Thanachayanont, K. Inpor, A. Olarnwanich and A. Jiraseree-Amornkun, "Molecularly Imprinted Polymers Near Field Communication sensors for pesticide detection," *2022 19th International Conference on Electrical Engineering/Electronics, Computer, Telecommunications and Information Technology (ECTI-CON)*, Prachuap Khiri Khan, Thailand, 2022, pp. 1-4, doi: 10.1109/ECTI-CON54298.2022.9795549.
28. Y. Peng, X. Tuo, X. Li, G. Zhang, T. Pang and N. Fu, "Progress in near-infrared spectroscopy for baijiu quality and safety detection," *2023 CAA Symposium on Fault Detection, Supervision and Safety for Technical Processes (SAFEPROCESS)*, Yibin, China, 2023, pp. 1-6, doi: 10.1109/SAFEPROCESS58597.2023.10295906.
29. S. Choudhary, V. Narayan, M. Faiz and S. Pramanik, "Fuzzy Approach-Based Stable Energy-Efficient AODV Routing Protocol in Mobile Ad hoc Networks", in *Software Defined Networking for Ad Hoc Networks*, M. M. Ghonge, S. Pramanik and A. D. Potgantwar, Eds., Springer, 2022, https://doi.org/10.1007/978-3-030-91149-2_6
30. K. Dushyant, G. Muskan, A. Gupta and S. Pramanik, "Utilizing Machine Learning and Deep Learning in Cyber security: An Innovative Approach", in *Cyber Security and Digital Forensics*, M. M. Ghonge, S. Pramanik, R. Mangrulkar, D. N. Le, Eds, Wiley, 2022, https://doi.org/10.1002/9781119795667.ch12

10

Simulating the Effects of Climate-Smart Agriculture Practices on Farm Resilience: A Mathematical Modeling Approach

Kiran K. S.[1], Meenakshi Dheer[2], Mukesh Laichattiwar[3], Devendra Pal Singh[4], Vaidehi Pareek[5] and Soma Bag[6*]

[1]Department of Physics, Faculty of Engineering and Technology, JAIN (Deemed-to-be University), Karnataka, India
[2]Department of Agriculture, Vivekananda Global University, Jaipur, Rajasthan, India
[3]Department of Entomology, College of Agriculture, Parul University, Limda, Dist- Vadodara, Gujarat, India
[4]College of Agriculture Sciences, Teerthanker Mahaveer University, Moradabad, Uttar Pradesh, India
[5]Symbiosis Law School, Nagpur Campus, Symbiosis International (Deemed University), Pune, India
[6]Asadtala Nivedita Kanya Vidyamath, Nandigram, West Bengal, India

Abstract
We looked at research that showed how much more adaptable better aggregated farming systems are for climatic variability and climate alteration than much detailed farming techniques. The authors only discovered a small amount of literature specifically addressing the issue, making the usage of proxy approaches necessary to increase the specimen size. Where appropriate, the authors employed inter-annual climatic variability for climate change, a wide range of agricultural variables for resilience, and farming approach abundance and variety as proxies of the occurrence of the kind of mutual linkages which characterize IAS. The authors discovered that (1) 39 publications highlighted the issue through

**Corresponding author*: bagsoma2016@gmail.com
Kiran K. S.: ORCID: https://orcid.org/0000-0003-2055-9365
Meenakshi Dheer: ORCID: https://orcid.org/0000-0001-7126-807X
Devendra Pal Singh: ORCID: https://orcid.org/0009-0000-7479-8631

Sabyasachi Pramanik, Niranjanamurthy M., Ankur Gupta and Ahmed J. Obaid (eds.) Mathematical Modeling in Agriculture, (189–210) © 2024 Scrivener Publishing LLC

statistical modeling, (2) that aggregation was intensely (n = 17/24) connected with enhanced climatic resilience, (3) that the outcomes were primarily based on contrasts of much various or enhanced agricultural frameworks with lesser diversified or enhanced agricultural frameworks, and (4) that production, income, gain, and production variance were all positively correlated with resilience. Our results provide intriguing, but far from definitive, proof that agricultural framework aggregation might increase resilience and emphasize the requirement for more study to determine if aggregation strategies may provide comparable results.

Keywords: Integrated agricultural system, smart agriculture

10.1 Introduction

Compared to traditional farming yield methods, integrated agricultural systems (IAS) [1] may reduce dependency on outer resources, according to growing research. Possibly because of these benefits, IAS (and connected ideas called varied techniques, mixed approaches, agroecological [2] frameworks and agroforestry frameworks) is being enhanced by worldwide institutions, agribusinesses and governments. Regarding proof about the effectiveness of IAS in a shifting environment, the question is whether more IAS increases the capacity of farming techniques to adapt to changing or unpredictable climate. However, many modern climate adaptation strategies and initiatives operate with the presumption that much IAS will have better climate alteration resistance. For instance, Nigeria's National Agricultural Resilience Framework15 and Kenya's National Climate Change Response Strategy14 both support integration and diversity as adaptive strategies.

Adaptive strategies, diversification and integration are used. Even in impact assessments of the Worldwide Fund for Farming Growth's Adaptation, farm-level richness—a potentially appropriate proxy for aggregation—is employed as a proxy for climate resilience.

This evaluation aims to inquire about the evidence on the feasibility of farming aggregation for climatic resilience with a view to guiding future climate change adaptation strategy. We address the following seven issues utilizing a review of research on IAS and climatic alteration:

1. What characteristics did study examples that explored the impact of IAS on climatic flexibility share?
2. Was it discovered that integration boosts climate resilience?
3. Which significant parallels and contrasts did studies that discovered a link between IAS and climate resistance have in common?

4. What metrics were used to measure integration and toughness? How helpful and similar were the several metrics for how IAS affected climate resilience?
5. What associations between specific techniques, proxies, and results did you find?
6. Integration and climate resilience generalizable relationships discovered?
7. What specific examples of methods via which aggregation impacted the climatic resilience of farming frameworks does the literature offer?

10.2 Definitions, Concepts, and Methods for the Analytical Framework

IAS: An Overview, Comparisons to Other Alternative Agricultural Systems, and Definition

Our study examines the data showing how agricultural system integration has improved agricultural systems' climatic adaptability. Our definition of agricultural systems generally refers to groups of agricultural operations carried out at various sizes (plots, farms, villages, landscapes, etc.), either concurrently or in sequential order. The authors define IAS as farming frameworks which intentionally combine the three workable elements of farming—crops, livestock, and aquaculture or forests. In order to complete this review, IAS is what we refer to as agricultural systems that have at least one interaction between two or more components of agricultural systems that is facilitative or synergistic. Intercropping may boost soil moisture, animal waste can be used as fertilizer [3] for crops, and the yield of one commodity can act as a safety net against poor output or lower demand. The level of integration in farming frameworks is sometimes unknown and may even be partly impossible to detect. In order to examine the possible advantages of integration, observable IAS components must be used as stand-ins for integration. A system has the potential to be more integrated the more varied, mixed, or diversified it is. Consequently, in research that does not specifically quantify the synergistic interactions of IAS, we use the following:

Climate Variability and Change

It may be difficult to identify the effects of climate change on agriculture owing to a shortage of longer-term data and alterations in farming operation, technology [4] or infrastructure. These factors make it difficult to do research on the implications (or averted repercussions) of climate change on economic activity. As with earlier studies on the effects of climate change on agriculture, we also take into account survey which substitutes a better random inter-annual climatic variability for climatic alteration while considering constant farm characteristics and/or location characteristics. This method establishes a connection between a random variation in the climate and financial output. These interactions may be interpreted as causative, but there is no evidence to suggest that they are good indicators of long-term climate change if the agents involved can alter through time and recognition. Still, a current research found that agricultural yields in the US Midwest were not significantly affected by long-term changes in mean temperature or inter-annual temperature variations (per unit of temperature exposure). In one highly recognized essay devoted to interpreting and reporting every framework category, the authors facilitate clarity for every framework. These definitions may not always conflict with one another. In fact, several kinds have definitions that are almost identical. The alternate agricultural systems shown in this table are not all possible. Additionally, it doesn't provide a thorough analysis of all depictions for every framework.

Resilience

According to our perspective, resilience is a phenomenon that results from several intentional actions as well as numerous unintended interactivities among agents, the agricultural framework, and its surrounding physical, social, economic, cultural, and institutional factors. Resilience is the ability of a framework to withstand a disruption or return to a reference condition after an interruption. The author's clarity incorporates resilience theory from other disciplines. Resilient frameworks are those which may withstand harm by shocks to framework component, operating and assistance furnishing. Socio-ecological framework resilience constructs upon the acuteness utilized in environment by further tracking learning, modification and skill at the management of alteration as proof of resilience. Current clarity of resilience has focused on learning, adaptation, and expertise at managing alteration.

According to international development literature, it is the ability to sustain a standard of well-being in the face of unforeseen shocks by, for instance, a person, family, or community. Tracking a desired result before and after a disruption is a need for the research of resilience. Few researches divide the resilience timeframe into various periods, such as the time when people are resilient to shocks and the time when they are recovering from shocks. In reality, though, data constraints on shock time and the occurrence of farming state measurements could make these differences impossible. A system's resilience may be measured using a variety of system components. A system's structure and behavior may be tracked to show its resilience function or the ability and agency of actors to sustain and/or enhance their standard of living.

Time series [5] observations are useful for empirical resilience research, and various researches monitor farming technique indicators which are often and easily accessible (such as farm revenue and crop production farming system outputs). These metrics, meanwhile, could not accurately reflect the well-being of households. A system could not have a consistent reference state owing to ongoing climatic and environmental change, which undermines understanding of resilience. Utilizing additional resilience metrics, such as adaptive capacity and learning functions may help solve this issue.

Library Research

We looked for any study addressing the connection between IAS and climatic resilience. Peer-reviewed publications are located using Google Scholar keyword searches carried out between November 2015 and January 2016. The database screening consisted of various aggregations of the terms climatic alteration, farming aggregation, IAS, varied, agribusiness, agronomy, mixed frameworks, variety, climatic consequences, expenditures, advantages, production, gain and conversion due to the originality and accuracy of our research survey. After acquiring a first set of articles, we also looked through the works that these papers mentioned as well as the works that cited them. Less research has been done on how integration affects smallholders in the tropics' resistance to climate change. For finding a suitable amount of references, we consequently defined each system component in a manner that struck a compromise between inclusiveness and clarity. We looked at research that examined agricultural [6] systems comprehensively, using the plot, farm, home, county, and country as units of study. Additionally, we included studies that used a variety of agricultural framework indicators which we believe to be reliable stand-ins for resilience.

Literature Acceptable for Meta-Analysis

The search of the literature turned up 97 publications and working papers, of which we believed that either (1) they would contribute original research looking into the relationships between aggregation and resilience under climatic alteration (or related proxies), or (2) they would facilitate an overview of findings on similar subject. Forty-two more prospective primary research papers were found after we filtered the review articles. The result was the creation of 124 primary papers of study that may have met criteria (1). A selection of 3740-68 research that was quantitative in character was then found. Because the goal of this article is to assist policymakers, we limited our meta-analysis to these studies. In a changing environment, consider the possible advantages and disadvantages of encouraging integration. Since this inquiry is quantitative in character, it calls for the analysis of the quantitative study. However, qualitative approaches provide more information about the subject. In addition to using qualitative techniques as part of the meta-analysis, we cited several qualitative publications in the review.

The following features of the 37 papers were compiled into a database: the paper's premise and purpose; its geographical urgency; its farming dimension of concentration; its integration proxy (variety and synergy); continual vs binary treatment of aggregation; data aggregation methodology; what farming results were followed; if numerous results were monitored; the spatial and temporal scale at which the dependent variable(s) were investigated; and data survey methodology, specifically analytical modeling,

Quantitative or Qualitative Analysis

A narrative description of the key findings identification approach (if some) is utilized to evaluate the relationship between farming aggregation and climatic resiliency; variables utilized to reflect climate; if the effects of rule or professional mediations were investigated. This database is shown in Supplement I. After that, we encoded the database's findings into a quantitative dataset (Supplement II), which the author then utilized to perform a meta-analysis having the goal of explaining connections between the design, methodology, findings, and geographic locations of the research examined.

10.3 Results

1. What features did study examples that explored the impact of IAS on climatic resilience share?

The investigation turned up a wide-ranging and expanding body of literature looking at connections between IAS and climatic resiliency (see Supplement I in a complete breakdown of study-by-study features).

The studies are current; all but three of the 37 were released after 2020. The review work was mostly concerned with tropical developing nations: 24 researches focused only on the tropics; however, because of studies concentrating on Japan, this category was not a pure subset of the developing world group, which was the subject of 26 investigations. The literature placed a lot of emphasis on smallholders: 35% of studies just looked at smallholders, whereas none only looked at largeholders. The production data was mostly derived from household surveys. All Regression analysis was used in 22 researches while survey data was used in 26 studies. Although some of them looked at county-level data, many of them used and compared household-level data. The following research was mostly modeling studies parameterized utilizing aggregate data [7] as well as model farm types and survey data. More often than temperature shocks and change, precipitation shocks and change were studied. In a significant portion (59%) of the research, aggregation (or a stand-in for aggregation) was examined as a continuous variable. Nine surveys monitored crop diversity singly, six monitored crop richness singly, and twenty monitored both. The impacts of synergy beyond richness or variety were only examined in five researches. Despite the fact that the studies looked at and used a variety of factors and methodologies, we were nevertheless able to identify areas of variation and overlap, as will be further addressed in the following sections.

2. Has it been discovered that integration boosts climate resilience?
Collectively, the publications under consideration provided evidence that more integration on farms is connected to more resilience to climatic alteration and unpredictability [8]. Of the nine publications that were evaluated, a pragmatic correlation between one measure of more integration and one measure of stronger resilience was shown or found in 33 (or 87%) participants. Six of these surveys moreover showed a detrimental federation. A research discovered a solely unfavorable connection between integration and resilience. The percentage of research demonstrating a constructive correlation in equatorial areas was similar to the percentage in nontropical areas (around 70%). When profit was used as a stand-in for resilience, every study found a positive correlation, but this was not always the case when yield or agricultural output was looked at. While 71% of the statistical articles agreed that increased integration improved resilience, 80% of deterministic modeling studies did not.

3. What techniques were used in the literature? What significant parallels and differences did studies that discovered a link between IAS and climate resistance have in common?

Thus, the authors provide a descriptive analysis of modeling research and mathematical [9] analysis researches which exhibit relationships between aggregation and resilience.

(a) Implementations using models

Richness and variety have been shown to boost the resilience of agricultural frameworks and agricultural areas in model simulations. Martin and Magne53 modeled the effects of adaptive management on four "typical" cattle frameworks in Germany. Richness and evenness were two of the adaptive methods that were simulated. They discovered that forage availability was improved by both richness and evenness, especially when paired with two additional adaptation strategies — high "disparity" forage assemblages in the ideal environment for growth. Similar results were produced by [10], who conducted implementations to depict the spectrum of climatic volatility and mid-stretch effects on two different diverse crop-livestock frameworks (smaller, food insecurity vs. big, food security). A vivid illustration of the mutual aspects of mixed crop-livestock frameworks is provided by the straight and incidental results of fertilization on productivity.

(b) Statistical investigation

Regression analysis studies have provided proof of the connections between integration and climate resistance. Specialized crop farms in South Africa were shown by [11] to be more susceptible to climatic effects than mixed horticultural farms, especially among large holders who were market-oriented. Reduced rainfall and higher temperatures were linked to lower agricultural income, the scientists said, among other significant associations between varied climatic conditions and revenue. To demonstrate how crop diversification reduces yield reductions in years with drought irregularities, [12] analyzed data from counties in southern Italy and farm families in Ethiopia. [13] carried out a microeconomic analysis, analyzed mixed crop results and compared them against two specialist farms; there are cattle farms (just crops and just cattle) in the face of climate change. Their goal was to evaluate whether, on the assumption that farmers' decisions were influenced by maximizing net income, a trend toward mixed farming might be anticipated. The author used information from 9570 agricultural analysis conducted through 15 Asian nations.

Using soils, water accessibility, power, temperature, rainfall, and control factors such nation dummies, researchers looked to see whether climate had any influence on the decisions made by farmers. Then, when the climate was changed, they investigated differences in the likelihood of selecting each system. They learned that the cumulative incomes of the aggregated field and the specialty crop field were extremely delicate to the environment, and humid and dry weather patterns enhanced the likelihood of mixed farming adoption. [14] said that despite mixed farming's stronger resistance to climatic alteration compared to technical frameworks, forecasts of its difficulty maybe amplified because of three significant specialized defects.

It consists of the notion which decisions about field usage are motivated by profitable consideration (in actuality, agriculturists' goals aren't every time rigidly tied to revenue increment), the inadequate depiction of fields (the enormous heterogeneity which exists across farms), and limitations to the establishment of animals into the framework, like the scarcity of land, were ignored within each farm type, and these limitations were ignored, along with the exclusion of important variables from the analysis (that prevented a correct judging of spatial and temporal changeability, particularly considering the wide geographic prospect and lengthy implementation time under survey).

4. Which metrics may we utilize as a stand-in for integration and resilience? How helpful and similar were the several metrics pertaining to IAS's impact on climatic resilience?

The authors looked at researches which used a variety of indicators that we thought may serve as proxies in farming framework integration and which monitored a wide range of outcomes that we collectively referred to as resilience proxies (see Supplement I). Thus, the authors go through trends in the examined literature's usage of proxies and indicators.

Integration was not regarded as a binary variable but rather as a continuous variable in the majority of the analyzed studies. Therefore, our assessment took into account research that contrasted, for instance, less vs. more integrated farming systems, rather than simply evaluating the execution of aggregated vs. special agricultural frameworks with regard to climatic resilience. By using this strategy, more studies were added to the pool and research that used units of counties, states, regions, and countries are all subject to study. Comparing specialized vs. IAS would have omitted this category of research since aggregate units of analysis are rarely never fully integrated. Numerous integration proxies' effects on climate resilience have been studied in certain research. For instance, [15] showed how both the initial degree of aggregation and the rate of development in integration contribute to climate resilience.

(a) Resilience markers

Surveys which predicted the impact of climatic variability alteration, and upsets on a broad range of outcome variables were covered in our evaluation. Income, yield, revenue, labor, profit variance over time, crop loss rates in space, food variety, and threat were among the resilience indicators reviewed. Sometimes, we discovered that surveys mainly concentrated on results which weren't they resilience proxies' sill which were also applicable for examining links between aggregation and climatic resilience. For instance, [16] employed crop collapse rates as a driver of area shift between feeding and livestock even though their major emphasis was on the change in agricultural system area. Their intermediate outcome offers perceptions on resilience. Numerous studies reported multiple outcomes, i.e., both the amount of products manufactured and the cost generated by selling of these products. Various effects are useful for cross-study contrasts but don't add much to surveys where they are deterministically connected. Additionally, certain resilience measures are better suited for prolonged modification design as opposed to shorter-term design.

5. What associations between specific techniques, proxies, and results did you find?

Different outcome variables under study were linked to various research methodologies. The majority of modeling studies monitored a variety of resilience proxies, including agricultural outcomes like yield and income. To demonstrate the range of study designs examined, this table compares the research methodologies of a sample of the assessed papers. We concentrate on the unit of analysis for each research, the categories of climatic upsets recorded, how the surveys showed resilience, and if the survey evaluated or established a nonlinear climatic-resilience link. Revenue and other indicators, such as well-being, labor productivity, and foodstuff security, were frequently monitored using statistical analysis. Although most studies monitored [17] various production-oriented proxies (such as yield and production) no research examined yield or production alone. Few studies examined long-term resilience or environmental quality. Only seven of the statistical studies followed results over many years, perhaps due to data shortages and/or problems controlling for variables that make it difficult to identify how the climate affects productivity over time (such as technological progress). But as a consequence, comparing the outcomes of statistical analysis and modeling may be challenging.

6. Were correlations between aggregation and climatic resilience discovered? The many different kinds of climatic effects that have been studied, the vast range of agricultural integration techniques, and the many potential resilience indicators start a debate about whether the evidence demonstrating that IAS may boost climate resilience is generally applicable. Regarding six themes, we address this question: (1) the analysis unit; (2) the control variables used; (3) the spatial and temporal scale; (4) the autonomous vs. induced nature of aggregation (5) the linearity of the relationship between aggregation and resilience and (6) the productivity of field systems vs. countryside means.

(a) Analytical unit
Using aggregates of the field or home data as the unit of evaluation, ten of the researches assessed in Supplement I looked at the way farming aggregation has diminished climatic resistance. While [18] argued that organic diversity across environmental, spatial, and temporal scales guaranteed ecosystem facilities on which farming relies on, [19] demonstrated that transforming topographies and cropping frameworks improved agro-ecosystem achievement and resilience in various places of the US.

Despite the fact that studies sometimes used theoretical or methodological techniques that presupposed generalizability, they weren't always comparable or generalizable. One set of research examined how farm-level integration affected agricultural resilience. The majority of the time, they featured surveys with large home samples and relied on a number of observations for statistical [20] confirmation of their findings. Decisions to participate in (higher degrees of) agricultural integration were made on an individual basis in the agricultural environments depicted by such research. A different set of research examined how integration at the landscape level affected resilience at the landscape level. They used modeling simulations and examined the combined effects of several individual choices. In these circumstances, for instance, farmers formed partnerships and decided on land usage or agricultural methods together. It follows that the conclusions of a particular research could not be applicable across various units of analysis due to the various processes and motives which underpin the affecting of aggregation. Given the growing body of research demonstrating the prospective ecological and socio-economic benefits of a landscape technique, it is notable that there are relatively few quantitative studies that have looked at the offering of landscape-level aggregation to climatic resilience.

[21] said, for instance, that "a more diversified and structured landscape mosaic would favor diverse habits and tropic networks, as well as greater flexibility of the whole system to cope with potential socioeconomic and climate change induced hazards and crises".

(b) Controlling factors

Inputs, management techniques, economic indicators (like off-farm revenue origins, grants and costs), healthcare and nourishment, regional matters (such as trade flows, institutions, and census density), access to data, and climate were all used as control variables. The most often used controls were inputs, agricultural indicators, and climatic variables. Depending on the unit of analysis utilized, different control variables were used. We discovered that healthcare [22] and nourishment, data, and geographical issues were seldom taken into account while contrasting research at the farm, family, and field levels to surveys done at the environment, village, nation, or local levels. However, both aggregate and disaggregate studies included agricultural indicators, administration and climate. Farm characteristics were restricted to collect research, whereas inputs and economic variables were only included in disparate investigations. Most research indicates an agricultural integration and climate resilience showed a positive correlation when easy-to-measure farm-level factors like farm size, yield, irrigation, and soil type, inputs like fertilizer, labor, and machinery, management factors like plantation/harvest dates, diversity, and financial factors like off-field earnings, assets, and costs were taken into account. It's interesting to note that 62% of the research which didn't account for profitable measures revealed a negative link between aggregation and resilience.

(c) Temporal and spatial scale

From a geographical and temporal standpoint, the findings from the examined literature have a narrow range of generalizability. Smallholders in equatorial nations were a key emphasis of the evaluated survey with regard to the former, as was previously indicated. This sample bias impact could be made worse by the fact that land contributes to be more concentrated in these nations. There are concerns about how evident the association between farming aggregation and climatic resilience is across these two classes given the variations between small and large-scale landholders in terms of their susceptibility to climate change and risk-coping mechanisms. This points to the necessity for more study on large holders and integration, as well as an examination of the degree to which small and big farmers may be compared and inferred from one another. In terms of the temporal scale, statistical evaluations often examined results at the level of a lone year in cross section or throughout time, but the greater bulk of modeling surveys provided results over many years. It is rare for statistical analyses of climate change's effects on agriculture to use the results of several farm systems as their unit of study. This is partly because longitudinal data are limited and it is difficult to account for variables that make it difficult to identify how the environment affects productivity over time (such as technological progress). But as a consequence, comparing the outcomes of statistical analysis and modeling may be challenging.

(d) Induced vs. autonomous integration

Only one of the studies that were analyzed looked on the advantages of aggregation for climatic resilience as promoted by policies or programs. The remaining research evaluated only autonomous integration, which is defined as aggregation which was seen and wasn't explained by a previous or ongoing rule or wasn't modeled deterministically. Therefore, it is advisable to interpret the findings of the literature review as proof which independent aggregation may increase climatic resilience. Despite not having been tested yet it is a reasonable expectation that resilience will also be conferred through integration fostered by policy actions.

(e) The linearity of the relationship between resilience and integration

The authors looked at some surveys, including those by [23], which demonstrated how the field's capacity to respond to external shocks changed nonlinearly. The research on socioecological systems highlights having as a widespread phenomenon: the variability of a system's reaction to a supplied inconvenience changes nonlinearly with sluggish or regulating framework characteristics. As a result, even within a study instance, the generalizability of conclusions about the impact of climatic alteration on agriculture may be constrained. However, tracking longer-term trajectories of framework traits that are known to influence how farm production responds to shocks would give a more precise assessment of the system's thresholds and resilience.

(f) The relationship between farming framework yield and rural maintenance.

It has long been recognized in the growth survey which farming framework yield is not always a reliable indicator of rural well-being or the reduction of penury. It's encouraging to note that just nine out of the research we analyzed (n = 9) tracked output, productivity, or profitability alone. Equally as many research (n = 9) monitored variations in profit and/or benefit. The rest trials (n = 21) monitored a variety of results. One of them monitored off-farm results (n = 8). Therefore, it should be understood that the reviewed research is monitoring results which are at least somewhat indicative of rural livings.

7. Which specific examples of methods via which aggregation impacted the climatic resilience of farming frameworks does the literature offer?
It may be used to assess the generalizability of findings and to guide actions that increase climate resilience via integration by comprehending the processes through which integration was discovered to do so. The evaluated research identified three mechanisms—the sampling result, the complementarily result, and the capability impact—by which aggregation can foster resilience. The complementary and sampling results are dependent on the connected environmental ideas of multiplicity-productivity and multiplicity-stability. The sampling result denotes the notion which in a certain location, there are a variety of organisms.

Given a more unpredictable environment the likelihood increases with the number of species that some will be able to prosper in their circumstances. The complementary effect is based on the notion that several species aid one another. The theory of complementarity in ecology is that it is high by default because each functional group has a unique resource niche. Even in agriculture, niches may exist. Agroecosystem components are chosen by people, and these components may have a range of functions. Because of this, [24] contend that the complementarity effect in agroecosystems must be divided into the disparity result (how dissimilar a system's species are from one another and whether they compete for the same resources) and the facility result (i.e., the degree to what the existence of one species eases or approves the existence of others via complementarity). According to the theory, the agroecosystem will be more resilient due to the complementarily process the more functional diversity it has. The capacity impact, the last mechanism, denotes to every human choices that strengthen resilience via aggregation.

The sample effect and the disparity effect received the greatest attention in the literature that was evaluated. Many of the simulation studies have these as their main areas of focus. By default, there was no facilitation. The research by [25] that moreover looked at assistance was an exception. Numerous more studies focused on the capacity impact. For instance, [26] shown that roughly 1% of the world's GDP may be influenced by farmers' freedom to choose a different crop mix. These transitions may result in more integration or specialization, but it shows that countries with the ability to switch and integrate have the highest chance of being robust. Nevertheless, [27]'s simulation included the assumption that producers would maximize profit, have full foresight into the future climate, and be able to diversify or specialize appropriately. The application of these outcomes to frameworks where manufacturers don't have ideal planning and don't pursue profits may be constrained by these inflexible assumptions. None of the statistical studies examined attempted to identify resilience mechanisms. Distinguishing the synergy of farming frameworks from the prospective synergy of farming frameworks would be a problem in this kind of study. Without this difference, it is impossible to separate the complementarity effect, which is determined by both variety and the sampling effect, from the latter.

To find a large enough body of research on the impact of IAS on resilience, prospective integration as a stand-in for integration—was required. Even so, it did not permit differentiation between synergy and potential synergy (i.e., extensive mixing). None of the four surveys we looked at which attempted to measure synergies made a distinction between them and diversity or richness. Finding trustworthy methods to represent synergistic agricultural activities that are separate from variety and/or richness is a significant area of future study. Monitoring feedback between ecological and agricultural systems is a significant scientific problem. It has been shown that IAS has the potential to be much more environmentally friendly than traditional systems. The research of the resilience techniques, which was the focus of this review, might eventually get mixed up with, enhanced by, or even confused by this impact. More than half of the studies, however, did not even monitor one non-climate environmental indicator, much less the effect of administration on the ecology.

10.4 Consequences for Political Implementations

IAS may theoretically be more robust than more specialized systems over various year periods under a fluctuating or altering environment, according to the deterministic modeling studies that have been examined collectively. But does more IAS really result in greater resilience? Statistical research included only one unfavorable correlation and several instances of good integration-resilience relationships. This is reassuring to us, but not definitive since only a small number of these associations have been shown to be causal. The major question is whether we can anticipate that IAS driven or directed by policy would have the same relationship to climate resilience that our assessment of the statistical literature has shown. Fostering IAS may not boost resilience and can also decrease it if independent IAS is an expression of the potential for resilience rather than an origin of resilience itself.

Sadly, only a handful of the research we analyzed attempted to differentiate between resilience resulting from IAS and resilience resulting from previous resilience ability. This crucial difference may be clarified by pilot policies and programs that distribute IAS promotion interventions in a random or as-good-as-random manner. Different origins of ambiguity in the proof depending on IAS and climatic resilience must also be addressed. To do this, such pilot policy actions must be accompanied by meticulous monitoring. Initially, these interferences can help in determining the magnitude of restoration to IAS from both a private and public standpoint. From a private standpoint, the authors should know if aggregation advantages agricultural frameworks significantly in comparison to the prices of climatic effects and various causes of livelihood variations. On the other hand, from the viewpoint of the general public, we want to comprehend how the expenses of the rule correspond to the resulting advantages to society. If these initiatives measure agricultural system indicators, second,

Participants' well-being and livelihoods (rather than just output and profitability) may aid in the selection of appropriate proxies for the resilience of rural livelihoods. Finally, strategies to promote the adoption of integration will need to be included in policies encouraging integration. When evaluating the net advantages which come from the various degrees of aggregation before and after exposure to climatic shock and unpredictability, aggregation acquiring determinants and the prices linked with them during the acquiring techniques itself should be taken into consideration. Collaborations between academics and decision-makers might be useful for impact assessment of adoption methods.

10.5 Advanced Research

The reviews' insights and shortcomings as a whole point to prospective directions for further study. First, there are numerous unknowns involved in studying the integration-resilience relationship, which is examined by unobserved factors. Inference is further complicated by concurrent changes in the socioeconomic environment, the managerial methods, and the climate. Therefore, to accurately attribute resilience to agricultural integration, implemented or quasi-experimental methods are required. On both the social and biophysical sides of the issue, we see room for more mechanistic methods. These could include methodologies like regionally dispersed long-run field trials, farm systems modeling to look back and future, and socio-behavioral experimental methods to investigate risk profiles of farmers. More empirical evidence on the cumulative, synergistic impacts of integration must come from such experiments in particular.

The investigation should take into account the intimate connections between agricultural and ecosystem function of general, short-term studies will fall short of properly contrasting integration with specialized systems. Estimates of agricultural benefits will be exaggerated if short-term returns to yield are prioritized without taking future ecological damage into consideration. Additionally, from the standpoint of climatic alteration adaptation, the adoption of IAS can lower ensuing adaptation prices, benefiting ecosystems and farmers in ways that short-term return metrics would not take into consideration. Longer-term surveys are essential for addressing this issue but are uncommon. Therefore, we encourage imagination when modeling probable longer-run outcomes that shorter-run data suggests.

Ultimately, commensurability is necessary when contrasting integration with alternatives. The capacity to perfectly contrast these advantages with various advantages and/or with trade-offs, which is, expenses incurred, is thus crucial to proving the value case for integration. Unfortunately, there aren't many studies that provide us enough information to balance the benefits and costs of integration in a compelling way. To help with such calculations, we encourage the collecting and sharing of additional agricultural and home data. Although there are difficult side effects, comparing cost-benefit ratios across location, cases, and time is still preferable to not comparing at all.

Two more improvements to the literature may be made using the extra data for cost-benefit analysis. It could allow for the deconstruction of the processes contributing to resilience and the testing of mechanistic hypotheses (e.g., a field might be aggregated based on one indicator and/or technique but not another). Additionally, it will raise the possibility of integrating aggregation proxies and resilience proxies across research. Cross-study comparison is now hindered since there are so many different methods to depict both integration and resistance. The harmonization of climatic data will moreover facilitate if such data and duplication data are supplied in a spatially specified manner.

Our analysis makes it abundantly evident that farm family and geographical heterogeneity both influence climate resilience related to IAS. The result is a significant data problem. To determine the advantages of agricultural integration, we require local data and contextual information. To follow results, we require precise household data that span a large enough spatiotemporal region to allow us to evaluate a variety of climatic and environmental factors. We believe that the continued development of this research is the greatest solution to this data dilemma. Data must be acquired with harmonization in mind. For example, [28] provide harmonized, geographically explicit homely analysis data across locations. A significant quantity of additional, comprehensive data on on-field aggregation might be collected with very small adjustments to nationally representative household surveys, drastically changing our ability to evaluate the benefits of IAS for enhancing climate resilience. First, having uniform IAS definitions and metrics would allow for comparisons across surveys. Second, in order to more precisely simulate the link between integration and resilience, it is crucial to collect information that goes beyond the existence of aggregation (for example, the quantity of crop remainder given to livestock as contrasted to a simple yes/no stating the existence of this action). We also stress the significance of such surveys being panel in nature (i.e., returning to the similar fields over a number of years), as this will enable us to (1) command for unnoticed field confounders and implement causal inference techniques, as well as (2) monitor various frameworks with various levels of IAS trajectories over time, both preceding and succeeding undergoing climatic shocks. Designing analysis tools which also allow for cross-nation and analysis contrasts of IAS while also being appropriate for use in various contexts is thus a pertinent technical research agenda as the growth community works to enhance tools to measure resilience in a climatic growth context.

10.6 Conclusions

We looked at research on the integration of agriculture and climate resilience. Our review discovered that adopting a comprehensive perspective on the depiction of climatic alteration, IAS, and climatic alteration resilience. Thirty-seven researches have looked at whether integrating agricultural systems has increased climatic alteration resistance. We discovered that the vast bulk of this research showed an association between IAS and increased farm level climate change resistance. This impact was especially noticeable in studies that followed farm results across periods of time long enough to account for changing climatic circumstances. Numerous studies also took important potentially confounding variables, such as socioeconomic and agronomic characteristics, into account. But the vast majority of statistical analyses demonstrating positive integration-resilience relationships required a clear method for determining how integration affects climate resilience. Several studies made trade-offs between resilience indices, especially during short time periods or under homogeneous climatic conditions. Despite the fact that our analysis shows a potential correlation agricultural integration and climate change, complete costs and benefits are seldom analyzed in research. These agricultural techniques limit our understanding of important integration results.

References

1. P. T. Khanh, T. H. Ngọc, and S. Pramanik, Future of Smart Agriculture Techniques and Applications, in *Advanced Technologies and AI-Equipped IoT Applications in High Tech Agriculture*, Ed., A. Khang, IGI Global, 2023.
2. T. H. Ngọc, P. T. Khanh and S. Pramanik, Smart Agriculture using a Soil Monitoring System, in *Advanced Technologies and AI-Equipped IoT Applications in High Tech Agriculture*, Ed., A. Khang, IGI Global, 2023.
3. B. Swaminathan, S. Palani, S. Vairavasundaram, K. Kotecha and V. Kumar, "IoT-Driven Artificial Intelligence Technique for Fertilizer Recommendation Model," in *IEEE Consumer Electronics Magazine*, vol. 12, no. 2, pp. 109-117, 1 March 2023, doi: 10.1109/MCE.2022.3151325.
4. S. Ahamad, S. Janani, V. Talukdar, T. Sharma, A. Sahu, S. Pramanik, and A. Gupta, Apollo Hospital's Proposed Use of Big Data Healthcare Analytics, in *Big Data Analytics Techniques for Market Intelligence*, IGI Global, 2023.
5. V. Veeraiah, V. Talukdar, Manikandan K., S. B. Talukdar, V. D. Solavande, S. Pramanik, and A. Gupta, Machine Learning Frameworks in Carpooling, in *Handbook of Research on AI and Machine Learning Applications in Customer Support and Analytics*, Eds., M dS. Hossain, R. C. Ho and G. Trajkovski, IGI Global, 2023.
6. S. Pramanik, "An Effective Secured Privacy-Protecting Data Aggregation Method in IoT", in *Achieving Full Realization and Mitigating the Challenges of the Internet of Things*, Eds., M. O. Odhiambo and W. Mwashita, IGI Global, 2022, DOI: 10.4018/978-1-7998-9312-7.ch008
7. V. Jain, M. Rastogi, J. V. N. Ramesh, A. Chauhan, P. Agarwal, S. Pramanik, and A. Gupta, FinTech and Artificial Intelligence in Relationship Banking and Computer Technology, in *AI, IoT, and Blockchain Breakthroughs in E-Governance*, Eds., K. Saini, A. Mummoorthy, R. Chandrika and N.S. Gowri Ganesh, IGI Global, 2023.
8. A. Bhattacharya, A. Ghosal, A. J. Obaid, S. Krit, V. K. Shukla, K. Mandal and S. Pramanik, "Unsupervised Summarization Approach with Computational Statistics of Microblog Data", in *Methodologies and Applications of Computational Statistics for Machine Learning*, D. Samanta, R. R. Althar, S. Pramanik and S. Dutta, Eds, IGI Global, 2021, pp. 23-37, doi: 10.4018/978-1-7998-7701-1.ch002
9. A. Mandal, S. Dutta, S. Pramanik, "Machine Intelligence of Pi from Geometrical Figures with Variable Parameters using SCILab", in *Methodologies*

and Applications of Computational Statistics for Machine Learning, D. Samanta, R. R. Althar, S. Pramanik and S. Dutta, Eds., IGI Global, 2021, pp. 38-63, doi: 10.4018/978-1-7998-7701-1.ch003

10. N. Mangra; F. Behmann; A. Thakur; A. Popescu; G. Suciu Jr; G. Giannattasio; R. Uppal; and W. Montlouis, "5G Enabled Agriculture Ecosystem: Food Supply Chain, Rural Development, and Climate Resiliency," in *5G Enabled Agriculture Ecosystem: Food Supply Chain, Rural Development, and Climate Resiliency*, vol., no., pp. 1-40, 30 Jan. 2023.
11. P. D. Rosero-Montalvo, C. A. Gordillo-Gordillo and W. Hernandez, "Smart Farming Robot for Detecting Environmental Conditions in a Greenhouse," in *IEEE Access*, vol. 11, pp. 57843-57853, 2023, doi: 10.1109/ACCESS.2023.3283986.
12. M. Car, B. A. Ferreira, J. Vuletic and M. Orsag, "Structured Ecological Cultivation with Autonomous Robots in Agriculture: Toward a Fully Autonomous Robotic Indoor Farming System," in *IEEE Robotics & Automation Magazine*, vol. 30, no. 4, pp. 77-87, Dec. 2023, doi: 10.1109/MRA.2023.3315934.
13. A. A. AlZubi and K. Galyna, "Artificial Intelligence and Internet of Things for Sustainable Farming and Smart Agriculture," in *IEEE Access*, vol. 11, pp. 78686-78692, 2023, doi: 10.1109/ACCESS.2023.3298215.
14. M. F. Celik, M. S. Isik, G. Taskin, E. Erten and G. Camps-Valls, "Explainable Artificial Intelligence for Cotton Yield Prediction with Multisource Data," in *IEEE Geoscience and Remote Sensing Letters*, vol. 20, pp. 1-5, 2023, Art no. 8500905, doi: 10.1109/LGRS.2023.3303643.
15. Waluyo, A. Widura, F. Hadiatna and D. Anugerah, "Fuzzy-Based Smart Farming and Consumed Energy Comparison Using the Internet of Things," in *IEEE Access*, vol. 11, pp. 69241-69251, 2023, doi: 10.1109/ACCESS.2023.3291616.
16. R. M. Saleem *et al.*, "Internet of Things Based Weekly Crop Pest Prediction by Using Deep Neural Network," in *IEEE Access*, vol. 11, pp. 85900-85913, 2023, doi: 10.1109/ACCESS.2023.3301504.
17. Y. Meslie, W. Enbeyle, B. K. Pandey, S. Pramanik, D. Pandey, P. Dadeech, A. Belay, and A. Saini, "Machine Intelligence-based Trend Analysis of COVID-19 for Total Daily Confirmed Cases in Asia and Africa", in *Methodologies and Applications of Computational Statistics for Machine Learning*, D. Samanta, R. R. Althar, S. Pramanik and S. Dutta, Eds, IGI Global, 2021, pp. 164-185, doi: 10.4018/978-1-7998-7701-1.ch009
18. R. Barreto, J. Cornejo, R. Palomares, J. Cornejo and M. Vargas, "Controlled Environment Agriculture and Bio-Automation Systems to Improve Plant Growth Methods in Space," *2023 IEEE Seventh Ecuador Technical Chapters Meeting (ECTM)*, Ambato, Ecuador, 2023, pp. 1-8, doi: 10.1109/ETCM58927.2023.10309029.

19. M. Konain, R. Janapati, S. M. Ahmed and M. M. Ahmed, "IoT based Solar-Powered Mushroom Farming for Sustainable Agriculture," *2023 International Conference on Sustainable Computing and Smart Systems (ICSCSS)*, Coimbatore, India, 2023, pp. 944-948, doi: 10.1109/ICSCSS57650.2023.10169539.
20. G. Taviti Naidu, KVB Ganesh, V. VidyaChellam, S. Praveenkumar, D. Dhabliya, S. Pramanik, and A. Gupta, Technological Innovation Driven by Big Data, in *Advanced Bioinspiration Methods for Healthcare Standards, Policies, and Reform*, Hadj Ahmed Bouarara, IGI Global, 2023, doi: 10.4018/978-1-6684-5656-9
21. D.-Y. Jeong, S.-K. Jo, I.-B. Lee, H. Shin and J.-G. Kim, "Digital Twin Application: Making a Virtual Pig House Toward Digital Livestock Farming," in *IEEE Access*, vol. 11, pp. 121592-121602, 2023, doi: 10.1109/ACCESS.2023.3313618.
22. V. Veeraiah, D. J. Shiju, J. V. N. Ramesh, Ganesh K. R., S. Pramanik, D. Pandey, and A. Gupta, Healthcare Cloud Services in Image Processing, in *Handbook of Research on Thrust Technologies' Effect on Image Processing*, IGI Global, 2023.
23. R. Fu, X. Ren, Y. Li, Y. Wu, H. Sun and M. A. Al-Absi, "Machine-Learning-Based UAV-Assisted Agricultural Information Security Architecture and Intrusion Detection," in *IEEE Internet of Things Journal*, vol. 10, no. 21, pp. 18589-18598, 1 Nov. 1, 2023, doi: 10.1109/JIOT.2023.3236322.
24. L. Lobefaro, M. V. R. Malladi, O. Vysotska, T. Guadagnino and C. Stachniss, "Estimating 4D Data Associations Towards Spatial-Temporal Mapping of Growing Plants for Agricultural Robots," *2023 IEEE/RSJ International Conference on Intelligent Robots and Systems (IROS)*, Detroit, MI, USA, 2023, pp. 4212-4218, doi: 10.1109/IROS55552.2023.10342449.
25. A. Aborujilah *et al.*, "IoT Integration in Agriculture: Advantages, Challenges, and Future Perspectives: Short survey," *2023 10th International Conference on Wireless Networks and Mobile Communications (WINCOM)*, Istanbul, Turkey, 2023, pp. 1-7, doi: 10.1109/WINCOM59760.2023.10322958.
26. Vignesh, L., Nishanth, J. C., H. Prasad H R, Jaythan Kumar A, and C. Pomu Chavan, "Smart Farm Android Application Using IoT and Machine Learning," 2023 *IEEE 8th International Conference for Convergence in Technology (I2CT)*, Lonavla, India, 2023, pp. 1-4, doi: 10.1109/I2CT57861.2023.10126447.
27. M. Jun, H. Jeong, C. Shin and T. Kim, "Development of Light-Emitting Grid Panel for Fish Growth in Indoor Aquaculture Farm," *OCEANS 2023 - Limerick*, Limerick, Ireland, 2023, pp. 1-4, doi: 10.1109/OCEANSLimerick52467.2023.10244359.
28. R. I. Mukhamediev *et al.*, "Coverage Path Planning Optimization of Heterogeneous UAVs Group for Precision Agriculture," in *IEEE Access*, vol. 11, pp. 5789-5803, 2023, doi: 10.1109/ACCESS.2023.3235207.

11

Modeling the Dynamics of Agrochemical Regulations and Impacts on Agricultural Productivity

Hannah Jessie Rani[1], Akanchha Singh[2], Aishwary Awasthi[3], Ashwani Rawat[4], Nuvita Kalra[5] and Ankur Gupta[6*]

[1]*Department of Electrical and Electronics Engineering, Faculty of Engineering and Technology, JAIN (Deemed-to-be University), Karnataka, India*
[2]*College of Agriculture Sciences, Teerthanker Mahaveer University, Moradabad, Uttar Pradesh, India*
[3]*Department of Mechanical Engineering, Sanskriti University, Mathura, Uttar Pradesh, India*
[4]*School of Agriculture, Dev Bhoomi Uttarakhand University, Dehradun, Uttarakhand, India*
[5]*Symbiosis Law School, Nagpur Campus, Symbiosis International (Deemed University), Pune, India*
[6]*Department of Computer Science and Engineering, Vaish College of Engineering, Rohtak, Haryana, India*

Abstract
In seven representative halophytic species, including the shoot framework of Halocnemum strobilaceum (Pall.), the effects of seasonal inequalities on lipid peroxidation were evaluated relating to malondialdehyde (MDA) constituent, and constituents of chlorophyll a and b, carotenoid, ascorbic acid constituent, proline constituent, glycine betaine constituent, and cumulative phenols constituent. Arthrocnemum macrostachyum Bieb., K. Suaeda vermiculata forssk. Koch. ex J. F. Gmel., Zygophyllum album L., Tammarix nilotica Bunge, Atriplex halimus L.,

**Corresponding author*: ankurdujana@gmail.com
Hannah Jessie Rani: ORCID: https://orcid.org/0000-0002-5449-104X
Akanchha Singh: ORCID: https://orcid.org/0009-0008-3360-9951
Aishwary Awasthi: ORCID: https://orcid.org/0000-0002-4985-2584
Ashwani Rawat: ORCID: https://orcid.org/0009-0002-0626-2135

Sabyasachi Pramanik, Niranjanamurthy M., Ankur Gupta and Ahmed J. Obaid (eds.) Mathematical Modeling in Agriculture, (211–222) © 2024 Scrivener Publishing LLC

and Limonium pruinosum (L.) Chaz. All of the species under study had a decline in carotenoid levels and an increase in glycine-betaine and lipid peroxidation throughout the summer. All species, with the exception of L, had a rise in ascorbic acid concentration throughout the winter except in Z, the level of total phenols rose over the summer in pruinosum album. Carotenoids and chl were shown to correlate positively. a, chl. The first four species belong to the Amaranthaceae family and include proline, ascorbic acid, cumulative phenols, and proline constituent with cumulative phenols. According to these findings, glycine betaine production is crucial for attenuating the oxidative harm caused by abiotic stress, particularly in Amaranthaceae.

***Keywords*:** Agrochemical, agriculture, productivity, modeling

11.1 Introduction

Reactive oxygen species (ROS) are produced by environmental stressors, such as salt, in plant cells. ROS cause the breakdown of proteins, carbohydrates, and DNA as well as the enhancement of lipid peroxidation, which results in cell death. According to [1], reactive oxygen specimens (ROS) comprise the loose radicals, superoxide radicals, hydroxyl radicals, per hydroxyl radicals, alkoxyl radicals, and non-radical forms such as H_2O_2 and singlet O_2. According to [2], malonodialdehyde (MDA) constituent, a byproduct of lipid peroxidation, was used as a marker of oxidative harm that causes disruptions in metabolic function and cellular aggregation. In order to defend their cells against ROS, plants have enzymatic and non-enzymatic antioxidant defense techniques. The specified effective cell approach to counteract the damaging results of hydroxyl radicals on cellular compositions is non-enzymatic antioxidants. Ascorbic acid, carotenoids, and phenolics are non-enzymatic antioxidants that directly interact with ROS to scavenge them. In addition to acting as an antioxidant, proline has been hypothesized to have osmoprotective properties. Through its synergistic interactions with other nutrients, ascorbic acid increases the activity of several enzymes and mitigates the harmful effects of the oxidative process.

In addition to acting as an antioxidant, it also acts as a signaling molecule, prompting the declaration of certain genes and proteins consisted in the manufacturing of antioxidant enzymes to be unregulated. This neutralizes the toxic oxidants. By improving the ROS detoxification system, glycine betaine was shown to reduce oxidative harm in alpha and tobacco suspension cells and suspended cells. Along with its buildup in particular plants under salt stress circumstances, it also plays a role in protecting the large-scale structural elements of plant cells, like membranes and protein complexes. Phenolics serve critical roles in the growth and development of plants, notably in the biosynthesis of lignin and pigments as well as in the reduction of peroxides, quenching of singlet oxygen, and neutralization of free radicals. Phenolic compounds serve as filters that absorb photons and lessen the excitation of chlorophyll when circumstances are adverse for the photosynthetic machinery in addition to their role as free radical scavengers.

The goal of the current research was to look at the impact in Halocnemum strobilaceum M., the effects of the moist and dried seasons on lipid peroxidation levels, photosynthetic pigment concentrations, ascorbic acid, proline, glycine, and cumulative phenolics are discussed. Arthrocnemum macrostachyum Bieb., K. Suaeda vermiculata forssk, Zygophyllum album, Atriplex halimus L., and Limonium pruinosum (L.) Chaz.

11.2 Resources and Techniques

Plant and soil components

Unprocessed substances (shoot specimens) in every three plants reproduce gathered arbitrarily for every specimens from seven chosen halophytes, viz., Halocnemum strobilaceum, Arthrocnemum macrostachyum, Suaeda vermiculata and Atriplex halimus, Zygophyllum album, Tamarix nilotica and Limonium pruinosum increasing naturally at downstream of Southwest Africa. During the winter (in February) and summer (in July) of 2021, the plants were labeled, and shoot specimens were taken from the similar person on a seasonal basis. The investigated site was gently dried, ground, and put through a 2 mm sieve at a depth of 0 to 30 cm. The location of the study site was 30° 27′ 12.7° N and 33° 48′ 2° E using a geographical compass. Laboratory tests were performed on fresh plant samples to identify lipids.

To assess the amount of glycinebetaine and total phenolics, specimens were scorched at 62 °C and crushed into a thin powder. They also included chlorophyll a and b, carotenes, ascorbic acid, and proline soil examination. Thermo Scientific USA measured electrical conductivity in soil water solution (1:5) and computed moisture content in accordance with Rowell (1994).

Levels of malondialdehyde (MDA)

According to [3] methods, the degree of lipid peroxidation in the examined specimens was calculated by the amount of MDA present. 2.6 ml of 0.2% TCA (Techno Pharmchem Bahadurgar Haryana) was used to homogenize

0.5 g of new samples before they were centrifuged for 12 min at 5 °C with 15,500 g (Sigma 1-16 K). Add 1.6 ml of 5% TBA (Bio Basic) to 0.5 ml of supernatant dissolved in 22% TCA, was mixed (Inc., Canada). Using a UV/VIS spectro-photometer, the immersion was calculated at 535 and 625 nm after centrifugation at 95 °C in a water bath for 25 minutes. From the MDA standard curve, the concentration of MDA was determined as ppm using the equation y = 0.0250 + 0.0038 (R2 = 0.98), and the findings were represented as (mol gl FW).

Fresh samples weighing 0.6 g were drawn out using a predetermined amount of 80% acetone (ADWIC, Egypt). Using the UV/VIS spectrophotometer ChromTech CT-2400, the absorbance of the filtrate was measured at 664, 645, and 453.5 nm. Equations were used to determine the outcomes in a quantitative manner.

Chl. a (μg/ml) = 10. 3E663 – 0.918E644

Chl. b (μg/ml) = 19. 7E644 – 3.87E663

According to [4] and represented as (mg100 gl FW), carotenoids (g/ml) = 4.3E452.5 (0.0254chl. a + 0.486 chl. b).

Analysis of ascorbic acid

10 grams of fresh specimens were weighed and blended with 50 ml of 0.38 mol/l metaphosphoric acid from Oxford Chemicals and 2 ml of sodium citrate. The 2, 7-dichlorophenolindophenol sodium solutions, 3.5 mol/l were immediately titrated with 10 ml of the filtrates. According to [5] estimation method, ascorbic acid was calculated and represented as (mg102 g 1 FW).

Determination of the proline

New specimens weighing half a gram were merged in 12 ml of 4% aqueous sulfosalicylic acid and the solution was then filtered. Acid-ninhdrin (Oxford Chemicals) was blended with 3 ml of the filtrate;

1.32 g of ninhydrine were dissolved in 40 ml of warm glacial acetic acid, 20 ml of 6 M phosphoric acid, and 3 ml of glacial acetic acid (SDFCL, India), then they were transferred to an ice bath and withdrew with 4 ml of toluene. According to [6], the toluene-holding color was sucked from the hydrated stage, and the immersion at 525 nm was measured utilizing a UV/VIS spectrophotometer called the ChromTech CT-2500. The amount of proline was measured using L-proline calibration curve was expressed as (g gl FW) utilizing equation

Y = 0.0243 0.0056 (R2 = 0.979).

Determine glycine betaine

According to [7] description, glycine-betaine concentration was calculated colorimetrically. 10 ml of distilled water were used to extract a half gram dry powdered sample. 1 ml of 2 N H2SO4 (ADWIC, Egypt) was blended to 1 ml of filtrate, and the mixture was thereby chilled in ice water. Half a milliliter of this blend was added to 0.3 milliliters of KI solution (16.7 grams of I and 25 grams of KI dissolved in 105 milliliters of H_2O), carefully mixed with a vortex mixer, and kept at 5 degrees Celsius for 17 hours, and centrifuged for 15 minutes at 10,250 rpm utilizing a Sigma 3K20 extractor.

Using a glass tube with a fine tip, the supernatant was aspirated. Using a UV/VIS spectrophotometer, the ChromTech CT-2500, the rest was blended in 9 ml of 1,2-dichloroethane, and the concentration was measured at 375 nm. Glycinebetaine was computed utilizing the equation

Y = 0.0045 0.1867 (R2 = 0.978) using the betaine standard curve from Sigma-Aldrich in Germany.

Determination of total phenols

Using the Folin-Denis indicator, as suggested by [8], the total phenolics were determined. A plant sample with a known weight was separated using 80% ethanol, and then 2 ml of the extraction was blended with 0.6 ml of the Folin reagent and well mixed before the mixture of 2 ml of soaked Na_2CO_3 and 4 ml of distilled H_2O. After one hour, catechol was used as a reference in a spectrophotometer to measure the blue color's absorbance at 725 nm using the equation

Y = 0.022 (R2 = 0.976). The outcomes were presented as (mg gl DW).

Statistical analysis and experimental planning

Three replications of a randomized complete block design were used in the experiment, which included two-way interactions among the four Amaranthaceae. According to [9], correlation coefficient evaluation was performed on all evaluated features in Amaranth species. The statistical [10] program MSTATC (Michigan State University) was used to do a mathematical evaluation of variance mean contrast. For the Amar-anthaceae group, means were separated using the F test, and for the two seasons of winter and summer, t-tests were used for all examined features with the exception of MDA data, which was transmitted by arc sin prior to inspection of variance. LSD estimates were computed with Gomez and Gomez's (1984) methodology.

11.3 Results

Climate information and soil characteristics for the sample region: while moisture constituent fell from 34.62% in winter to 14.59% in summer, electrical conductivity rose from 1.82 dS m-1 to 3.63 dS m-1. Typical temperature climate information for the sample region as well as certain chemical parameters of the soil supporting the researched species was calculated.

Oxidative stress among the species under study

MDA generation was used to evaluate lipid peroxidation. MDA concentrations significantly rose over the summer in all the species that were examined. MDA in Amaranth family species showed that Atriplex halimus and Suaeda vermiculata had the greatest levels, whereas Arthrocnemum macrostachyum and Halocnemum strobilaceum had lower levels with no appreciable change of Antioxidants in the species under study. Photosynthetic pigments; chlorophyll are not subject to lipid peroxidation. a, chl. In all of the examined halophytes, with the exception of chl, b and carotenoids as well as the a/b ratio tended to decline throughout the summer. carotenoid and b in S.

Vermicuata levels and the a/b ratio were raised in both Z. music & T. nilotica produced no discernible effect. Amaranth specimens include A. Halimus achieved the largest chl estimate chl and a. b and S. The maximum carotenoid value was found in vermi-culata, whereas H. both strobilaceum and A. The photosynthetic pigments with the lowest levels were found in macrostachyum. Ascorbic acid significantly dropped over the summer in Amaranth species, Z, as enhanced in L. nilotica. pruinosum. For the varieties of amaranth, A the maximum ascorbic constituent was found in halimus (12.41 mg/10 0 g/1 fr. Wt), whereas A. The least level was achieved with macrostachyum (3.09 mg10 0 g1 FW).

In the summer, the proline concentration in plants belonging to the Amaranth family varied from 2.73 g/gFW in H. strobilaceum to 17.46 g/gFW in S. vermiculata. Controlled by stress (drought and/or salinity), the mean proline in Amaranth minimized insignificantly; still, it rose in A. macrostachyum and S. vermiculata. Z. album, T. nilotica, and L. pruinosum are non-Amaranth specimens, and in the summer, their proline contents dropped significantly by 89.1%, 93.8%, and 78.7%, respectively. In all the plant species that were tested, glycine betaine tended to rise in the summer. Glycine betaine was abundant in the investigated Amaranth family, with H. strobilaceum having the greatest content (345.55 mole g1

DW), followed by A. macrostachyum (286.07 mole gl DW), S. vermiculata (250.83 mole gl DW), and A. halimus (172.65 mole gl DW). S. vermiculata among the Amaranth specimens had the greatest cumulative phenolic content (21.59 mg gl DW), whereas H. strobilaceum had the lowest (4.39 mg gl DW). It mostly rose in the Amarantheceae over the summer, but fell in H. strobilaceum, A. macrostachyum, and A. halimus. While it decreased in Z. album in the summer, it rose in T. nilotica and L. pruinosum. The results provide an analysis of the simple connection between several characteristics recorded in the shoots of the examined Amaranthaceae species. Some antioxidants, such as chl. b, carotenoids, proline, cumulative phenols, and proline with chl. a, chl. b, and glycine betaine, have a weakly positive link with lipid peroxidation. A slight negative connection between lipid peroxidation with chl. a and ascorbic acid, ascorbic acid with proline and glycine betaine, and glycine betaine with total phenols was found, on the other hand. Among the Amaranthaceae species, the glycinebetaine had an intermediately unfavorable correlation having lipid peroxidation (0.435), chl. a (0.462), and carotenoids (0.474). Strong positive relationships between ascorbic acid and chl. a (0.76) and chl. b (0.753) and between ascorbic acid and carotenoids (0.628, proline, and total phenolics) were also shown in the table. Proline and total phenolics were shown to have a substantial association (0.899) in Amaranth plants.

11.4 Discussion

Summer may be seen as a time of drought and/or salinity stress based on soil analysis. According to results, the pressure of increasing soil salinity, lower soil moisture content, and higher temperatures was the cause of the rise in lipid peroxidation over the summer. Similar findings have also been observed for Salvia macrosiphon Boiss, Cynodon dactylon (L.) Pers., and Salsola tetrandra Forssk. Additionally, MDA was induced by salt stress in Vicia faba L. and water stress in Periploca angustifolia Labill. Both H. strobilaceum and A. macrostachyum had lower levels of lipid peroxidation, which suggests that both species have more effective systems for scavenging free radicals, protecting them against lipid peroxidation at the cellular level.

In Salsola tetrandra, the dry season. Similar to what Kinai *et al.* (2008) observed, ROS generation in thylakoids is the primary cause of water stress' capacity to lower the concentration of chlorophylls and carotenoids in plant tissue. Chlorophyll levels may have dropped because of slower production or quicker pigment degradation. According to Kumar *et al.* (2017), Nerium oleander L. leaves contain significantly less carotenoids under salt and water stress. The reduction in the a/b ratio under stress was consistent with what [11] discovered in three of the four Brassica species when there was a water shortage. The lesser carotenoid levels in H. strobilaceum and A. machrostachyum, two members of the Amaranth family, suggest that these two species are genuine halophytes with reduced lipid per oxidation and aren't required to synthesize much antioxidant as a defensive technique.

In addition to its function as an antioxidant, ascorbic acid may be thought of as a cofactor for certain hydroxylase enzymes and violaxanthin deepoxidase. The findings of [12] who reported a reduction in ascorbic acid constituent in the mangrove plant Bruguiera parviflora under salt stress are consistent with the loss of ascorbic acid constituent during summer. With the exception of L. pruinosum, it's possible that intracellular ascorbic acid in the investigated halophytes is least fruitful at counteracting the negative consequences of abiotic stress.

According to [13], neither water stress nor salinity stress caused a major alteration in the proline constituent of Nerium oleander leaves, which is relevant to the insignificant alteration in the proline constituent of Amaranth specimens. The findings of [14] who assured that proline has a reversal result under drought and salinity condition in Salsola tetrandra were consistent with the reduced proline constituent under dry condition in Z. album, T. nilotica, and L. pruinosa. According to [15], proline was recommended as a significant cytoplasmic osmoticum and trustworthy biochemical marker for salt stress. It is known to scavenge free radicals and buffer cellular redox ability further to its role as a stabilizer of sub-cellular structures.

Data support the claim that glycine betaine production is one of Amaranthaceae's key benefits. It is obvious that larger levels of glycine betaine are associated with better plant salinity tolerance in H. strobilaceum and A. machrostachyum. In general, increased glycine betaine buildup in the summer strongly shows that this substance is important for preventing oxidative damage to plant cell processes under abiotic stress. Positive correlations was reported between salt content and aggregation of glycine betaine in Atriplex grif- fithii Moq., Suaeda fruticosa (L.) Forssk and Halopyrum mucronatum (L.) Stapf, Nerium oleander, Medicago sativa L. In addition to safeguarding membrane and enzymes, it balances PS11 protein pigment complexes under demanding circumstances. Their key defensive technique of drought tolerance and/or oxidative stress is shown by the enhanced cumulative phenolics concentration during drought stress. This concurs with findings presented by other researchers. However, discovered a decline in cumulative phenolics in the grapevine's leaves and roots during drought stress. In Mentha species, there is a significant positive association between phenolic concentration and antioxidative activity.

The connection between carotenoids and the five antioxidants—chl. (a, b), ascorbic acid, proline, and cumulative phenols—is positive, suggesting that two antioxidants may work together to combat free radicals. Furthermore, phenolics, other compounds in plants are important for the biosynthesis of lignin and pigments. In Mentha species, [16] found a significant positive connection between total chlorophylls and carotenoids. Proline metabolism can be a viable method for rerouting metabolites in place of its gathering toward cumulative phenols, as shown by the high link between proline and cumulative phenolics. Proline could be connected to the pentose-phosphate pathway framework in this regard, modifying the synthesis of NADPH and sugar phosphates for the anabolic pathway, thereby concurrently facilitating the minimizing counterparts required for mitochondrial oxidative phosphorylation in the formation of proline as an option to NADPH for the TCA cycle. Proline and phenolic compounds have recently been linked positively in several plant species, including Cyrtanthus contractus and Cyrtanthus guthrieae L. Bol., Cannabis sativa L. leaf and root served with NaCl, and Parthenium hysterophorus L. leaves treated with cadmium.

11.5 Conclusion

As a result of the increased buildup of glycine betaine, Halocnemum strobilaceum and Arthrocnemum ma-chrostachyum had the lowest levels of lipid peroxidation. To maintain equilibrium, plant cells in the drier season have developed more glycine betaine production to counteract the negative consequences of more lipid peroxidation. Additionally, proline is not reliant on the outcome to abiotic stress in the surveyed species, but phenolic biosynthesis plays a major responsibility in the resistance to salt and/or drought stress. By decreasing photosynthesis and storing phenolic chemicals, the examined species may be able to prevent oxidative damage caused by a biotic stress. Furthermore, L. pruinosum's oXidative process is significantly reduced by ascorbic acid.

References

1. S. Das *et al.*, "Antimicrobial Efficacy of Argon Cold Atmospheric Pressure Plasma Jet on Clinical Isolates of Multidrug-Resistant ESKAPE Bacteria," in *IEEE Transactions on Radiation and Plasma Medical Sciences*, vol. 7, no. 4, pp. 421-428, April 2023, doi: 10.1109/TRPMS.2023.3235358.
2. M. A. Acho, R. O. Olsehinde, A. G. Oluwafemi, C. O. Nwonuma and R. O. Arise, "Free Radical-Scavenging and Oxidative Hepatic Injury-Alleviating Properties of Azadirachta Indica Seed Protein Isolate and Hydrolysates," *2023 International Conference on Science, Engineering and Business for Sustainable Development Goals (SEB-SDG)*, Omu-Aran, Nigeria, 2023, pp. 1-15, doi: 10.1109/SEB-SDG57117.2023.10124554.
3. P. Maksimchuk *et al.*, "Oxide Nanocrystals with Variable Valence Ions for Hydroxyl Radical Neutralization," 2023 *IEEE 13th International Conference Nanomaterials: Applications & Properties (NAP)*, Bratislava, Slovakia, 2023, pp. NRA05-1-NRA05-5, doi: 10.1109/NAP59739.2023.10310932.
4. J. Lopatin, "Estimation of Foliar Carotenoid Content Using Spectroscopy Wavelet-Based Vegetation Indices," in *IEEE Geoscience and Remote Sensing Letters*, vol. 20, pp. 1-5, 2023, Art no. 2500405, doi: 10.1109/LGRS.2023.3237010.
5. J. -C. Chou *et al.*, "Potentiometric Nonenzymatic Ascorbic Acid Sensor for Static and Dynamic Measurements," in *IEEE Transactions on Instrumentation and Measurement*, vol. 72, pp. 1-10, 2023, Art no. 4001310, doi: 10.1109/TIM.2022.3223074.
6. S. R. Anton *et al.*, "Automated Detection of Corneal Edema with Deep Learning-Assisted Second Harmonic Generation Microscopy," in *IEEE Journal of Selected Topics in Quantum Electronics*, vol. 29, no. 6: *Photonic Signal Processing*, pp. 1-10, Nov.-Dec. 2023, Art no. 7201010, doi: 10.1109/JSTQE.2023.3258687.
7. M. Cervera-Díaz, M. F. León-Chávez, C. Santiesteban-Toca and C. Lozoya, "Greenhouse smart irrigation based on soil moisture and vegetation index measurements," *2023 IEEE Conference on Technologies for Sustainability (SusTech)*, Portland, OR, USA, 2023, pp. 20-24, doi: 10.1109/SusTech57309.2023.10129559.
8. A. Loumani *et al.*, "Physicochemical Study of the Conservation of Hot Pepper Post-Harvested in Adrar, Algeria by Convective Solar Drying," *2023 IEEE International Conference on Artificial Intelligence & Green Energy (ICAIGE)*, Sousse, Tunisia, 2023, pp. 1-4, doi: 10.1109/ICAIGE58321.2023.10346495.
9. Ashisha G. R., Anitha Mary X., T. George, Martin Sagayam K, Unai Fernandez-Gamiz, Hatıra Günerhan, M. N. Uddin and S. Pramanik, Analysis of Diabetes Disease Using Machine Learning Techniques: A Review, *Journal of Information Technology Management*, 2023.

10. A. D. Andrushia, T. M. Neebha, A. T. Patricia, K. M. Sagayam, and S. Pramanik, Capsule Network based Disease Classification for VitisVinifera Leaves, *Neural Computing and Applications*, doi: 10.1007/s00521-023-09058-y, 2023.
11. P. Ravichandran et al., "Utilization of hyperspectral imaging to characterize herbicide phytotoxicity in oat and mustard," *2023 11th International Conference on Agro-Geoinformatics*, Wuhan, China, 2023, pp. 1-6, doi: 10.1109/Agro-Geoinformatics59224.2023.10233445.
12. C. Sharma, A. Dey, H. Khatun, J. Das and U. Sarma, "Design and Development of a Gas Sensor Array to Detect Salinity Stress in Khasi Mandarin Orange Plants," in *IEEE Transactions on Instrumentation and Measurement*, vol. 72, pp. 1-9, 2023, Art no. 2006409, doi: 10.1109/TIM.2023.3300476.
13. D. Wieland *et al.*, "A common hard-failure mechanism in GaN HEMTs in accelerated switching and single-pulse short-circuit tests," *2023 IEEE International Reliability Physics Symposium (IRPS)*, Monterey, CA, USA, 2023, pp. 1-6, doi: 10.1109/IRPS48203.2023.10117943.
14. T. J. Pawar *et al.*, "Synthesis and Insecticidal Activity of Nitroguanidines Featuring R-and S-Proline," *2023 IEEE International Conference on Engineering Veracruz (ICEV)*, Boca del Río, Veracruz, Mexico, 2023, pp. 1-5, doi: 10.1109/ICEV59168.2023.10329640.
15. C. Sharma, A. Dey, H. Khatun, J. Das and U. Sarma, "Design and Development of a Gas Sensor Array to Detect Salinity Stress in Khasi Mandarin Orange Plants," in *IEEE Transactions on Instrumentation and Measurement*, vol. 72, pp. 1-9, 2023, Art no. 2006409, doi: 10.1109/TIM.2023.3300476.
16. A. L. J. T. Velasquez *et al.*, "Vision-Based Chlorophyll-a Measurement for Iceberg Lettuce Using Levenberg-Marquardt-Optimized Shallow Neural Network," *TENCON 2023 - 2023 IEEE Region 10 Conference (TENCON)*, Chiang Mai, Thailand, 2023, pp. 1181-1186, doi: 10.1109/TENCON58879.2023.10322332.

12

Optimizing Energy Consumption in Greenhouse Production: A Mathematical Modeling Approach

Beemkumar Nagappan[1], Arun Gupta[2], Sachin Gupta[3], Diksha Nautiyal[4], Aarti Kalnawat[5] and Dharmesh Dhabliya[6*]

[1]Department of Mechanical Engineering, Faculty of Engineering and Technology, JAIN (Deemed-to-be University), Karnataka, India
[2]Department of Mechanical Engineering, Teerthanker Mahaveer University, Moradabad, Uttar Pradesh, India
[3]Department of Management, Sanskriti University, Mathura, Uttar Pradesh, India
[4]School of Agriculture, Dev Bhoomi Uttarakhand University, Dehradun, Uttrakhand, India
[5]Symbiosis Law School, Nagpur Campus, Symbiosis International (Deemed University), Pune, India
[6]Department of Information Technology, Vishwakarma Institute of Information Technology, Pune, Maharashtra, India

Abstract
The management of groundwater quality and quantity is severely problematic in Delhi, the national capital region. The deepness of the water table and the quality of the groundwater in Delhi vary from location to location. The most typical groundwater contamination is dissolved nitrate. For many years, rising nitrate levels in local groundwater and waterways have been of concern. Agricultural practices and animals are often the primary causes of high groundwater nitrate concentrations. Through percolation, these activities contaminate groundwater as well as surface water channels. The fertilizers' worse water quality causes issues

**Corresponding author*: dharmesh.dhabliya@viit.ac.in
Beemkumar Nagappan: ORCID: https://orcid.org/0000-0003-3868-0382
Arun Gupta: ORCID: https://orcid.org/0000-0001-7445-3780
Sachin Gupta: ORCID: https://orcid.org/0000-0002-4900-0082
Diksha Nautiyal: ORCID: https://orcid.org/0009-0007-2169-3036

Sabyasachi Pramanik, Niranjanamurthy M., Ankur Gupta and Ahmed J. Obaid (eds.) Mathematical Modeling in Agriculture, (223–248) © 2024 Scrivener Publishing LLC

on the economic and ecological fronts. Since the turn of the century, increased fertilizer usage and the development of high-yield crops have increased nitrate levels in groundwater. Initially, the Drinking Water Directive represented the primary concern, which was related to human health. The Nitrates Directive, a more recent piece of regulation, made mention of nitrate's contribution to eutrophication. A research study was conducted to ascertain the nitrate pollutant of groundwater in Kolkata utilizing statistical models as a result of the study's findings on the significance of groundwater quality in urban areas and how it deteriorates owing to numerous pollutants, led by nitrogen. Groundwater models use statistical equations depending on some simplification guesses to represent the flow and transport mechanisms of groundwater. These presumptions often relate to the flow control, the groundwater's shape, the heterogeneity or anisotropy of the sedimentation inside the groundwater, the transportation paths for contaminants, and chemical reactions. The rate and regulation of water flow via aquifers and enclosing units in the sub-surface are determined using surface-water flow models. Simulations are the name given to these computations. The convective dispersion equation of nitrogen transportation in the specimen has been solved using the finite difference approach of statistical modeling. In the computation of nitrogen flow and transport, the model with the fewest restrictions and presumptions was used. Greater insights about nitrate behavior are provided by simulations utilizing homogeneous nitrogen to nitrate conversion ratio. The findings of the mass balance equation that forecasts the average nitrate concentration are compared to the outcomes of the transport simulation. Due to the numerous modeling assumptions used, modeling mistakes are often found in simulations. The use of the best hydro-geological judgment is required to reduce modeling mistakes. The significant change in the nitrate content is shown by hydro-geochemical data. These facts serve as the foundation for the simulation's conversion of nitrogen to nitrate ratio.

Keywords: Greenhouse production, mathematical modeling, agriculture, statistics, hydro-geology, simulation, 3D module, transportation framework

12.1 Introduction

An essential tool in environmental engineering is the groundwater model. In the survey of surface-water flow and pollutant transportation, the use of surface-water frameworks is essential. Numerous hydro-geologic situations have been investigated using groundwater models. For assessing the danger of surface-water quality, groundwater models are most recently utilized to forecast the transit of pollutants. Models are often intellectual summaries which analyze physical frameworks via the usage of statistical [1] equations. Models are approximate representations of physical systems or techniques. Sustainable alternative scenarios may be projected, tested, and assessed by numerically [2] or mathematically modeling [3] a reduced form of a hydro-geological system. How sharply the equations resemble the actual framework under study determines how helpful or applicable a model is. It is crucial to show a full grasp of the physical framework and the presumptions used for the derivation of the mathematical equations for assessing the utility or applicability of a framework.

Surface-water frameworks use mathematical equations and some simplifying assumptions to explain the flow and transport mechanisms of groundwater. These presumptions often relate to sediment anisotropy or flow direction, aquifer geometry, aquifer heterogeneity or bedrock, pollutant transportation processes, and chemical reactions. A framework should be treated as estimation [4] and not an actual replication of physical circumstances because of uncertainty in the values of the data included for the model. Although they are an approximation, groundwater models are a helpful investigative tool that surface-water hydrologists and researchers can employ for a variety of purposes.

Water quality, learning about the significant facets of the saturated region, implementing water flow and contaminant migration in the saturated region, consisting of river-groundwater relationship, evaluating the environmental consequence of alterations to the surface-water system, arranging optimized [5] tracking networks, and creating surface-water protected zones are just a few applications of already existing groundwater models. Models of surface-water flow are utilized to track the speed and direction of surface-water flow across aquifers and confining regions in the earth's crust. Simulations [6] are the name given to these computations [7]. An in-depth grasp of the hydro-geologic features of the model site is necessary for the modeling of groundwater flow. A thorough characterization of the following should be included in the hydro-geologic investigation:

- (Hydro-Geologic Framework) Subsurface conditions and aquifer and confining zone thickness.
- Groundwater and pollutant migration are governed by hydrologic borders, often known as boundary conditions.
- The confining zones' and aquifers' hydraulic qualities.

An analysis of the horizontal and vertical distribution of hydraulic head over the simulated region under starting circumstances, equilibrium and transitional situations can be done when hydraulic head may or may not change over time. Stresses include the distribution and intensity of surface-water recharge, groundwater pumping or injection, leakage into or out of bodies of surface water, etc. Depending on the circumstances at the site, these pressures may be continuous (not changing over time) or may fluctuate over time (transient).

So far, only academic and research institutions in India have conducted modeling studies. The majority of experts continue to utilize lumped frameworks for planning surface-water recharge and growth. The scattered nature of the surface-water regime and the sub-surface investigations are entirely ignored by these models. They cannot take into consideration the stream-aquifer interaction or the dependency of lateral recharge on the water table pattern alone since they are dependent on very conservative ideas like safe productions [8]. As a result, perennial yield and acceptable pumping (withdrawals above vertical recharge) are not possible. In India, the major goals of modeling research were (i) surface-water recharge, (ii) stream-aquifer interactivity, (iii) water table dynamics, and (iv) saltwater intrusion, etc. In order to accurately assess the use or utility of these models, it is crucial to comprehend broad elements of both surface-water flow and the transportation frameworks.

Subsurface Water Pollution

Nitrates: The most frequent pollutant in groundwater is dissolved nitrate (in the form of NO3-). High levels may produce carcinogens, lead to blue baby illness in kids, and encourage eutrophication in sources of surface-water. Fertilizers, sewage, landfills, air pollution, and industry are fewer origins of nitrates.

Pathogens: These are viruses and bacteria that may spread via water and cause diseases including cholera, typhoid, dysentery, polio, and hepatitis. Sewage, septic tanks, landfills, and animals are some of the sources.

Heavy Metals: Lead, Cadmium, Mercury, Copper, Chromium, and Nickel are examples of trace metals or heavy metals. These heavy metals may be harmful and cancer-causing. The discharge of industrial and mining performances, and the outgrowth or ash pond dumping of thermal power plants' fly ash, is sources of trace metals disposal of industrial solid waste and groundwater contamination caused by rain.

Inorganic Materials: Inorganic mixed salt contaminants, like SO_4, chloride, etc., together with sodium, potassium, expanding higher mixed materials; aggregation of bicarbonates, carbonates, together with calcium, and magnesium, swelling higher hardness of water; changing potable water into hard water; causing abdominal issues in humans if they use surface-water as a drinking source;

Organic substances: They consist of semi-volatile and volatile chemicals including PCBs, insecticides, and petroleum derivations. Amid the origins of organic compounds consist of industrial releases, discharges, sewage garbage, street drainage frameworks, and farming operations.

The most difficult environmental issue is water pollution, and nitrate is one of the most frequent and pervasive groundwater contaminants. Increased nitrate concentrations in groundwater are often caused by dispersed contamination from livestock and agricultural operations. Through seepage and percolation, these agricultural [9] practices contaminate groundwater as well as surface water systems. The difficulties caused by fertilizers for the environment and the economy are exacerbated. The advent of high-yield crops and agricultural technology throughout the previous century affected fertilizer consumption, which raised groundwater nitrate concentrations.

Although the proof supporting nitrates as an influence on these illnesses is debatable, high nitrate concentrations in drinking water were connected to human health problems such as methemoglobinemia in babies and stomach cancer in adults. Nitrate concentrations in ecosystems that are too high may have detrimental effects on the environment by eutrophicating nearby surface water bodies. Given the significance of groundwater quality in metropolitan areas and the degradation caused by numerous pollutants, nitrate being the most significant, a research study was conducted to ascertain the level of nitrate pollution in Delhi's groundwater supply.

Goals of the Study

- To research various environmental model types
- Creation of groundwater pollution models
- To create a model for Delhi's groundwater pollution due to nitrates.

12.2 Literature Review

It might be difficult to determine the exact definition of the word "model" since it is used so often and has so many meanings. The simplest straightforward definition of a model, according to [10], is a depiction of an actual system or process. The use of groundwater models as investigative and predicative tools may be done in order to anticipate future circumstances or to determine how a proposed action would affect current groundwater regime conditions. In the regulatory mode, they may also be used as screening or general techniques to produce management standards and guidelines.

Craner (2006) created a steady-state statistical surface-water flow model utilizing MODPATH and MODFLOW to comprehend the Southern Willamette's nitrate transport channels, groundwater age, and flow direction. According to the research, it could take many decades in certain places before groundwater nitrate levels can be measurably declining. A soil nitrogen dynamic framework was created by [11] to calculate the amount of nitrate that will leak into the groundwater. A model for the luck and transportation of nitrate in surface-water was created using these estimations. The approach takes into account both point and non-point nitrogen origins via various land usage groups. The approach was used for the Sumas-Blaine aquifer in Nevada State, US, which has a substantial concentration of berry and dairy crops. To assess the ultimate effects of present land usage uses and the effectiveness of suggested protective measures on nitrate pollution in the aquifer, simulations were run using the created framework.

Groundwater modeling may benefit from the data management and spatial survey capacities offered by GIS. The construction of a theoretical framework for any surface-water flow and contaminant transportation issue requires the use of GIS. Automatic data collecting, systematic framework parameter task, the formation of geographical mathematics, and the visual presentation of framework locations are all features that are offered by it and may all help to enhance and ease modeling.

For the Walloon area of Belgium, [12] created a GIS database with tools for hydrogeological modeling and surface-water risk assessments. Between the geographical database plan and the surface-water numeric framework interface GMS (Groundwater Modeling System), a "loose-coupling" tool was developed. The hydrogeological information saved in the database may be simply utilized inside various groundwater numerical models after temporal and geographic searches. Another helpful approach for collecting geographically scattered data for groundwater modeling is remote sensing. Recognition of collapses and dikes, and the depth of magnetic anomalies are all possible using airborne geophysical surveys characteristics. This knowledge is useful for creating accurate conceptual models of aquifers. According to [13], for unconfined aquifer (water-table) systems, it is necessary to assume that flow is horizontal and equipotential lines are vertical, which means that the horizontal hydraulic gradient is equal to the water table's slope and that the repository coefficient equals to the specific production (Sy).

The relationships between water levels, hydraulic heads, fluid pressures, and flow velocities are neither easy nor obvious when the water characteristics are heterogeneous and (or) transitory. In these situations, fluid pressures, fluid densities, and the inherent porosity of the porous medium are used to formulate and solve the flow equation.

Despite the fact that the ratio of the dispersion coefficient to the scale of the measurement aims to reduce with bigger scales, the majority of published values of the dispersion coefficient lie within the range of 0.02 to 2.0 times the scale of the measurement. Large-scale geographic differences in hydraulic characteristics lead to field-scale dispersion, also known as macrodispersion. As a result, it is problematic to describe transport in geological systems using relatively high levels of dispersivity together with homogeneous hydraulic characteristics (Kij). The conventional practice of dividing a transient flow farm by a mean steady-state flow farm misses some of the velocity fluctuation by default and must be made up for by employing higher levels of dispersivity, particularly transverse dispersivity. Numerous systematic outcomes for the flow equation have been created, but the majority of their uses are restricted to well hydraulics issues requiring radial symmetry. Equal aggregated compositions of the partial differential equations are evaluated using finite-element techniques using presumptive functions of the dependent variables and parameters. The use of finite-element techniques to solve groundwater issues is thoroughly analyzed and reviewed in [14].

Second-order differential equations are a category of equation that may be used to categorize equations that describe surface-water flow and solute transportation. Second-order differential equations may be broadly categorized as parabolic, elliptic, or hyperbolic. The chemical character of reactions and their relationship to the formulation of statistical problems are discussed and categorized by [15]. In their 1987 study, Bahr and Rubin compared the formulations of local and kinetic equilibrium for solute transport impacted by surface reactions. [16] outlined the direct technique, variation technique, weighted residual technique, and energy balance technique as the four techniques of formulating a problem using the finite element technique.

12.3 The Creation of Mathematical Models a Range of Models

There are several kinds of models [17] that may be used to solve the statistical equations which approximate the processes of surface-water fate, flow, and transportation. Analytical models are perfect answers to equations that represent relatively basic flow or transport circumstances, while numerical models are only approximate solutions to equations that describe very complicated situations. Instead of simulating all of the flow and transport processes, most of the frameworks can additionally replicate one or much of the techniques which control surface-water flow or pollutant relocation. The advective movement of pollutants is simulated by particle tracking [18] models like MODPATH, but other flow and transport mechanisms are not taken into consideration. It is crucial to ascertain if the framework equations shows the primary physical and chemical techniques taking place at the location before choosing a framework for usage there. In hydro-geological and remedial studies, any framework, whether it is a logical framework or a sophisticated numerical framework, can be beneficial and applicable.

Research Models

The most basic models are analytical ones, which are created by making several simplification assumptions. These are the precise answers to certain groundwater flow or transport equations that are often greatly simplified. These equations are condensed versions of much intricate 3-D equations [19] for surface-water flow or solute transportation prior to the creation and widespread usage of computers. Three-dimensional equations had to be made simpler when the human brain got involved since they were difficult to solve. Due to these simplifications, the solute transportation equation was reduced to one or two dimensions and the surface-water flow to a single dimension. The model equations for homogeneous and isotropic aquifers, simple uniform aquifer geometry, and uniform hydraulic and chemical reaction characteristics were altered as a consequence. Although some groundwater flow models (such as analytical element models) are two dimensional, analytical frameworks are usually steady-state and 1-D. Fewer pollutant transportation frameworks also guess 1-D surface-water flow conditions and 1-, 2-, or 3-D transportation states.

It isn't always feasible to take into account for location characteristics which change over time or space due to the simplifications included into analytical models. These situations include shifting groundwater flow or direction, shifting hydraulic head or chemical reaction characteristics, shifting hydraulic stresses, and shifting hydro-geologic or chemical boundary states.

The best situations for using analytical models are:

- Creating data collecting strategies before starting fieldwork, Initial site evaluations when a higher degree of precision [20] is not required.
- An impartial review of simulation results from a numerical model.

- Locations where the analytical framework's simple assumptions are aided by the field data.

Mathematical models

The more difficult equations describing surface-water flow and solute transportation are resolved using numerical models. These difficult equations often include one-dimensional numerical models as well as multi-dimensional models for chemical processes, solute transport, and groundwater movement. These numerical models approximate the differential equations representing surface-water flow or solute transportation using techniques like finite differences, the matrix [21] method, or finite elements. The approximations demand that the framework's domain and time be specified in detail. The framework domain is stated in this discrete process by a network of grid cells, and the simulation [22] time is denoted by a time interval. The graph [23] and curve show how a parameter continuously varies throughout the model space or time domain. The curve is approximated by the bars as a time interval step-by-step representation.

The accuracy [24] of any numerical model is dependent on the precision of the I/P data, the extent of the discrete in both space and time (the smaller the discrete steps, the less the probable fault), and the numerical technique utilized to clarify the model equations. Mathematical or numerical models, as opposed to analytical models, may reflect a complicated, multi-layered hydro-geologic network. To do this, the network is divided into distinct cells or components. Numerical models can be utilized to implement very basic flow and transportation circumstances that can be readily approximated utilizing an analytical framework. These applications are in addition to sophisticated 3-D surface-water flow and solute transportation issues. However, numerical models are often employed to represent issues that analytical models are unable to adequately capture.

12.4 Formulation of a Model

The objectives of the model, which specify why constructing a groundwater model is necessary, must be specified. These modeling goals will have a significant influence on the amount of modeling work required.

Characterization of Hydro-geology

To comprehend the significance of pertinent flow or solute-transport processes, it is required to accurately characterize the hydro-geological conditions at a location. A detailed field characterization must be performed due to the rise in attempts to use natural attenuation as a corrective measure. It takes more site-based farm-work than merely a starting evaluation to reach this level of characterization, including additional tracking wells, surface-water specimens, an enhancement in the quantity of laboratory analyses, and taking much field characteristics into account. Without accurate farm characterization, it is impossible to choose a desirable framework or create a framework that can be consistently calibrated. In the worst scenario, this characterization requires the following hydro-geological and geochemical data:

- Subsurface geology is described by regional geologic data.
- Data about topography, such as the height of surface waters.
- Using data from observed stream discharge (base flow) and existing surface-water bodies.
- Drawings showing cross sections of geological characteristics from well logs and soil borings.
- Diagrams of well construction and soil boring logs.
- Accurate hydraulic head data.
- Calculations of hydraulic conductivity based on slug test and/or aquifer data.
- Groundwater sources and sinks, including their locations and estimated flow rates.
- Detection of hazardous substances and other pollutants in a contaminated plume.
- Pollutant plume's vertical and horizontal extent.
- History, location, and the pace at which contaminating sources or sinks are being bulk loaded or removed.
- Identification of downgradient receptors and the direction and rate of pollutant movement.
- Sediment organic carbon content.
- Geochemical field characteristics that are appropriate (such as pH and dissolved oxygen) e.g., electron acceptors and degradation byproducts) are suitable criteria as geochemical indicators.

Conceptualization of a Model

In order to explain groundwater flow and the transfer of pollutants at a site, data characterizing site characteristics must be gathered and collected in a methodical manner. Framework conception helps in selecting the modeling methodology and modeling software to be used. The following questions should be asked while creating a conceptual model; however, they are not the only ones:

- Is there enough data to describe the site's hydro-geological states?
- How many different directions does groundwater flow?
- Can the conveyance of contaminants or the flow of groundwater be classified as 1-, 2-, or 3-dimensional?
- What is the aquifer system's composition, if it consists of more than a single aquifer, and how significant is the vertical flow between aquifers?
- Does rainfall or leakage from a river, drain, lake, or infiltration pond replenish the aquifer?
- Does the surface-water leak into a river or lake, run into a drain, or be extracted by a well in order to leave the aquifer system?
- Does it seem that the hydro-geological properties of the aquifer stay mostly consistent through time and space, or does geologic evidence suggest significant fluctuation at the location?
- Do the boundary conditions surrounding the outside of the model domain have a clear definition and a hydro-geological or geochemical foundation?
- Do the circumstances at the source of contaminants or groundwater flow stay stable throughout time?
- Are there geochemical reactions occurring in the surface-water on site, and are the mechanisms realized?

Software for Modeling Selection

After the conceptual model is created and the hydro-geological characterization of the site is finished, a computer [24] modeling program is chosen. The chosen model should be able to simulate the site's circumstances. A few broad principles should be followed when determining if a model is appropriate:

- Where to utilize analytical models.
- Site data demonstrate the relative simplicity of processes involving groundwater flow or pollutant transfer.

- It is necessary to do an initial hydro-geological state assessment or to screen potential corrective measures.
- Use numerical or mathematical models in the following situations:
- Site data demonstrate the relative complexity of processes involving groundwater flow or pollutant transfer.
- Hydraulic or chemical sources and sinks, hydro geological or geochemical conditions, and groundwater flow directions are all spatially and temporally variable.

The following scenarios call for the employment of a one-dimensional groundwater flow or transport model:

- Preliminary evaluations in when the level of aquifer heterogeneity or anisotropy is unknown.
- Locations where a possible receptor is right next to a source of contamination.
- The following justifies the development of two-dimensional models:
- Issues involving one or more groundwater sinks or sources (such as injection or pumping wells, drains, rivers, etc.).
- Sites with two-dimensional groundwater flow (such as radial flow to a well, single aquifer systems with a low vertical hydraulic head, or gradients in pollutant concentration, for example).
- locations where the aquifer's hydraulic characteristics differ significantly.
- Difficulties with contamination migration when it is necessary to estimate the lateral or vertical spread of the contaminated plume and consider the effects of transverse dispersion.

When the hydro-geologic conditions are well understood, three-dimensional flow and transport models should be utilized.

There are many aquifers present, the vertical flow of pollutants or groundwater is significant.

The justification for choosing proper model software must be included in the model survey report. The framework developer is in charge of choosing the model software to utilize at a location whatever logical groundwater flow or outcome if the framework code has previously been evaluated, validated, and documented, and transfer model software. However, it is advised that the model developer get in touch with the GMP as soon as the investigation begins to go through the best model software options. The model documentation report must be sent to the GMP together with a copy of the software and the program documentation if the GMP is not already using it. When field data specify that surface-water flow or transportation mechanisms are reasonably straightforward, for instance, analytical models may be used. Similarly, depending on the hydro-geological characterization and model conceptualization, 1-D [25], 2-D [26], or 3-D surface-water flow and transportation frameworks must be used. Figure 12.1 shows the flowchart of the model creation method.

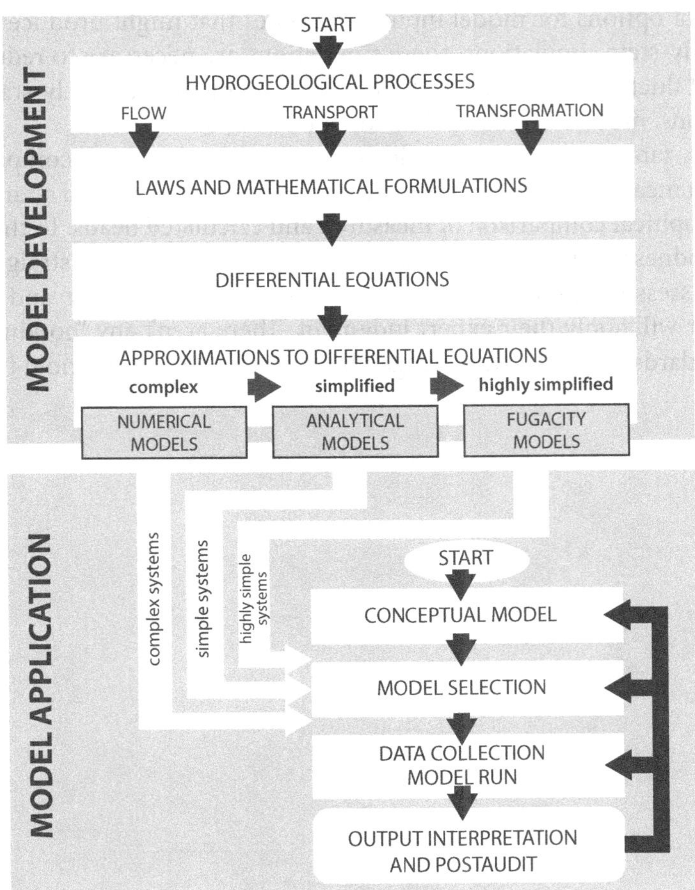

Figure 12.1 Flowchart of the Model Creation method.

Design of the Model (Input Parameters)

Model calibration involves adjusting the parameters that go into the model in an effort to get the site circumstances to meet certain standards. Field conditions at a place must be accurately defined. A framework which is measured to a set of circumstances that is not reflective of the real field conditions may come from improper site characterization. Measuring to steady-state and transient situations is a typical step in the calibration process. For the field circumstances, steady-state models show no alterations over time in the hydraulic head or pollutant gathering, which is being imitated. Transient calibration involves a change in hydraulic head or pollutant concentration over time (such as an aquifer test, an aquifer pressured by a well-field, or a migrating contamination plume). Since there are a wide variety of options for model input data values that might produce identical steady-state simulations, these simulations are necessary to reduce the range of fluctuation in the input data. Without modeling steady-state flow conditions, models can be calibrated, but it's not as simple.

Maps, tables, or graphs should be used to display these comparisons between measured and calculated heads. Figure 12.2 displays a straightforward graphical comparison of measured and calculated heads. In this case, the "goodness-of-fit" is better the closer the heads fall on the straight line. When assessing the calibration findings, each model creator and model reviewer will apply their expert judgment. There aren't any "goodness-of-fit" standards that are acknowledged by everyone and always hold true. To

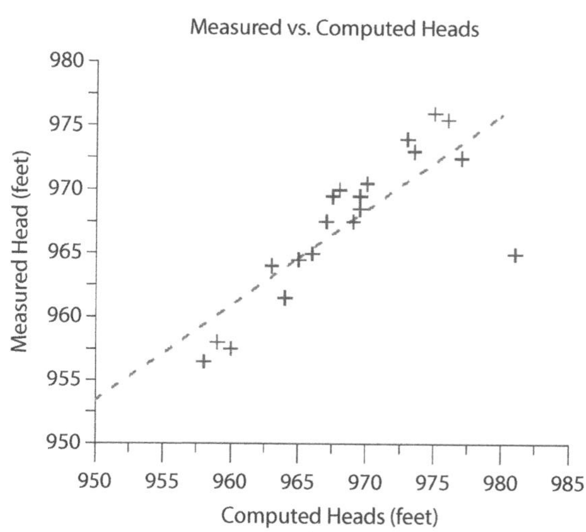

Figure 12.2 A contrast of the hydraulic heads computed and measured.

minimize the discrepancy between model simulations and observed field circumstances, the model creator must make every effort. The residual (difference between simulated and real-world circumstances) for a decent model should typically be less than 12% of the variability in the farm data across the model domain. The plot in Figure 12.3 depicts residuals at tracking wells.

The model developer should also avoid the bias of scaling down the dispersion of field data when altering the model's input data. A model that seems to be calibrated as a consequence of this "over calibration" approach is really based on a data set that is unsupported by field data. Even models that haven't been calibrated might provide meaningful findings for first analyses. Uncalibrated models may be used as a screening tool to assess the relative efficacy of potential remedial treatment options or as a guide for data gathering operations for hydro-geological research. Sensitivity analysis is the process of changing a model's input parameters throughout a tolerable range (the range of uncertainty in the parameter's value) and tracking how the model responds relatively to those changes. The observed change in hydraulic head, the transfer of contaminants, or the flow rate is often recorded in these data. Future field characterization would be necessary for data for which the model is somewhat sensitive, but not for model-insensitive data.

The sensitivity analysis's goal is to show how sensitive model simulations are to variations in the values of the model's input data. It is also shown how sensitive one model parameter is in relation to other ones.

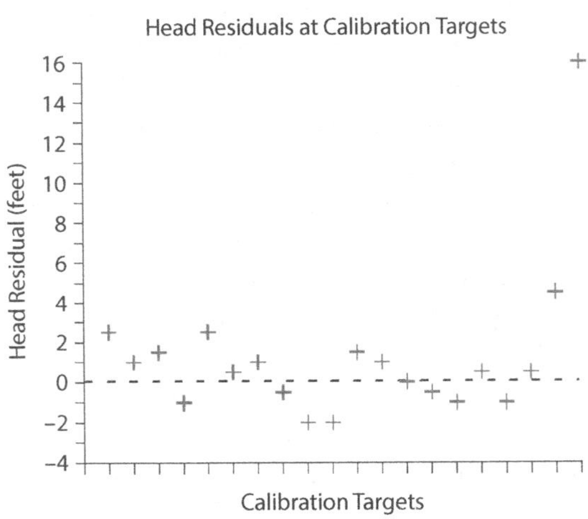

Figure 12.3 Remaining at calibration targets from a comparison of measured and calculated heads.

Sensitivity analyses are also helpful in deciding how to proceed with next data gathering efforts.

Verification of Models

A calibrated model employs chosen estimates of sources, sinks, boundary conditions, and hydro-geologic parameters to match past field circumstances. The framework can need to be additionally calibrated or improved as a consequence of the verification procedure. Predictive simulations may begin after the model has accurately re-created observed changes in field circumstances.

Dynamic Simulations

Future groundwater flow or pollutant transport conditions may be predicted using a model. The model may be used to compare various remediation options. Any model forecast, however, is just an estimate due to flaws and uncertainties in the study of surface-water flow and solute transportation. Since there is uncertainty in the input data and parameter values and the underlying assumptions, all model predictions must be stated as a range of potential results.

The succeeding instances show in a simplified way how model estimations can be portrayed to highlight the variety of potential results arising in framework input data uncertainty. The range in estimated head at calibration targets at a certain time that results from changing a framework parameter throughout its range of uncertainty is shown in Figure 12.4.

In reaction to shifting stressors on the aquifer system, hydraulic heads are anticipated for a future time period in Figure 12.5. The black line represents the findings that were anticipated using the calibrated flow model. The hydraulic heads predicted using slightly varied model input parameters are shown by the red and blue lines. Once again, it is important to show the expected head range so that prudent or acceptable judgments may be made about the groundwater resource.

An example of how to define a wellhead protection area (WHPA) for a public water supply is shown in Figure 12.6. The range of hydraulic conductivity values recorded from an aquifer test at a municipal well was used to mimic these WHPAs. The WHPA is broader and shorter when defined using the low hydraulic conductivity value (shown in brown) than when defined using the high hydraulic conductivity value (shown in gold). It's crucial to estimate the designated WHPA conservatively in order to

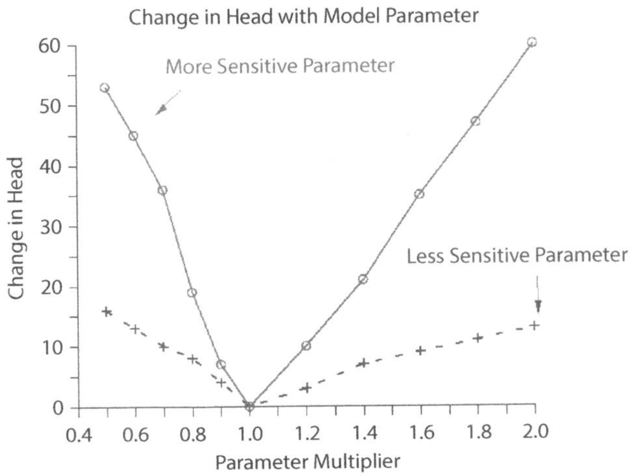

Figure 12.4 A simulated change in hydraulic head due to a parameter value change.

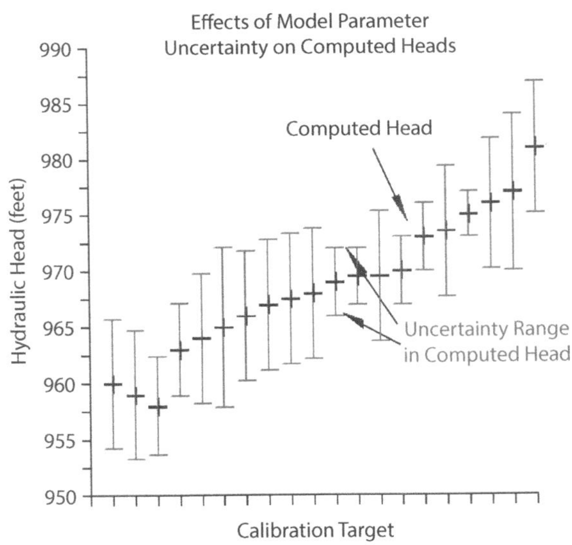

Figure 12.5 The simulated uncertainty in the calibration objectives for hydraulic heads.

safeguard the public water supply. For instance, the final WHPA that is advised is a combination of the two simulated WHPAs from before.

The simulation of a pollutant concentration down-gradient from a source location is shown in this last example (Figure 12.7). The simulation's

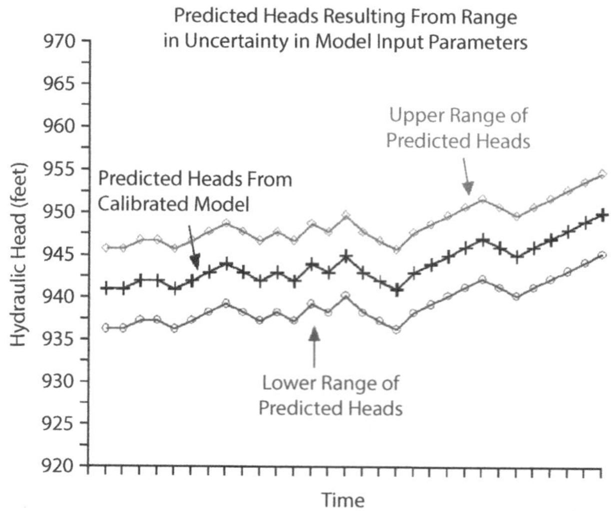

Figure 12.6 Predicted range in hydraulic heads.

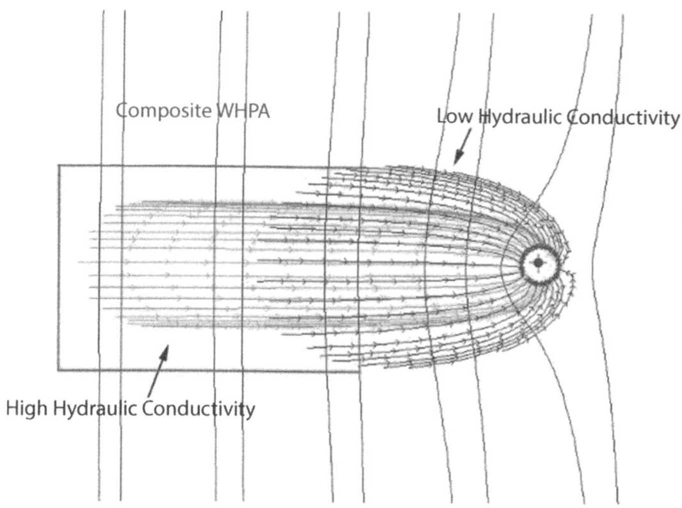

Figure 12.7 Simulated wellhead protection zones utilizing a variety of hydraulic conductivities.

goal was to calculate the pollutant concentration at the compliance point, which is located around 1000 feet down gradient from the source location. The amount of organic carbon [27] fraction employed in each simulation differs between the two simulations shown in the graphic. The amount of attenuation of the pollutant plume increases as the content of organic

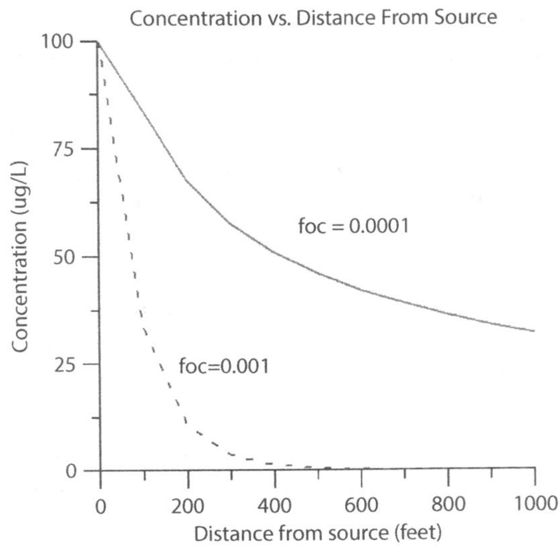

Figure 12.8 Simulated pollutant concentrations.

carbon rises. It is advisable to display the simulation results as a range of values since the organic carbon content of sediments is seldom determined with a high degree of precision and a single number is often utilized in fate and transport models. Figure 12.8 shows the simulated pollutant concentrations.

Performance Evaluation Strategy

Additionally, groundwater models are utilized to forecast pollutant concentrations and their movement routes. Even if they are tiny in scale, predictive model mistakes build up over time and may lead to glaring inaccuracies in solutions that are projected ahead. To compare predicted field conditions with actual field conditions, performance monitoring is necessary.

Performance [28] monitoring is necessary to quantify the hydro-geologic system's real behavior and demonstrate conformity with ecological state. Simulations of groundwater models are approximations and should not be used in lieu of field data collection. The following are just a few examples of applications for pollutant transport and groundwater modeling that call for performance monitoring:

- Hydraulic containment systems for which a remedial action plan has set certain geochemical and hydraulic head parameters, which gauge the efficacy of the corrective procedure.

- The chemistry of the contaminated plume discharging to a surface-water body must fulfill the necessary standards in groundwater/surface-water interface (GSI) mixing zones.
- A contaminant transport model forecasts that pollutants particular to a place would exceed regulatory limits before reaching a defined point in natural attenuation treatments.

The extent of performance monitoring necessary at a location relies on the simulated circumstances or activities and, if relevant, the level of danger to the down gradient receptors. For instance, hydraulic containment of pollutant plumes by pump-and-treat systems would need intensive hydraulic head monitoring and maybe less regular groundwater quality monitoring. In order to show the desired hydraulic response, hydraulic-head monitoring stations should be at evenly spaced places. The hydraulic containment system's capture zone would need less frequent chemical monitoring, but sites down gradient or lateral to it would require more regular chemical monitoring. Sites where contaminant biodegradation has been simulated need for intensive chemical parameter monitoring and less frequent hydraulic head monitoring. In order to assess the movement or mass removal of pollutants and determine the groundwater quality at the selected sites of compliance, chemical monitoring would be necessary at a sufficient number of locations.

12.5 Modeling of Groundwater Quality

An assumption about how a system or process works is called a conceptual model. This claim may be quantified as a set of mathematical equations known as a mathematical model. Since physical characteristics and geological features are well known they are employed as constants in mathematical models, while specifics of the system's condition or potential are used as variables. Mathematical models [29] are abstractions that function as equations. The word "model" has several meanings and is used so often that it might be challenging to determine its exact meaning. The simplest definition of a model is as a visual depiction of a system or process that really exists at the location. Most of the groundwater models now in use are deterministic mathematical models.

PDEs must often be solved in order to use deterministic groundwater models. All spatial objects must adhere to the universal rule of conservation of mass and energy. Deterministic mathematical models clarify cause-and-effect relationships by relying on the conservation of mass, momentum, and energy. The fundamental premise is that the system's response to any set of stresses can be predicted, even if the magnitude of the new stresses is outside the range of previously observed stresses, given a thorough understanding of the processes and knowledge of the parameters that influence how stresses on a system affect subsequent responses in that system. Analytical models may often provide precise answers, but they need more precise data collection methods and physical parameter uncertainty levels in order to be used.

All geologic systems exhibit heterogeneity of the medium or variation in the aquifer characteristics, which has been demonstrated to have a significant impact on groundwater flow and solute transport. Some deterministic models interpret the characteristics of porous media as lumped parameters (basically, as a black box), despite the fact that homogeneous systems are not existent in the actual world. However, this prevents the depiction of heterogeneous hydraulic behavior in the model. Applying distributed-parameter models yields more predictable outcomes because they provide accurate depiction of more realistic distributions of system attributes. Through the discrete of space and time, numerical techniques provide approximations of the solutions to the governing equation (or equations). The changeable internal attributes, limits, and stresses of the system are approximations within the discrete issue domain. The inflexible idealized conditions of analytical models or lumped-parameter models may be relaxed by deterministic, distributed-parameter, numerical models, making them more realistic and adaptable for replicating field circumstances.

The ideas of the predominant controlling processes dictate the quantity and kinds of equations that must be solved. The dependent variables of the equations are the measurements of the state of the system and are theoretically determined by the solution of the equations. The coefficients of the equations are the parameters that are measures of the qualities, boundaries, and stresses of the system. The computer code that results from the implementation of a numerical technique to resolve one or more partial differential equations is also referred to as a generic model. The resultant computer program is a site-specific model when the grid dimensions, boundary conditions, and other characteristics (such as hydraulic conductivity and storability) are defined in an application of a general model to represent a specific geographic location. Examples of applications to simplified issues are generally used to show how well generic models can solve the governing equations.

12.6 Conclusion

It is necessary to increase the consumption of nitrate by denitrifying bacteria and plants. Utilization of the research for the contaminations resulting from irrigational investigations, the current research on the modeling of groundwater for transport of nitrate and ammonia pollutants was carried out. This may aid in managing the application of fertilizers and other chemicals in agricultural settings. The findings of this research will be helpful to people, communities, and regulators who are all interested in keeping drinking water sources safe. When devising mitigation strategies for existing polluted wells and changing how future wells are placed to avoid contamination, knowledge of the processes regulating nitrate contamination would be helpful. Goals of the research are that the future of this groundwater modeling project is quite promising. The nature of the flow and transport of contaminants in groundwater at the fields has to be thoroughly studied. It is important to thoroughly investigate the relationship between the intermixing of contaminants and groundwater caused by leaks from surface sources. By applying the research to other sources of contaminants and changing the types of pollutants, it may be expanded into a much bigger context. This is a really helpful tool for managing our water supplies, which are limited and running out. We can have a more compelling understanding of nature by simulating these numerical models.

References

1. Ashisha G. R., Anitha Mary X., T. George, Martin Sagayam K, Unai Fernandez-Gamiz, Hatıra Günerhan, M. N. Uddin and S. Pramanik, Analysis of Diabetes disease using Machine Learning Techniques: A Review, *Journal of Information Technology Management*, 2023.
2. A. D. Andrushia, T. M. Neebha, A. T. Patricia, K. M. Sagayam, and S. Pramanik, Capsule Network based Disease Classification for VitisVinifera Leaves, *Neural Computing and Applications*, doi: 10.1007/s00521-023-09058-y, 2023.
3. A. Roy and S. Pramanik, A Review of the Hydrogen Fuel Path to Emission Reduction in the Surface Transport Industry, *International Journal of Hydrogen Energy*, https://doi.org/10.1016/j.ijhydene.2023.07.010, 2023
4. A. M. Ray, S. Pramanik, B. Das, A. Khanna, Hybrid Cryptography and Steganography Method to Provide Safe Data Transmission in IoT, *ICDAM*, 2023.
5. S. Pramanik, M. G. Galety, D. Samanta, N. P Joseph, Data Mining Approaches for Decision Support Systems, *3rd International Conference on Emerging Technologies in Data Mining and Information Security*, 2022.

6. V. VidyaChellam, V. Veeraiah, A. Khanna, T. H. Sheikh, S. Pramanik and D. Dhabliya, A Machine Vision-based Approach for Tuberculosis Identification in Chest X-Rays Images of Patients, *ICICC*, 2023, Springer doi.org/10.1007/978-981-99-3315-0_3
7. S. Praveenkumar, V. Veeraiah, S. Pramanik, S. M. Basha, A. V. Lira Neto, V. H. C. De Albuquerque and A. Gupta, Prediction of Patients' Incurable Diseases Utilizing Deep Learning Approaches, *ICICC*, 2023, Springer, doi.org/10.1007/978-981-99-3315-0_4
8. T. H. Ngọc, P. T. Khanh and S. Pramanik, Smart Agriculture using a Soil Monitoring System, in *Advanced Technologies and AI-Equipped IoT Applications in High Tech Agriculture*, Ed., A. Khang, IGI Global, 2023.
9. P. T. Khanh, T. H. Ngọc, and S. Pramanik, Future of Smart Agriculture Techniques and Applications, in *Advanced Technologies and AI-Equipped IoT Applications in High Tech Agriculture*, Ed., A. Khang, IGI Global, 2023.
10. E. M. Pechlivani, G. Gkogkos, N. Giakoumoglou, I. Hadjigeorgiou and D. Tzovaras, "Towards Sustainable Farming: A Robust Decision Support System's Architecture for Agriculture 4.0," 2023 *24th International Conference on Digital Signal Processing (DSP), Rhodes (Rodos)*, Greece, 2023, pp. 1-5, doi: 10.1109/DSP58604.2023.10167922.
11. K. Sudharson, B. Alekhya, G. Abinaya, C. Rohini, S. Arthi and D. Dhinakaran, "Efficient Soil Condition Monitoring with IoT Enabled Intelligent Farming Solution," *2023 IEEE International Students' Conference on Electrical, Electronics and Computer Science (SCEECS)*, Bhopal, India, 2023, pp. 1-6, doi: 10.1109/SCEECS57921.2023.10063050.
12. H. Jang, K. Yu and S. Park, "Managing 3D GIS Data for Indoor Environment Using Property Graph Database," in *IEEE Access*, vol. 11, pp. 37216-37228, 2023, doi: 10.1109/ACCESS.2023.3266519.
13. O. H. Kombo, S. Kumaran and E. Ndashimye, "A Decision Support Tool Based on WSN and ML Models for Adaptation and Management of Groundwater Resources in Sub-Saharan Africa," 2023 *IEEE AFRICON*, Nairobi, Kenya, 2023, pp. 1-5, doi: 10.1109/AFRICON55910.2023.10293319.
14. J. Kim, S. -Y. Lin, T. Singh and R. P. Singh, "InSAR Time Series Analysis to Evaluate Subsidence Risk of Monumental Chandigarh City (India) and Surroundings," in *IEEE Transactions on Geoscience and Remote Sensing*, vol. 61, pp. 1-15, 2023, Art no. 4505715, doi: 10.1109/TGRS.2023.3305863.
15. Y. Wang *et al.*, "Multiscale Feature Fusion and Semi-Supervised Temporal-Spatial Learning for Performance Monitoring in the Flotation Industrial Process," in *IEEE Transactions on Cybernetics*, doi: 10.1109/TCYB.2023.3295852.
16. Z. Liu, H. Yu, K. Xu and X. Miao, "RMCW: An Improved Residual Network with Multi-Channel Weighting for Machinery Fault Diagnosis," in *IEEE Access*, vol. 11, pp. 124472-124483, 2023, doi: 10.1109/ACCESS.2023.3328906.
17. Reepu, S. Kumar, M. G. Chaudhary, K. G. Gupta, S. Pramanik and A. Gupta, Information Security and Privacy in IoT, in *Handbook of Research*

in Advancements in AI and IoT Convergence Technologies, Eds., J. Zhao, V. V. Kumar, R. Natarajan and T. R. Mahesh, IGI Global, 2023.

18. S. Pramanik, A. J. Obaid, N M, and S. K. Bandyopadhyay, Applications of Big Data in Clinical Applications, Al-Kadhum 2nd International Conference on Modern Applications of Information and Communication Technology, *AIP Conference Proceedings* 2591, 030086 (2023), https://doi.org/10.1063/5.0119414
19. V. Jain, M. Rastogi, J. V. N. Ramesh, A. Chauhan, P. Agarwal, S. Pramanik, A. Gupta, FinTech and Artificial Intelligence in Relationship Banking and Computer Technology, in *AI, IoT, and Blockchain Breakthroughs in E-Governance*, Eds., K. Saini, A. Mummoorthy, R. Chandrika and N. S. Gowri Ganesh, IGI Global, 2023.
20. D. Pradhan, P. K. Sahu, N. S. Goje, H. Myo, M. M. Ghonge, M., Tun, R, Rajeswari and S. Pramanik, "Security, Privacy, Risk, and Safety Toward 5G Green Network (5G-GN)", in *Cyber Security and Network Security*, Eds., Wiley, 2022, doi: 10.1002/9781119812555.ch9
21. R. Anand, J. Singh, D. Pandey, B. K. Pandey, V. K. Nassa and S. Pramanik, "Modern Technique for Interactive Communication in LEACH-Based Ad Hoc Wireless Sensor Network", in *Software Defined Networking for Ad Hoc Networks*, M. M. Ghonge, S. Pramanik and A. D. Potgantwar, Eds., Springer, 2022, https://doi.org/10.1007/978-3-030-91149-2_3
22. K. Dushyant, G. Muskan, A. Gupta and S. Pramanik, "Utilizing Machine Learning and Deep Learning in Cyber security: An Innovative Approach", in *CyberSecurityandDigitalForensics*, M.M.Ghonge, S.Pramanik, R.Mangrulkar, D. N. Le, Eds., Wiley, 2022, https://doi.org/10.1002/9781119795667.ch12
23. A. Gupta, A. Verma and S. Pramanik, Security Aspects in Advanced Image Processing Techniques for COVID-19, in *An Interdisciplinary Approach to Modern Network Security*, S. Pramanik, A. Sharma, S. Bhatia and D. N. Le, Eds., CRC Press, 2022.
24. D. K.aushik, M. Garg, Annu, A. Gupta and S. Pramanik, Application of Machine Learning and Deep Learning in Cyber security: An Innovative Approach, in *Cybersecurity and Digital Forensics: Challenges and Future Trends*, M. Ghonge, S. Pramanik, R. Mangrulkar and D. N. Le, Eds., Wiley, 2022, doi: 10.1002/9781119795667.ch12
25. A. Bhattacharya, A. Ghosal, A. J. Obaid, S. Krit, V. K. Shukla, K. Mandal and S. Pramanik, "Unsupervised Summarization Approach with Computational Statistics of Microblog Data", in *Methodologies and Applications of Computational Statistics for Machine Learning*, D. Samanta, R. R. Althar, S. Pramanik and S. Dutta, Eds., IGI Global, 2021, pp. 23-37, doi: 10.4018/978-1-7998-7701-1.ch002
26. Y. Meslie, W. Enbeyle, B. K.Pandey, S. Pramanik, D.Pandey, P.Dadeech, A. Belay, and A. Saini, "Machine Intelligence-based Trend Analysis of COVID-19 for Total Daily Confirmed Cases in Asia and Africa", in *Methodologies and Applications of Computational Statistics for Machine Learning*, D. Samanta, R.

R. Althar, S. Pramanik and S. Dutta, Eds., IGI Global, 2021, pp. 164-185, doi: 10.4018/978-1-7998-7701-1.ch009

27. G. Taviti Naidu, KVB Ganesh, V. VidyaChellam, S Praveenkumar, D. Dhabliya, S. Pramanik, A. Gupta, Technological Innovation Driven by Big Data, in *Advanced Bioinspiration Methods for Healthcare Standards, Policies, and Reform*, Hadj Ahmed Bouarara, IGI Global, 2023, doi: 10.4018/978-1-6684-5656-9
28. S. Pramanik, Intelligent Farming Utilizing a Soil Tracking Device, in *Convergence of Cloud Computing, AI and Agricultural Science*, Eds., A. K. Sharma, N. Chanderwal and R. Khan, IGI Global, 2023.
29. V. Talukdar, A. Z. Kurniullah, P. Keshwani, H. Khan, S. Pramanik, A. Gupta, D. Pandey, Load Balancing Techniques in Cloud Computing, in *Emerging Trends in Cloud Computing Analytics, Scalability, and Service Models*, IGI Global, 2023.

13

Analyzing the Economic and Legal Impacts of Intellectual Property Rights on Plant Breeding Innovations: A Mathematical Modeling Study

Gopalakrishna K.[1], Bhirgu Raj Maurya[2], Rajeev Kumar[3], Sushila Arya[4], Himanshi Bhatia[5] and Ankur Gupta[6]*

[1]Department of Electronics and Communication Engineering, Faculty of Engineering and Technology, JAIN (Deemed-to-be University), Karnataka, India
[2]College of Law, Teerthanker Mahaveer University, Moradabad, Uttar Pradesh, India
[3]Department of Agriculture, Sanskriti University, Mathura, Uttar Pradesh, India
[4]School of Agriculture, Dev Bhoomi Uttarakhand University, Dehradun, Uttrakhand, India
[5]Symbiosis Law School, Nagpur Campus, Symbiosis International (Deemed University), Pune, India
[6]Department of Computer Science and Engineering, Vaish College of Engineering, Rohtak, Haryana, India

Abstract
Knowing the ways plants react to density in monocultures and the ways in which the presence of weeds impacts crop productivity has been the focus of several researches on plant competition. The evolution and present knowledge of plant competitiveness are covered in this botanical briefing, with a focus on the theory of plant competition, the creation and use of statistical frameworks to study crop-weed competitiveness, as well as the dynamics of weeds in crops. It is believed that by combining the findings of earlier studies in this way, researchers would have

**Corresponding author*: ankurdujana@gmail.com
Gopalakrishna K.: ORCID: https://orcid.org/0000-0003-1039-1722
Bhirgu Raj Maurya: ORCID: https://orcid.org/0009-0000-6220-1295
Rajeev Kumar: ORCID: https://orcid.org/0009-0009-4143-5776
Sushila Arya: ORCID: https://orcid.org/0000-0001-7492-4757

Sabyasachi Pramanik, Niranjanamurthy M., Ankur Gupta and Ahmed J. Obaid (eds.) Mathematical Modeling in Agriculture, (249–264) © 2024 Scrivener Publishing LLC

a framework in which to think about new lines of inquiry for ongoing research on competitive surveys and its implementation to aggregated weed management.

Keywords: Intellectual property, modeling, plant breed, legal impact

13.1 Introduction

The efficacy of weed management strategies is shown by the 95% decrease in weed seed quantities in cultivable farms in the 21st century. The significant use of herbicides, adjustments to crop rotations, and a variety of cultivation techniques are chiefly responsible for this effectiveness in weed management. Still, the continued usage of a variety of agro-chemicals in current years has increased the quantity of herbicide-resisting weed specimens, changed the weed flora of cultivable farms from one that was blended weeds to one that was influenced by grass weeds, and raised concerns about the use of pesticides in the environment and its effect on public health.

As a result, techniques of weed management which permit a decrease in the usage of herbicides are becoming more and more popular. This may be seen in the rise in interest in intercropping, organic farming, and non-chemical weed management approaches. Recent arguments have also been made that using genetically adjusted herbicide-tolerating crops with glyphosate and glufosinate herbicides can permit a much more adaptable, knowledge-dependent approach to weed management, allowing for a huge weed growth before the time in the season than is achievable in contemporary frameworks to assist bio-diversity. To avoid reducing yields, it is necessary that the influence of deferred command on the competitiveness equilibrium between weeds [1] and crop [2] be properly realized if such methods are to be used.

Thus, a thorough grasp of the dynamics of competition will increase the effectiveness of utilizing agronomic methods to control weed populations. As our knowledge of the process of plant competition has grown empirically, mathematical models have also been created to characterize it. The predominant weed control strategy has been mirrored in the model [3] framework. An empirical modeling method and the estimate of production loss because of weed competitiveness were the main topics of earlier study. There has been a hike in the usage and growth of more mechanistic-dependent and dynamic population frameworks for weeds as a consequence of much current scrutiny in controlling competitiveness, which has been fueled by better understanding of the environmental and biology of involving specimens. The mathematical model [4] is a crucial and important component of the survey of plant competitiveness and is used both as a research tool and as a strategy for prediction [5].

This botanical briefing aims to review the growth of statistical frameworks of plant competitiveness, mainly in their implementation to the management of farming weeds, and to give an abstract of our present grasp of the techniques of competitiveness at each plant level. The purpose of this article is to provide a general overview of the huge range of methodologies which was used in the research on plant competitiveness as it pertains to weed control. The measurement of intra-specific and inter-specific competitiveness in the crop-weed environment as well as the effects of competition on plant performance within the dynamics of rising crop and weed populations, are the main topics of this article. It is believed that by compiling the findings of earlier research in this way, researchers would have a framework in which to think about possible avenues for further investigation into competitiveness surveys and their relevance to weed control.

13.2 Competition Postulates

Despite the fact that there are various definitions of competitiveness, they may generally be categorized into two groups: those which emphasize techniques and resource acquiring and those which emphasize the decline in fitness provided by a cooperative need for a resource with a restricted availability. The mechanistic modeling method in crop-weed competition research emphasizes the significance of resource acquisition and preemption in the competitive process. The creation of phenomenological frameworks of crop-weed competitiveness, the measurement of production loss, and the dynamics of weeds in agro-environments, on the other hand, have been impacted by definitions of competition that place a greater emphasis on fitness.

13.3 Transparent Competition

A number of publications in the 1970s and 1980s by various Korean researchers are largely responsible for the current knowledge of intra-specific competitiveness in plant populations. The journals recognized three main outcomes of intra-specific competitiveness in monocultures: density-dependent mortality (self-thinning), size hierarchy development, and competitiveness-density result that causes the average dimension of survived plants to decrease with enhancing density. When plants in an agricultural setting seldom achieve the aggregations of weights and densities where self-thinning will be anticipated to prosper, the process of self-thinning is not discussed here.

13.3.1 Effect of Competitiveness-Density

The past groundbreaking research on farming of plants sparked a great deal of interest in the statistical analysis of the connection between plant attainment and density. According to [6], a reciprocal equation of the form

$$x = x_m(1 + bM)^{-a} \tag{13.1}$$

where x is the average plant weight, M is the plant density, x_m is the average dry weight of a remote plant at a given time, and a and b are parameters, best describes the decrease in plant performance with density as an outcome of competitiveness. While parameter b shows whether the yield-density connection is over-turning ($a > 1$), asymptotic ($a = 1$), or monotonically rising ($a<1$) with density, parameter b relates to the density at which intra-specific competitiveness affects production. The region of resources needed to create a yield of wm in an isolated plant may be thought of as parameter a, hence the parameters x_m and are often positively connected. This model, or one that is close to it, is crucial to explaining density-based techniques which arise from competitiveness in plant growths. It was effectively utilized to characterize production-density connections in a variety of plant monocultures.

13.3.2 Changes to the Population's Size Structure

The most often used metrics for assessing the significance of plant competition are the dimension form of a population, which may be quantified using the skewness, coefficient of variation [7], and Gini coefficient of histograms [8]. The earliest investigations on variations in biomass frequency distribution with time and density in plant monocultures were conducted by Koyama and Kira in 1956. They provided data showing that frequency distributions of herbs and tree species were initially symmetrical and gradually shifted towards a positive skew as they grew. They emphasized that this could be accounted for by pointing out that exponential growth and a symmetrical distribution of relative growth rates are what led to this unavoidable outcome. Therefore, the emergence of log-normal weight frequency distributions is not evidence of plant competition.

However, there is evidence that competition, as well as the symmetry of competitiveness, both influence the formation of frequency distributions and play a significant role in creating the plant-to-plant variability in relative growth rates which impact frequency distributions of weight. If a small quantity of people use an extremely huge part of the resources at hand, it results in asymmetrical competition that stunts the development of smaller neighbors. Each plant grows according to its size under symmetrical competition. Asymmetrical competition often results in increased biomass disparity within a population. The spatial arrangement of plants, the resource's feature, its spatial heterogeneity, its episodic availability, and the plant's physical and morphological responses to steps of resource substitution, however, interact in a complex way. Due to these intricate interconnections, it is not feasible to draw the conclusion that the emergence of size disparity is evidence of asymmetrical competition.

Numerous population models have described the emergence of a size hierarchy, and many variables have been identified as being crucial in determining an individual's position within a size hierarchy. However, frameworks of the formation of dimension hierarchies was not used to this aim despite the clear economic relevance of diversity in plant dimension, particularly in vegetable crops which must fulfill tight size restrictions to be marketable. Instead of using experimentally obtained connections, frameworks to forecast the impact of farming operations on alterations in population dimension structure have rather depended on predicting alterations in the parameter amounts of frequency distribution graphs.

13.4 Concurrence Inter-Specific

Studies in agronomy that attempt to assess competitiveness between two specimens often choose weed and crop specimens, as well as two crops planted as an intercrop, to a much smaller level. Competition in mixtures has been studied using a range of experimental methods and statistical analysis. Replacement series, additive, and neighborhood designs are the main topics of discussion here. The replacement series strategy entails keeping a consistent overall stand density while cultivating two species in a variety of ratios, including monoculture. Although widely regarded as a trailblazer in the analysis of competitiveness within mixtures, the model has come under fire for its dependence on total stand density and its inability to distinguish between the outcomes of intra and inter-specific competitiveness, particularly when environmental conditions are changing. Despite these flaws, the replacement series still has supporters, although a limited number. However, the authors will argue that it is a bad plan for the study of competitiveness in farming settings where knowing how the impacts of competitiveness change with density is a major problem.

13.4.1 Adding Damage

Experiments using additive designs vary in both the density and percentage of species in mixes. The defined "partial additive design" enables the density of a single species to maintain a constant value while the density of the second specimens is altered throughout a span of densities in its easy and often used formation in a farming environment. Due to the lack of data on the impact of the crop on the weed, the data collected by this structure may only supply a restricted image of the interactions between specimens. Thus, this blueprint is especially preferred for the research of crop-weed competitiveness. Using the average leaf area index to weight densities, one may account for variations in emergence periods.

The hyperbolic model's success is partly attributable to its capacity to meet the four essential biological realities of crop-weed competition, according to [9]:

1. The absence of weeds will result in no yield loss;
2. At lower weed densities, the impacts of growing weed numbers would be cumulative;
3. Crop production loss cannot surpass 100%;
4. At higher weed growths, there is a non-linear link between crop production and weed growth. The framework is at times blamed as it provides less insight into the basic mechanisms affecting competition results and because it is difficult to extrapolate to a wide variety of species or environments. However, the framework is easily parameterized, and given data collected from a variety of locations at various periods, it is easy to make generalizations regarding the elements that most significantly affect production loss.

If the partly cumulative framework normally entails increasing a species while altering the density of the other, an extreme variation entails developing a species that compete with one another and those that do not. Due to this, a variety of indices have been created to measure competition by, basically, comparing a plant's performance in a monoculture with that in a mixture. Although there has been some support for the usage of these easy complex indices in combination with the implemented framework, they only enable a rudimentary understanding of the competitiveness technique to be obtained. Moreover, the indices are sensitive to density and can incorrectly ascribe changes in inter-specific competition to impacts on relative performance in mixtures.

The additive series is at the other extreme of the spectrum and it entails reproducing all possible density pairings for two specimens across a broad spectrum of densities. When analyzed utilizing a two-species regression [10] model, it enables the assessment of both intra- and inter-specific competitiveness. More broadly, the lone species model (eqn 1) for this kind of study may be expanded to two or more species utilizing the relationship version of the hyperbolic model, which has been widely utilized to explain the harm done to crop production.

$$X_L = jM_w \,(1 + jb^{-1} M_w)^{-1} \tag{13.2}$$

where i represents the % production loss per weed plant at lower weed densities, X_L represents the proportion of production lost, M_w represents weed density, and a represents the maximum production loss at higher weed densities. Moreover, the equation can be updated to find variations in weed and crop competitiveness brought on by variations in emergence time.

The four fundamental biological truths of crop-weed competition that Cousens (1985) identified are as follows: (1) crop yield loss can never be greater than 100%; (2) the effects of increasing weed numbers will be additive at low weed densities; (3) crop yield loss can never exceed 100%, and (4) there is a non-linear response of crop yield to weed density at high weed densities. These truths are partially explained by the hyperbolic model's popularity. The model has been subject to criticism on occasion due to its limited ability to elucidate the fundamental mechanisms that determine competition outcomes and its challenges in generalizing to a wide variety of species or environments. Nonetheless, the framework is easily parameterized, and generalizations about the elements most important in predicting yield loss may be made using data collected from various places at various periods.

An extreme variation of partial additive design includes growing a species both with and without interspecies rivalry, while the conventional use entails flourishing a single species at a fixed density and thereby altering the other. As a result, many indices have been created to measure competition by simply contrasting a plant's performance in a mixed and monoculture. Although some researchers support the usage of these straightforward composite indices in combination with this implemented framework, their results only provide a rough image of the competitiveness technique. Again, the indices are density-favorable and can mistakenly link shifts in interspecies competition to the impact of environmental factors on relative performance in mixes.

13.4.2 Neighborhood Function

[11] invented the neighborhood technique of analyzing plant competition, which links a single chosen species' performance to the population of nearby species in a certain area. This plan is predicated on the idea that a chosen plant's performance is influenced by the quantity, biomass, cover, gathering or separation of nearby plants. The data needs for neighborhood models, however, could be very resource-intensive and might result in comparable outcomes to mean density models with reduced spatial explicitness. Despite their limited usage in this context in agricultural research, they are particularly useful when it comes to quantifying competition under various plant spatial layouts.

13.4.3 Innovative Design and Analysis

There is growing agreement that the types of conclusions which can be legitimately concluded from a research are primarily found by the implemented framework utilized. This is true even though the debate over the best implemented plan and technique of measuring competing strength in blended species draws considerable attention. As the targets, goals, and possibilities differ from research to compute from species to species, there is no ideal design for competitive trials. However, the vast majority of research, which has shown that competition's impacts on performance in plant mixes can be explained by straightforward hyperbolic models, suggests that the issue with quantifying plant competitiveness is one of regression. Regression [12] analysis is often the most reliable way for analyzing competition in real-world settings, according to theory. In order to better understand how shifting states and resources influence the technique of competitiveness in the farming ecology and the ensuing effects of competitiveness on crop production and weed accomplishment, this makes the case for compared techniques to the research of plant competitiveness under both controlled ecology and farm states.

13.5 Dynamic Plant Growth and Competition Models

Most models of plant competitiveness fall into one of two categories: phenomenological, which solely describes the results of competition, or mechanistic, which represents the physiological mechanisms underpinning plant development. The conclusions that may be formed about the competitive process from studies of competition that merely take into account ultimate yield are inherently constrained. To comprehend the shifting dynamics of species interactions and clarify the competitive processes affecting individual development over time, quantitative growth measurements obtained during a growing period are required. Additionally, dynamic growing investigation enables the separation of ontogeny—a person's phenotypic growth trajectory—from the impacts of the environment on growth. Mechanical frameworks, and particularly eco-physiological frameworks that take into account how plants' physiological systems react to their surroundings, often include a large number of factors. The usage of mechanical frameworks as a tool for decision-taking is problematic since parameter estimations might be difficult to calculate and, as a result, include significant inaccuracy. However, there has been a general rise in their usage recently as a consequence of the growth of more economical mechanical plant development frameworks.

In many eco-physiological models, light interception is the most developed component. This is because resource capture, especially the interception of solar radiation, plays a significant role in defining a species' capacity for competition. For instance, [13] suggested a framework, which was later improved by [14] and associates that accounted for the spatial location of leaves and roots by splitting the canopy and root area into various horizontal layer. Then, implemented development was divided between the two species based on how much each species contributed to the overall leaf area. [15] used a less thorough method to modeling development under a multi-species canopy, estimating the photosynthetic capacity of leaf located in sunny and shaded parts of the canopy. [16] offered a less complex framework that replicated competitiveness for light by segmenting the canopy into a limited quantity of layer.

The level of specificity utilized to estimate the local accessibility and interception of light varies between the aforementioned models. By utilizing the mechanical, spatially direct, stochastic forest implementation framework SORTIE, it was interestingly shown by [17] that the level of facts utilized to express the regional availableness of light had unexpectedly lesser impact on the performance of its estimations at the level of forest growth. Even yet, a lesser precise measurement of light does affect the development and death rates that are anticipated for each individual tree.

While the Conductance Model suggests a straightforward mechanical technique to implement the development of same and dissimilar height species in monoculture and blends as a function of various resources, ecophysiological frameworks were accomplished to comprise of resources, like N, in addition to light. The growth of a single response to a diverse resource contribution has also been described using cellular automata models. However, weed control program design has not yet been significantly influenced by dynamic frameworks of plant development and competitiveness. Despite the clear potential for applying these frameworks to crop-weed competitiveness through, for instance, altering the crop's canopy to subdue weeds and delaying the application of herbicides to permit weed development in the initial phases of crop growth to foster biodiversity; these strategies have not yet been widely used. The rigorous studies are the issue.

13.5.1 Dynamic Population

Understanding the different density-based models is necessary to comprehend the growth dynamics of a species via time and processes that have an independent and density-dependent effect on a population's birth and death rates. A hyperbolic equation of the type predicts the population dimension of species i(Ni) at time t + 1 as a function of the population dimensions at time t in population frameworks, which are dependent on populations of plants at either blooming or fertilization.

$$M_j(l+1) = \partial_j M_j(l)(1+\sum \beta_{ij} M_j)^{-a} \tag{13.3}$$

The parameters of equation (13.2) are the competition coefficients, which describe the per capita impacts of intra-specific (ii) and interspecific (ij) contenders, and the finite rate of rise, Z, which is described as the maximum average rate of census hike from lower densities. This approach enables the dynamics of the specimens to be described utilizing parameters which may be absolutely inferred from data on adds the number of plants. It can also be easily modified to incorporate a seed bank. A variety of weed species, including Avena fatua, Chenopodium album and Vulpia bromoider, have been studied using population frameworks of the type shown by eqn (13.2). Measuring the intensity of intra- and interspecific competition is a crucial part of all the models mentioned above. This was always done by adjusting densities and using regression scrutiny to calculate the competitive parameters in eqn (13.2), utilizing either population development rate or a plant performance indicator (like seed output) as the dependent variable. Sensitivity study clarifies critical life cycle characteristics that affect population size and points out parts of the cycle where controls could be successful. For Chenopodium album, it permitted [18] to draw the conclusion that better systems for understanding germination behavior will be more beneficial for estimating the results of altering management on longer-term quantity of the weed than managing the competitiveness capacity of the crop, that will only be fruitful if huge alterations in competitive outcome may be estimated.

13.6 Aspects Impacting the Result of Competitiveness

Finding the ecological and life-history features that impart competitive ability are required if a crop's inherent potential to control weeds is to be used. It has been suggested that a variety of life-history characteristics affect a species' or an individual plant's capacity for competition. Many of these features, like biomass partitioning, are morphological and exhibit significant phenotypic flexibility, which a plant might use to its advantage in a hostile environment. A number of other features have also been noted as possible predictors of competitive aptitude, including emergence time, plant size, seed size, seedling size, and seed size. The capacity of a single plant to acquire resources is affected or reflected by characteristics in one way or another.

While life history characteristics alone might provide some insight about a person's or a species' capacity for competition, it is becoming more widely accepted that relative differences between two competitive species have a significant role in determining the outcome of competition. The comparative time of emergence, comparative leaf size, and relative biomass are the life history characteristics that are most frequently taken into account when comparing the relative differences between different species examined the competitiveness interactions between seven species and discovered that competitive coefficients are closely related to species variations in their maximum sizes, root distribution, emergence times, and seed sizes. The difference in the largest dimension of plants cultivated separately was the strongest predictor, however; relationships between the other features and the competition coefficients are caused by influences on the maximum size. The research also demonstrated transitivity (estimated hierarchies of competition between species) and coefficient reciprocity (inverse correlations between the interspecific coefficients in species pairings). Theoretically, short-term competition for finite resources underlies expectations that coefficients would behave in this way.

According to [19], competition evaluated in trials using plant weight as the dependent variable isn't always equivalent to the intensity of rivalry in a population dynamic sense. Therefore, caution should be used when drawing conclusions regarding the result of competitiveness from research on plant performance singly. In terms of population dynamics, the result of competition is influenced by both the size of the competitive coefficients and the limited rate of population growth.

13.7 Crop-Weed Competition Models Applied in Practical Situations

The yield loss of crops caused by different weed densities has been widely studied using crop-weed competition models. In one of the most straightforward adaptations of this strategy, herbicide-weed resource curves and information of crop-weed competition have been coupled to implement the outcomes of herbicide usage on crop output and offer a basic economic analysis of herbicide handling. Using bio-economic models has allowed for a more thorough economic study. These are made up of a number of sub-models that generally describe the economic system, crop-weed competitiveness, and the life cycles of the crop and weed.

A different strategy has been to use crop-weed models to determine the lowest, or "threshold," weed density that justifies weed treatment. The weed density at which the expense of chemical management is equal to the net benefit on crop output is known as the threshold obtained by lessening the competition from weeds. [20] and others have criticized this modeling technique, citing the lack of data, uncertainty, and the non-random distribution of weeds. [21] made the case in a recent study that weed management decision frameworks need to be assessed from the viewpoints of biological correctness, grade of suggestions and usability. The researchers also claimed that developing affordable ways to estimate weed populations is necessary for future usage, showing that using models results in more accurate judgments and that there is reliable, longer-term financing for the models' upkeep and support.

While improving the cost-effectiveness of using herbicides has been the main driver behind many threshold models, aggregated weed control frameworks that implement an aggregation of various chemical, cultural, and biological weed management techniques facilitate a better longer-term solution to weed management. Decision aids like WEEDSIM and WEEDCAM implement the longer-term effects of a blend of various control choices on the ecology supplied an early prediction of the weed, seed-bank or seedling collection by using a choice of sub-frameworks to demonstrate biological techniques, consisting of crop-weed competitiveness. The usage of statistical frameworks in aggregated crop security was noticeably under-used, although presenting a possibly useful tool in evaluating the ecological and the economic implications of weed control measures.

The core of these frameworks is the analysis of competitiveness added with the limited pace of census growth. Their range of application includes, for example, explaining the disappearance of a once-common weed, foreseeing methods that provide ecological control of weeds, comparing the effects of large-scale alterations in farm control for the dynamics of weeds, and for thus, a variety of issues that cannot be addressed without quantifying plant competition are being addressed via the use of weed population models. The implemented plans and analyses which are utilized to measure competitiveness are crucial for this reason.

13.8 Conclusion

For a small number of species that are economically significant, crop-weed models that include competition have been very successful in modeling how the technique of competitiveness impacts crop production and the way planned weed control actions effect weed levels. If model parameters are to be employed more in predictive circumstances and to gather data for a variety of weeds and crops, we must, however, improve our knowledge of the spatial and temporal variability in those parameters. Contrarily, mechanical frameworks have, as yet, had very little effectiveness in understanding crop-weed competitiveness and minimal success in predicting crop usefulness in the process of weed control. The issue with these frameworks is that they are limited by their inborn requirement for precise information pertaining to important physiological techniques and require highly comprehensive investigations in order to be adequately parameterized. Due to the need for a more broad examination of competitive and physiological processes, the creation of more concise models would be advantageous and provide for insight beyond that of the specific species. Understanding how weed control choices made throughout the growth season of crops impact (a) the crop's production due to competitiveness for resources, and (b) the biodiversity and quantities of weeds in the present and subsequent crops is necessary for investigating integrated weed management. Here, phenomenological and mechanical models both have a role to play. When the latter, despite not consisting of these complex particulars, allows investigation of planned control settlements on the excess of weeds via crop rotation, the former provides various features of the links between plant characteristics and the ecology to permit investigation of within-season control conclusions on crop production.

References

1. P. T. Khanh, T. H. Ngọc, and S. Pramanik, Future of Smart Agriculture Techniques and Applications, in *Advanced Technologies and AI-Equipped IoT Applications in High Tech Agriculture*, Ed., A. Khang, IGI Global, 2023.
2. T. H. Ngọc, P. T. Khanh and S. Pramanik, Smart Agriculture Using a Soil Monitoring System, in *Advanced Technologies and AI-Equipped IoT Applications in High Tech Agriculture*, Ed., A. Khang, IGI Global, 2023.
3. Reepu, S. Kumar, M. G. Chaudhary, K. G. Gupta, S. Pramanik and A. Gupta, Information Security and Privacy in IoT, in *Handbook of Research in Advancements in AI and IoT Convergence Technologies*, Eds., J. Zhao, V. V. Kumar, R. Natarajan and T. R. Mahesh, IGI Global, 2023.
4. V. Veeraiah, V. Talukdar, Manikandan K., S. B. Talukdar, V. D. Solavande, S. Pramanik and A. Gupta, Machine Learning Frameworks in Carpooling, in *Handbook of Research on AI and Machine Learning Applications in Customer Support and Analytics*, Eds., Md S. Hossain, R. C. Ho and G. Trajkovski, IGI Global, 2023.
5. V. Jain, M. Rastogi, J. V. N. Ramesh, A. Chauhan, P. Agarwal, S. Pramanik, and A. Gupta, FinTech and Artificial Intelligence in Relationship Banking and Computer Technology, in *AI, IoT, and Blockchain Breakthroughs in E-Governance*, Eds., K. Saini, A. Mummoorthy, R. Chandrika and N. S. Gowri Ganesh, IGI Global, 2023.
6. K. Dushyant, G. Muskan, A. Gupta and S. Pramanik, "Utilizing Machine Learning and Deep Learning in Cyber security: An Innovative Approach", in *Cyber Security and Digital Forensics*, M. M. Ghonge, S. Pramanik, R. Mangrulkar, D. N. Le, Eds., Wiley, 2022, https://doi.org/10.1002/9781119795667.ch12
7. B. K. Pandey, D. Pandey, S. Wairya, G. Agarwal, P. Dadeech, S. R. Dogiwal and S. Pramanik, "Application of Integrated Steganography and Image Compressing Techniques for Confidential Information Transmission", in *Cyber Security and Network Security*, https://doi.org/10.1002/9781119812555.ch8, Wiley, 2022.
8. D. Kaushik, M. Garg, Annu, A. Gupta and S. Pramanik, Application of Machine Learning and Deep Learning in Cyber Security: An Innovative Approach, in *Cybersecurity and Digital Forensics: Challenges and Future Trends*, M. Ghonge, S. Pramanik, R. Mangrulkar and D. N. Le, Eds., Wiley, 2022, doi: 10.1002/9781119795667.ch12
9. B. Chen, W. Peng, X. Cao and J. Röning, "Hyperbolic Uncertainty Aware Semantic Segmentation," in *IEEE Transactions on Intelligent Transportation Systems*, doi: 10.1109/TITS.2023.3312290.
10. B. K. Pandey, D. Pandey, V. K. Nassa, A. S. George, S. Pramanik and P. Dadheech, Applications for the Text Extraction Method of Complex Degraded Images, in *The Impact of Thrust Technologies on Image Processing*, Nova Publishers, 2023.

11. A. M. Ray, S. Pramanik, B. Das, A. Khanna, Hybrid Cryptography and Steganography Method to Provide Safe Data Transmission in IoT, *ICDAM*, 2023.
12. G. Taviti Naidu, K. V. B. Ganesh, V. Vidya Chellam, S Praveenkumar, D. Dhabliya, S. Pramanik, A. Gupta, Technological Innovation Driven by Big Data, in *Advanced Bioinspiration Methods for Healthcare Standards, Policies, and Reform*, Hadj Ahmed Bouarara, IGI Global, 2023, doi: 10.4018/978-1-6684-5656-9
13. L. Lei, Z. Li, G. Yang and H. Yang, "High-Throughput Extraction of the Distributions of Leaf Base and Inclination Angles of Maize in the Field," in *IEEE Transactions on Geoscience and Remote Sensing*, vol. 61, pp. 1-28, 2023, Art no. 4411228, doi: 10.1109/TGRS.2023.3332869.
14. R. R. Chandan, S. Soni, A. Raj, V. Veeraiah, D. Dhabliya, S. Pramanik, A. Gupta, Genetic Algorithm and Machine Learning, in *Advanced Bioinspiration Methods for Healthcare Standards, Policies, and Reform*, Hadj Ahmed Bouarara, IGI Global, 2023, doi: 10.4018/978-1-6684-5656-9
15. Y. Zhang, X. Han and J. Yang, "Estimation of Leaf Area Index Over Heterogeneous Regions Using the Vegetation Type Information and PROSAIL Model," in *IEEE Journal of Selected Topics in Applied Earth Observations and Remote Sensing*, vol. 16, pp. 5405-5415, 2023, doi: 10.1109/JSTARS.2023.3283535.
16. X. Huang, F. Cheng, J. Wang, P. Duan and J. Wang, "Forest Canopy Height Extraction Method Based on ICESat-2/ATLAS Data," in *IEEE Transactions on Geoscience and Remote Sensing*, vol. 61, pp. 1-14, 2023, Art no. 5700814, doi: 10.1109/TGRS.2023.3240848.
17. H. Hao, X. Zhang, Y. Chi, R. Gao, A. Xie and M. Xu, "Cooperative Carrier Aircraft Support Operation Scheduling via Multi-Agent Reinforcement Learning," *2023 24th IEEE International Conference on Mobile Data Management (MDM)*, Singapore, Singapore, 2023, pp. 297-302, doi: 10.1109/MDM58254.2023.00055.
18. D. Mondal, A. Ratnaparkhi, A. Deshpande, V. Deshpande, A. P. Kshirsagar and S. Pramanik, Applications, Modern Trends and Challenges of Multiscale Modelling in Smart Cities, in *Data-Driven Mathematical Modeling in Smart Cities*, IGI Global, 2023, doi: 10.4018/978-1-6684-6408-3.ch001
19. B. K. Pandey, D. Pandey, V. K. Nassa, A. S. Hameed, A. S. George, P. Dadheech and S. Pramanik, A Review of Various Text Extraction Algorithms for Images, in The Impact of Thrust Technologies on Image Processing, Nova Publishers, 2023.
20. P. K. Mall, S. Pramanik, S. Srivastava, M. Faiz, S. Sriramulu and M. N. Kumar, FuzzyNet-Based Modelling Smart Traffic System in Smart Cities Using Deep Learning Models, in *Data-Driven Mathematical Modeling in Smart Cities*, IGI Global, 2023, doi: 10.4018/978-1-6684-6408-3.ch005
21. A. Namdev, V. Veeraiah, S. Dhamodaran, S. Islam, S. Pramanik and A. Gupta, Cloud Computing Applications in Biomedicine, in *Emerging Trends in Cloud Computing Analytics, Scalability, and Service Models*, IGI Global, 2023.

14

Simulating the Effects of Land Use Regulations on Agricultural Land Values: A Mathematical Modeling Study

Ashwani Rawat[1], Ramachandran T.[2], Yogesh Chandra Gupta[3], Manoj Kumar Mishra[4], Gabriela Michael[5] and Sabyasachi Pramanik[6*]

[1]*School of Agriculture, Dev Bhoomi Uttarakhand University, Dehradun, Uttarakhand, India*
[2]*Department of Mechanical Engineering, Faculty of Engineering and Technology, JAIN (Deemed-to-be University), Karnataka, India*
[3]*College of Law, Teerthanker Mahaveer University, Moradabad, Uttar Pradesh, India*
[4]*Department of Agriculture, Sanskriti University, Mathura, Uttar Pradesh, India*
[5]*Symbiosis Law School, Nagpur Campus, Symbiosis International (Deemed University), Pune, India*
[6]*Department of Computer Science & Engineering, Haldia Institute of Technology, Haldia, West Bengal, India*

Abstract

We assess the status of agricultural systems science at the moment, paying close attention to the strengths and weaknesses of farming system frameworks. We talk about the framework's current condition in relation to five distinct use cases that span field, farm, landscape, regional, and global spatial scales and ask queries about the past, present, and future. Major developments that are pertinent to a variety of applications for agricultural system models at different geographical and temporal dimensions have been developed as a result of contributions from many disciplines. Although the elements of present agricultural systems models

**Corresponding author*: sabyalnt@gmail.com
Ashwani Rawat: ORCID: https://orcid.org/0009-0002-0626-2135
Ramachandran T.: ORCID: https://orcid.org/0000-0002-6991-0403
Yogesh Chandra Gupta: ORCID: https://orcid.org/0009-0003-3699-537X
Manoj Kumar Mishra: ORCID: https://orcid.org/0009-0007-7321-9838

Sabyasachi Pramanik, Niranjanamurthy M., Ankur Gupta and Ahmed J. Obaid (eds.) Mathematical Modeling in Agriculture, (265–294) © 2024 Scrivener Publishing LLC

are required for the Use Cases, we discovered that each of them has shortcomings and needs to be enhanced. All of the Use Cases had two shortcomings that we were able to pinpoint: 1) a lack of data for creating, assessing, and using farming system frameworks, and 2) insufficient knowledge systems for successfully communicating framework outcomes to the community. We contend that these restrictions obstruct development more severely than conceptual theory gaps or the methodologies for applying system models that are currently available. A culture shift among agricultural researchers is also necessary to ascertain that data for directing the spectrum of Use Cases are accessible in future framework enhancements and implementations. New efforts on open data display pledge for tackling the data challenge. We get to the conclusion that several policies and various frameworks are required for framework implementations for various reasons. The Use Cases facilitate a helpful model for analyzing the strengths and weaknesses of current frameworks and data.

***Keywords*:** Models of integrated agriculture systems, crop models, financial frameworks, models of cattle, use cases, data on agriculture

14.1 Introduction

A broad variety of disciplinary contributions have helped shape farming frameworks research as the authors know it over the past 45 years. In general, throughout this same time span, as plenty of researchers, designer and analysts graduate from colleges having expertise in systems modeling, statistical [1] techniques and IT tools, respect for and receipt of farming frameworks science has incremented. Additionally, throughout this time, there has been a commensurate rise in the need for agricultural systems science to solve societal issues that go beyond farming [2]. Related queries include how to organize frameworks more effectively for better and much methodical yield, what adjustments must be made in an agricultural framework for increased advantages without hampering the ecology, what regulations must be put in place to support farming systems as they develop to solve larger community objectives, and what systems must be in place to adjust to the ongoing alterations which farming features, such as climate alteration, shifts in consumer request for farming products, variable energy costs, and a host of various factors. Models of farming frameworks are under pressure to go past only consisting of sustainability and economic concerns. New Sustainable Development Goals have a strong agenda and call for frameworks of dietary grade of food that go past mass productions and multifunctionality.

A confluence of research and technologies which utilize data and cognitive sciences would probably be advantageous for developing sustainable solutions that meet many aims. The authors will ideally have a virtual lab with models, data, statistical tools and IT tools to perform research evaluating the effects and tradeoffs of various technologies, policies, or scenarios for evaluating these many aspects of agricultural and food frameworks. Users will be able to design scenarios, specify studies encompassing various social, political, and resource conditions and various geographical and temporal dimensions, and provide outputs appropriate for interpretation and utilization by decision takers using the virtual lab. That virtual laboratory is obviously a myth. But where are we now in terms of this perfect circumstance? In order to answer that question, this paper will examine the condition of farming frameworks science and its capacities for the two key domains of farming frameworks, framework applications—smallholder farming in developing nations and commercial farming in industrial nations—represented by the Use Cases represented by [3]. This study expands on past analyses of certain components. [4] examine the applications of NextGen in worldwide alteration in research, another important field of agricultural systems model applications.

14.2 Models of Component Agricultural Systems

Here, we talk about models as parts of models for integrated agricultural systems, concentrating on how applicable the models are to certain Use Cases. The proficiencies and limitations of numerous data and analysis tools for the various Use Cases, and what is required for the next-generation of frameworks and cutting-edge frameworks, are covered by [5]. Grassland models and agricultural systems crop modeling research publications were recently published, summarizing model capabilities and applications. For instance, [6] conducted a survey of crop framework creators and users to evaluate the status of crop frameworks for utilization in climate change analysis and decision-taking. They underlined the need for further model development as well as for data that was both more plentiful and of higher quality. [7] evaluated crop models in terms of their suitability for carrying out integrated evaluations of climate change effects and highlighted significant flaws in the majority of crop models. [8] spoke about changes in tools and uses across time. For the last 30 years in 43 countries [9] examined the accuracy of the CERES maize, wheat, and rice models in comparison to observed data. They claimed that model performance for the variables studied was excellent when site-specific inputs were used (e.g., a mean relative error of 13% for grain production).

Both frameworks depict crop or field agro-ecosystem development and production answers to climate [10], soil, plant species features and control. Models of cropping and meadow systems share these fundamental properties. However, there are particular difficulties with several features of grassland/rangeland modeling. The necessity which grassland frameworks include numerous interacting species with perennial and woody kinds of grasses is the cause of many of these difficulties. The models are compelled to take long-term residual effects into account since plants persist over a number of years. Dependence on nutrients found in the soil or human-caused troubles such as fire strengthen the need for a long-term outlook in meadow modeling. Since a majority of the biophysical techniques are comparable (e.g., in terms of photosynthesis, development, absorption of water and nutrients from the soil, etc.), modeling grasslands takes extra elements into account.

Yield, whether it is grain, or pasture biomass production, is the most often modeled response variable for cropping systems. For determinate annual crops, this yield is collected all at once; however, for indefinite crops and meadows it can be produced more than once. Though mathematical models can be helpful for estimating these organic productions in acknowledgement to a variety of weather status, nutrient concentrations, irrigation quantities, etc., they don't estimate outcomes to nonlinearities and threshold results outside the span of features in the data utilized to produce them. Vigorous cropping and pasture system frameworks, on the other hand, may replicate these biological yields as well as other reactions that are crucial for analysts to understand, including crop water usage, N absorption, nitrate filtering, soil corrosion, soil C content, greenhouse gas discharges, and remaining soil nutrition. Additionally, dynamic frameworks may be utilized to predict reactions in contexts where no previous experiments have been conducted, as well as for times, locations, and situations. They may be utilized to implement trials and predict results, enabling users to compare the environmental and economic costs and benefits of various systems. The results of simulation trials may be used to predict how an organism will react to numerous genetic, climatic, and soil-related elements. In an implemented design of climatic consequence and conversion, "hybrid" agricultural system models may simulate policy-relevant "treatment effects" by fusing dynamic crop implementations having proper economic frameworks.

Crop development and productivity in pastures and farms are influenced by a variety of variables. To help crop models focus on the few essential components, early pioneers in crop modeling categorize the scenario in which crops are produced. The three crop output levels and the variables affecting each are summarized in Figure 14.1. Crop output that is entirely governed by defining elements such as CO_2, radiation, temperature, and crop traits (such as genetic regulation of physiology and canopy design), is referred to as potential production. The division of biomass development into grain and different plant components is another aspect of potential production models, and defining elements that impact these are modeled.

However, under extremely intensive management (providing sufficient water and nutrients and managing pests, weeds, and diseases), yield comes close to the potential level for the particular CO2, temperature, radiation and canopy architecture utilized. This powerful yield level is seldom reached in actual yield circumstances. For instance, in certain places, crops produced in greenhouses [11] or on highly maintained field yields that are at or close to developing phases. The following yield circumstance is known as water or nutrient-restricted production (Figure 14.1). The distinguishing

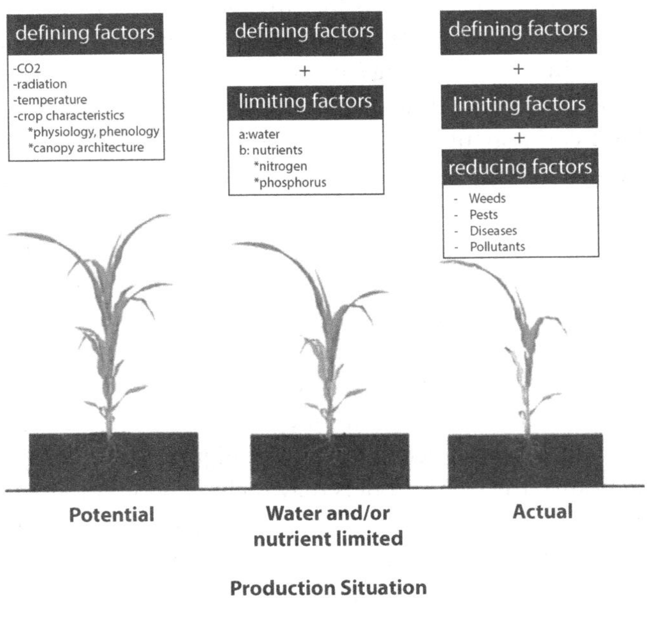

Figure 14.1 A diagram of the production scenario is used to identify the aspects that cropping system models incorporate and omit in order to assist in directing their development and educating users about the models' suitability to handle various concerns.

characteristics are still significant at this stage, but the amount of water or nutrients required to reach maximum development capability may also be constrained. To implement the time-changeable accessibility of water and nutrients, the intake of these resources, and minimizations in growth when they are not sufficient to meet future growing requests, crop frameworks which implement water or nutrient-shortages should consist of soil water and nutrition elemental methods. Due to the worldwide significance of these resources in determining yield, component modules which replicate soil H_2O, N, and C dynamics are included in the majority of cropping and grassland system models. The majority of models do not replicate reactions to phosphorus, potassium, or micronutrients, despite the fact that some of them do incorporate phosphorus. In addition to the complexities of the physical and chemical techniques which take place in soils, frameworks which characterize soil water, N, and C dynamics are also complex due to the management techniques utilized for these resources (such as water harvesting, drip irrigation, categories of applied inorganic/organic fertilizer etc.).

So, other elements (insects, illnesses, weeds, and pollutants) that may hinder development and yield are included in real production (Figure 14.1). While some crop models can simulate damage from diseases and insects, the modeling of these minimizing elements has lagged behind other advances in crop modeling (in a review of current development). The majority of organizations that simulate cropping and grassland systems exclude these components. As a result, few models now in use replicate reactions to damage caused by pests or diseases or to their control utilizing agrochemicals, resistant crop cultivars, or other methods. For certain applications, this poses a serious constraint. Models of cropping systems' crop, soil, atmosphere, and management [12] components.

The powerful production level (Figure 14.1), besides the water and N-restricted yield level, is often included in dynamic crop models. Different models, however, incorporate such aspects in a variety of ways. The Cropping System Model, which integrates the DSSAT models CERES, CROPGRO, and is shown schematically in Figure 14.2. Using the notion of coupling points, the CSM models may include pest and disease damage into certain crops as well as soil water, nitrogen, carbon, and phosphorus dynamics. As a result, carry-over alterations in soil H_2O, N, and phosphorus are implemented to depict long-term alterations in soil resources in answer to various control practices. It can also simulate many seasons.

Additional agricultural and grassland system models that have been developed, notably CROPSYST [13], are examples of models with similar

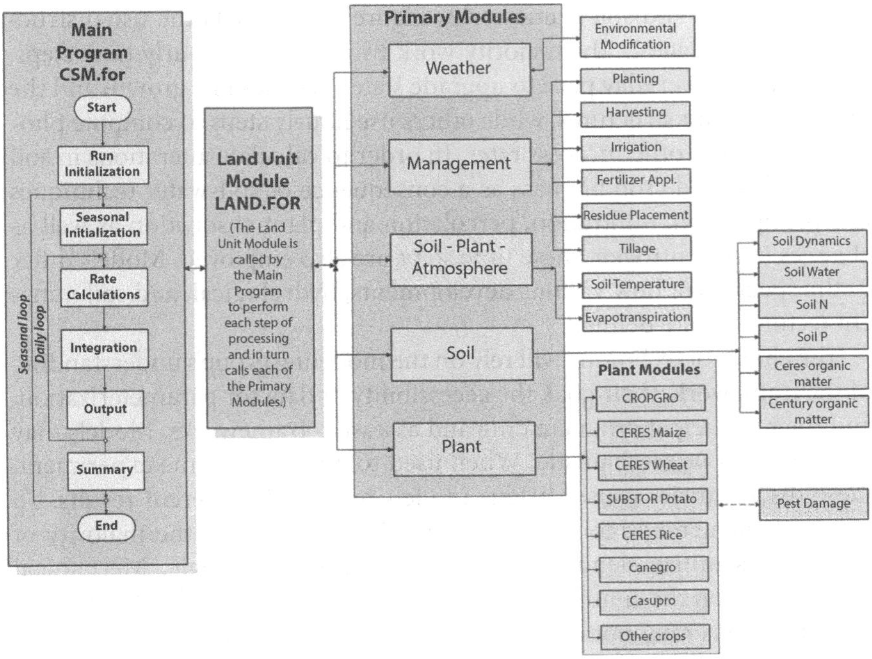

Figure 14.2 The DSSAT Cropping System Model includes elements for land, soil, crops, climate and control.

components and capabilities, even though most frameworks don't implement the results of pests and diseases without being externally linked with time-series. Intercropping may be simulated by certain models, such as APSIM. Modules from a set of frameworks cannot be used with different frameworks, which is a regrettable aspect of the existing crop and grassland models. Active pest and easing harmful modules from DSSAT CSM to APSIM is doable but there are limitations. For instance, APSIM's intercropping facilities are firmly ingrained in the system model and may not be easily shifted to different frameworks like DSSAT CSM requiring the non-trivial work of developing module "wrappers" to manage inter-model communications.

Techniques

The emphasis was on simulating a lone "point" in space (think of this as a farm) across time to examine variability in crop replies to soil, control and weather. Most "cropping system" frameworks have arisen as progress of elemental crop and soil frameworks. Figure 14.3 depicts the usual structure of this lineage 2. The majority work using daily or hourly time steps. Some employ everyday pace to upgrade state variables like growth and the biomass of plant structures, while others use hourly steps to compute photosynthesis and other process rates. In order to calculate alterations in soil H_2O, soil N, and crop biomass as a consequence of soil-water techniques like precipitation, infiltration, percolation and plant absorption as well as changes in soil nitrogen, these time steps are also employed. Models differ in the specifics of how various developments, hydrological, and soil nutrient techniques are defined.

The choice of technique will rely on the modeling group's understanding of the framework, their goal, the accessibility of data for parameterization, and their participation in creating and assessing frameworks. Models may be functional or mechanistic. When used to mimic the same experiment, these discrepancies cause various models to provide different results. To divide implemented biomass into multiple plant organs, the majority of frameworks utilize slender functional equations and logic. Mechanical frameworks, on the contrary, employ the potential energy of soil water and "instantaneous rate" notions from soil physics to represent the quantity of water deposited in a soil which is accessible to plants. Functional frameworks likewise predominantly utilize "capacity" concepts to do this. The gap between the upper and lower bound of soil water-holding capacity determines the quantity of water that is accessible to plants in capacity-based functional models. Although infiltration and runoff techniques can be estimated with small time gaps, in this sort of soil water model, water transport and its availability for crop development is represented by functional equations on a daily time step. Some modeling systems, such as APSIM, may function with either capacity-dependent or energy-dependent soil water modules. Ideally, an agro-ecosystem implemented engine or platform can be able to run with element modules configured to various "mechanism" levels. In spite of a few frameworks that contain input data on plant genetics, they are limited in number and are not yet used widely. The fact that most models are not genetically based is one of the reasons why it is so common to calibrate frameworks utilizing farm data in order to get genotype-based parameters.

The most well-known instance of this agricultural frameworks "platform" is APSIM. It is located at the intersection of the models of crop-soil frameworks shown by Figure 14.2 as well as the models for whole-farm optimization that are covered elsewhere in this study. These "agricultural systems" characteristics of the APSIM technique are fully described by [14] (Figure 14.3).

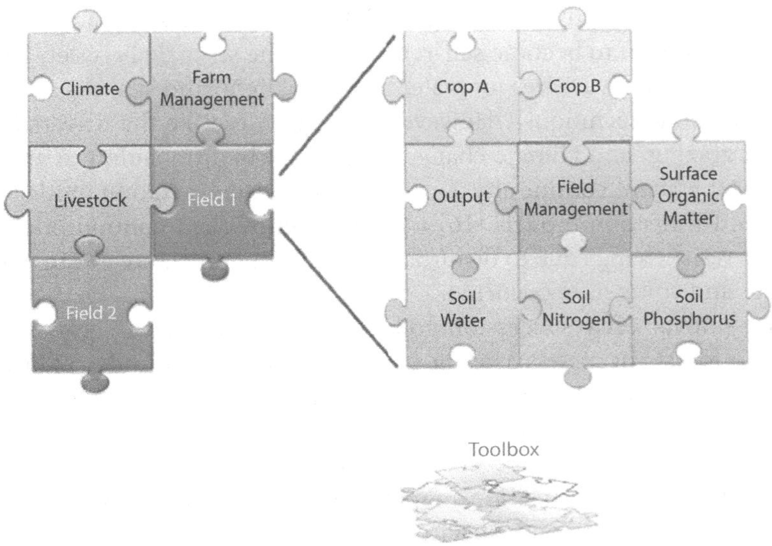

Figure 14.3 A theoretical framework for simulating "agricultural systems".

Additional things to keep in mind while simulating grasslands

According to [15], meadows are often mixed categories made up of a range of grasses and forbs, consisting of legumes and occasionally woody plants. Contrary to croplands, the variety of species often prevents the adoption of a lone-species parameterization as different species have varying capacities for vying for available resources including water, food (most often nitrogen), and light. In general, grassland models use one of the following to simulate plant action and competitiveness among herbaceous plants: (1) a collection of species, each with its own unique parameterization; (2) combinations of warm-season grasses, legumes, etc.) into parameters for plant functional classes (3) community-averaged parameterizations. These combined techniques need more work for parameterization but allow for depiction of alterations in plant group formation over time, such as those brought on by climatic change, competition between plant populations, and death.

The function of ecosystems may be significantly impacted by trees, which are dynamic elements of natural grazing areas across the globe. Because trees behave differently to diverse causes (such fire, grazing, and CO2 concentration) and rely on features of plant populations (like seed banks), representing tree/grass competition is difficult. Co-occurring population and biophysical changes might cause changes in plant community composition to become self-reinforcing. The struggle between herbaceous and woody varieties for water, nitrogen, light, and space is modeled using dynamic techniques. However, they also describe the drivers (such as fire, grazing, and climate change) that result in a transition to another quasi-stable plant community. Dynamic rangeland vegetation models and state-and-transition models [16] identify a set of plant communities that tend to resist change due to disturbance. Wide-ranging increases in woody species and cover have occurred.

The shift of early succession communities dominated by grasses and forbs to forests occurs as a result of spread phenomena under many, but not all, environmental conditions. It has been difficult to study woody invasion and comprehend the significance of competing forces, in part due to the sluggish rates of the processes causing changes. The drivers of transitions in state-and-transition models reflect these modest shifts, which raises questions about our capacity to accurately depict long-term changes in the tree-grass balance. Plant ecologists have been researching ecological succession from the early research done before 1945. There should be more collaboration between ecologists and agricultural modelers.

The removal of photosynthesizing tissues, altered light transmission through the canopy, influenced nutrient cycling, altered plant al-location patterns, and differentially impacted species mortality and recruitment rates in grasslands; these are just a few of the ways that grazing animals of all kinds affect plant productivity. The makeup of plant communities may alter as a result of such changes to groupings of plants (species, functional groups, etc.). While grassland models that include species or plant functional types can depict grazing-induced changes in the competitive balance, such models that represent plants with a set of community-wide parameters typically rely on a combination of grazing response curves and LAI (Leaf Area Index)-driven reductions in production potential. Grazing (or cutting for hay) reduces part of the plant's productivity in grasslands and rangelands; hence models must be able to predict growth for plants with an uncertain future. Independent of the season or climate the quantity of biomass or leaf area changes. After a grazing episode, there may also be notable variations in development rates across and also within species.

Crop models in their shortened version

The variables to which models react differ across frameworks and change as modelers work to have them more thorough and broadly appropriate. However, the few analysts who wish to use them may not have all the necessary inputs, or they may need to include a crop framework into an economic model or another model to study reactions at different scales. Minimized form crop frameworks, which have a lot less input needs, run quickly, and give responses required for certain operations, have been developed by some researchers using more thorough crop models. For instance, [17] generated the parameters and variables required to run the far less complex nuances field scale crop model using the APSIM cropping system model. For various soil properties and management inputs, the minimized form synopsis framework reacts to N and P levels. Similar methods were utilized by [18] to demonstrate that minimized maize, peanut, and cotton frameworks parameterized from the DSSAT CSM framework faithfully replicated DSSAT findings across time and space. Researchers may create scenario-based crop models using reduced form data.

In using larger-scale evaluations which can include socioeconomic, livestock, and ecological sustainability elements, particular summary frameworks that roughly replicate the responses of a more complete model are available. Minimized-form crop frameworks may be seen as the "production function" which serves as the basis for economic yield frameworks, and they may be combined with economic frameworks to form "hybrid" frameworks for policy analytics and impact evaluation. A comparable procedure of developing a summary model was shown by [19].

Livestock management

The complicated nature of livestock systems necessitates modeling at several scales, including those of the individual animal, the herd, and the interactivity between the flock and its surroundings via the consumption of feed, usage of the environment's resources, and so on. In the past, a variety of models have been employed to explain the various parts of livestock systems. The Hurley Pasture Framework, PHYGROW, Century, APSIM, SAVANNA, and DSSAT are a few examples of these.

Models for animal performance

Animal performance is a key factor influencing output, profitability, and efficiency in livestock operations. Consequently, the models of livestock that forecast the production of milk and meat from animals are the most often utilized. During the 1950s, when the initial feed needs for cattle were created, there were precursors to performance models. Since then, many ideas have been consistently constructed and improved around the US and Europe. The feed industry's workhorse for formulating rations is nutrient requirements models, which are connected to linear programming frameworks for least-cost rations. These models often combine mechanistic concepts of mammalian energetic and protein metabolism with statistical regressions generated from experimental data. Models of flock dynamics track changes in animal populations and herd composition over time. Herds are often divided into cohorts of varying ages, weights, and sexes in herd dynamics models. These cohorts are distinguished by various mortality, reproductive and rates of replacement and sale. At certain reproductive rates, adult females give birth to children that grow or die, join the next cohort, are sold, and then the cycle repeats.

The finest of these frameworks contain interactions between animal nutrition and reproduction to compel reproductive and mortality parameters stochastically [20]. This aspect is crucial since feed availability and supplementing tactics have a big influence on the performance and reproduction of herds. Herd dynamics models may be used for a variety of purposes, such as determining the best stocking rates and conveying facilities, evaluating the effects of reproductive technologies and mortality minimizations, and forecasting the elimination of biogas from crop or pastoral frameworks. Livestock epidemiologists often use these models to determine the effects of illnesses on flock mortality and bitterness. They were utilized in linear programs applications to examine the best sales policies, herd numbers, and other issues, as well as with dynamic programming [21] to optimize substitution choices in commercial dairy flocks. In order to assist in the choice of management techniques in livestock systems, biological implementation models are seldom employed as input-output coefficient generators for linear program frameworks.

Models for aggregated livestock frameworks

These simulations depict the entire animal farms and their essential elements (Figure 14.4). The necessity to create whole-system frameworks employing implementation and optimization [22] approaches to describe various elements and their interactivities is justified by the complexity of certain livestock systems. For instance, it is impossible to describe grazing management techniques without also taking nutritional and herd management into account, as flock dynamics or feed enhancement operations dictate grazing frequency, fodder usage, and ultimately animal performance. So, adjustable frameworks depicting pasture development, formation and quality, every animal performance to evaluate nutritional techniques, and census dynamics characterizing the behavior of populations are all included in implementations of the biology of livestock companies. Stocking rates,

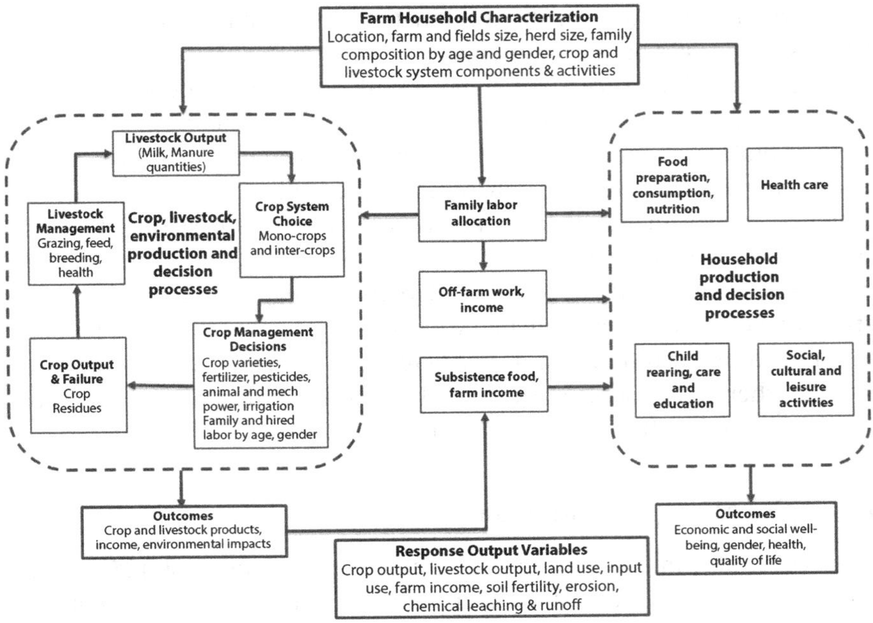

Figure 14.4 Diagram of a farming system that highlights the relationships and components of the home and production systems.

animal sales, mortality or substitution rates, and birth gaps are examples of administrative approaches at the herd level that ultimately find animal quantities and physical state classes.

Modeling of plant and animal pests and illnesses

For more than a century, biologists have developed statistical frameworks to explain the population dynamics of agricultural weeds, pests, and illnesses. [23] address recent advances in modeling these components; here, we place a greater emphasis on generic ideas and the level of development. The variety of modeling techniques that make up the state of science today may be categorized in a variety of ways. By threat and production type is the first and most evident one. As a result, models exist that explain the dynamics of weeds, illnesses, and pests which threaten agricultural crops, animals, and fish utilized in aquaculture. While hazards like pests and viruses have been known since the dawn of time, our understanding of the intricate microbial communities that live on crop surfaces, in the soil surrounding plants, and in the rumen and gut is still in its infancy. There are currently no models for the combination of advantageous and harmful species that makes up these systems.

Mechanical (or process-dependent) and non-process-dependent pest and disease frameworks may be broadly divided. The former uses an explicit biological approach, while the latter relies only on statistics. The chosen modeling strategy heavily relies on the application being used. For instance, a farm manager could be curious about the best time to use a preventive pesticide to combat a frequent insect problem. For this reason, a statistical [24] model including independent organic factors, like crop stage, and based on environmental variables, like temperature and precipitation, may be the most accurate predictor of future insect population density. In other circumstances, the model may include details about the pest itself, such as data from pheromone or other traps.

Examples of output from many runs

The usage of climate-coordinating frameworks to forecast [25] future pest issues is a separate statistical application. An organism's present distribution is predicted using a number of factors, including the climate. Following that, scenarios created from global climate models are used to map the "climate envelope" in order to forecast how the organism will be distributed following climate change. A lot has been written about the advantages and disadvantages of this strategy recently, with the assumption that organisms can travel readily to track climate being especially challenged. Implementation of contemporary spline and neural-net estimation approaches, as well as the usage of PCs and mobile phones, have all contributed to significant recent advancements in mathematical frameworks of pest dynamics.

Mechanistic models take into account fewer details of the biological aspects of the relevant crop and pest species. The frameworks might be quite simple, describing a pest census by a lone state variable like density, or they can be extremely complicated, with each bug having a variety of properties. If frameworks of in-between complexity have much organic information still are built such that simple evaluation frameworks may be retrieved as restricting instances to aid explanation, the simplest frameworks trade realism for mathematical tractability and generic in- views. The extent to which stochastic processes are explicitly included (typically crucial in hygienic frameworks) and if a census is treated as homogeneous or geographically variable are further differences between pest and disease frameworks.

The integration of crop frameworks with those of pests and diseases is a crucial topic. The challenges involved and current large-scale programs that have sought to close this gap are reviewed by [26]. Additionally, they offer a road map for enhancing pest and disease modeling that emphasizes enhancing data sources for parameterization and evidence, strengthening the combination of crop to adversary frameworks, and developing a group of researchers who may work together to allocate skills and create community tools.

Similar to statistical models, mechanical frameworks can be utilized to forecast pest and disease concerns in the near future. Biological insights may be more effective than mathematical frameworks that were described with crop frameworks, if they can fill in for missing data or when they can help with estimation by proposing a framework form which straightforward mathematical fitting will miss. Think about how an insect might react to the normal temperature. With enough weather and population data, a trend may be deduced that high temperatures can promote development and breeding and, as a consequence, lead to an increase in pests. As an alternative, the physiological reaction of the insect might be framed that would increase the framework's anticipating ability or enable the prediction of insect dynamics in systems with little data (or under future climatic conditions). There are many schools of physiological framing. But, to our knowledge, there has never been a thorough comparison of several methods or statistical approaches to the same issue.

The spread of illness in commercial animal herds has been one area where biological insights have proved effective. Understanding how animals interact and, more crucially, how they are relocated might help policymakers make key decisions. Modern Bayesian [27] parameter estimate is combined with data on animal migrations between particular farms in modern livestock frameworks. However, obtaining the kind of data required for such methods is either impractically costly or politically unpalatable for many nations.

Modeling was shown to be helpful in determining potential pest concerns and in directing broad policy formulation. The quantity of subsequent instances of a disease which are anticipated to arise when a single case manifests in a susceptible census is known as the basic epidemiological number (R0). Speculation in control measures and vaccine advancement has been prioritized using R0 calculations for common human illnesses. These days, there are complex mathematical techniques accessible for estimating R0 in the existence of stochastic outcomes, for regionally extended populations, and for highly structured populations.

Frameworks of the growth of pesticide hindrance and the existence of plant illnesses are perhaps the most complex implementations of census genetics to weed, pest, and illness challenges in farming. Generic or phenotypic evolutionary models may be widely characterized. In spite of phenotypic frameworks was investigated in farming, genetic [28] evolutionary models still make up the great bulk of evolutionary models. In an attempt to limit the pace at which resistance spreads, portions of farms were restricted without being sprayed or without being planted with transgenic crops which indicate an insecticide. The genetic underpinnings of plant-pathogen interactivities was decided for a quantity of key frameworks, allowing for a thorough survey of strain dynamics and the potential for disease propagation to be reduced down by the selective application of a variety of different crop types. Modern high-throughput DNA measuring methods are generating an abundance of data, which is utilized in cutting-edge work on genetic frameworks of weeds, pests, and illnesses to simulate how innovative genetic interference can be utilized to control pest census. A few of the more refined pest tracking software now contain particular economic variables with factors like commodity pricing that can be changed dynamically (usually dependent on mathematical process in place of process frameworks). Depending on the state of the market, the farmer may decide to take alternative pest management actions.

More broadly, reducing chemical inputs via "integrated pest (or disease) management" is an objective shared by many of those striving to improve the sustainability of agriculture. Although building the models necessary to support this kind of study is difficult, some of the more sophisticated ones combine both economic and biological processes. For a variety of dimensions and objectives, several methods were evolved to predict the economic effects of actions and rules. Here, we list the most significant methods that have been created together with their key drawbacks.

Prescriptive farm management advice is based on linear economic optimization models of farm frameworks which were created in the 1950s and 1960s. A complicated collection of linear inequality constraints which describe the productivity options accessible to an agriculturist define these models. The best manufacturing options are chosen using the simplex optimization technique. This method has the drawback that it can only explore extreme points in the multidimensional choice variable space and cannot explore intermediate solutions. The need for complex constraint structures in linear programming models, which limit the availability of other solutions and make implementation challenging for applications like the adoption and consequence of newer technologies, is a significant drawback.

Models for economic production

For models of a lone crop's yield function and lone-equation and simultaneous system frameworks which simulate input demand and output supply nature, econometric techniques have been developed and used. Primal representations and statistical estimate were the main topics of early research, but in the 1970s and later, many efforts switched to dual representations. Models that are both static and dynamic have been constructed. Direct data on the physical amounts of inputs and outputs seen in implemented plots, or, in later phases, thorough farming yield researches, are used to estimate single crop production functions.

For the last 45 years, sophisticated simulation models have been employed to explain dynamic agricultural systems. [29] theory of system dynamics, which employs storage and flow variables to characterize the system, was often used as a foundation for early examples. The fundamental idea that a complete and complicated feedback system is stable and repeatable, however, has never been proved. In order to reflect a combination of consistent behavioral interactions (i.e., that the amounts of produce given by farmers may be sold at a price that covers the expenses of production), structural simulation models might be beneficial.

Some agent-based models use mixed-integer optimization techniques and have more formal dynamic and calibration structures. The technique, however, is prone to the same problems with empirical verification and repeatability as previous large structural simulation models because of its generality. The population-based modeling strategy used by [30] is a more economical, all-purpose strategy intended to depict the variety of agricultural systems. It combines biophysical and economic simulation models to assess how changes in technology, government regulations, and the environment may affect sustainability Models for optimization are calibrated

A number of smaller-scale models, referred to as village or household models, were produced as a result of these macroeconomic models (such as computable general equilibrium, or CGE, models). General equilibrium village models take into consideration all economic flows as well as payments made inside the village to various employees and landowners. They also cover the flow of money into and out of the village boundaries. This is especially helpful in rural farming economies in underdeveloped countries when a large portion of the work is performed by relatives for little to no remuneration. Village-level equilibrium models also have the benefit of taking into consideration the value derived from local food production for sustenance. A social accounting matrix that takes into account flows both within and outside the economy serves as the foundation for these CGE models. A constant elasticity of substitute production, supply, or transformation function that is calibrated against exogenously calculated elasticity is another common method. Due to their more open-ended design and the rigorous assumptions needed for their solution, CGE models have the drawback of being data and computationally heavy. General equilibrium models are more difficult to include intricate process models into than more sophisticated partial equilibrium models that include bioeconomics. Recent researches on the use of models that mix bio-physical and economic models to depict agricultural systems are described by [31].

[32] also examine models of the economic agriculture system. The models they categorize as farm, landscape, regional, and national models are bio-economic models. Crops, livestock, and socioeconomic components interact in intricate ways in each of these systems at various sizes. For instance, Figure 14.5 illustrates the elements that must be included in farm-scale system models. These elements and operations include a farm's agricultural and animal production businesses, home production and decision-making processes, and interactions between the household and farm production systems.

The first way uses agronomic models, which translate the impacts of climate change into crop yields, to connect optimization models of agricultural output to climate models. Adams *et al.*'s (1990) national-scale modeling of this kind for the United States is a pioneering example, and the models' outputs were often utilized. Since then, many optimization models across smaller geographic regions have used this similar methodology under the influence of downscaled climatic data. Using formal economic models to predict climate change impacts has the benefit of including adaptation and mitigation effects as a consequence of changes in agricultural output that are economically advantageous under both altered growing conditions and a carbon price.

An alternative econometric method for assessing how climate change affects economic factors and agricultural crop yields such as property prices and financial rewards is to calculate statistical models using behavior that has been observed. Then, using information from projected future climate data, these statistical models are simulated. This method's justification is that it enables the model to include true adaptive behavior. This kind of approach, meanwhile, also has some serious flaws. For instance, it cannot capture changes in socioeconomic situations or be used to distinguish between technological adaptations and climatic impacts. It also cannot describe the effects of CO2 fertilization on agricultural output. Several scholars have modeled the impacts of climate on yields and other factors using statistical econometric techniques.

Accordingly, connected individual modeling systems for hydro-economic models were created. There are two approaches employed. In the first, hydrologic policy models (such as the Calvin water allocation model) are integrated with benefit-response functions from economic models. A second hydro-economic modeling strategy involves fitting a simpler function statistically to the output of intricate simulations across a variety of hydrologic and climatological factors in order to characterize the response in the hydrologic models. An economic policy model may then include these response functions. This method has often been used to examine the best use of common property resources utilized in agriculture, such as groundwater. [33] provided an economic viewpoint overview of the ideas and applications of hydro-economic models, whereas [34] provided a hydrologic perspective study.

Although there is a long history of modeling watersheds and environmental quality, few of these efforts have included agricultural and livestock models and many of these attempts have not taken into account objectives connected to agrarian systems. The interconnectivity of agricultural and ecological processes throughout the landscape is discussed from at least two distinct angles. The first viewpoint is that production systems, markets, and commerce are influenced by human systems, including those on the farm, in communities, and in administrative and political settings where agricultural systems interact via decisions and policies. The opposing viewpoint contends that the interdependence of hydrological and biophysical processes determines the fundamental behavior of agricultural and cultural systems throughout the landscape. This viewpoint results in an importance placed on comprehending the physical, chemical, and biological processes that take place in watersheds. Although both viewpoints are significant, agricultural models seldom take both into account when making judgments or developing models. Watershed hydrology models have several applications, particularly when employing the SWAT model as stated by [35], which primarily focuses on environmental quality and water resource challenges.

The aggregate of model findings from smaller spatial resolutions to a wider region is sometimes referred to as "scaling up" of model results. Hydrologists often describe the regions of interest as watersheds. As opposed to this, when regions are defined by economists, they often take the form of administrative and political entities (such as cities, districts, or whole nations) or socioeconomic strata (such as small and big farms). To assess practices and policies related to environmental quality, several agricultural system models have been created recently. With a focus on vertical fluxes of energy, water, carbon, nitrogen, and nutrients across the

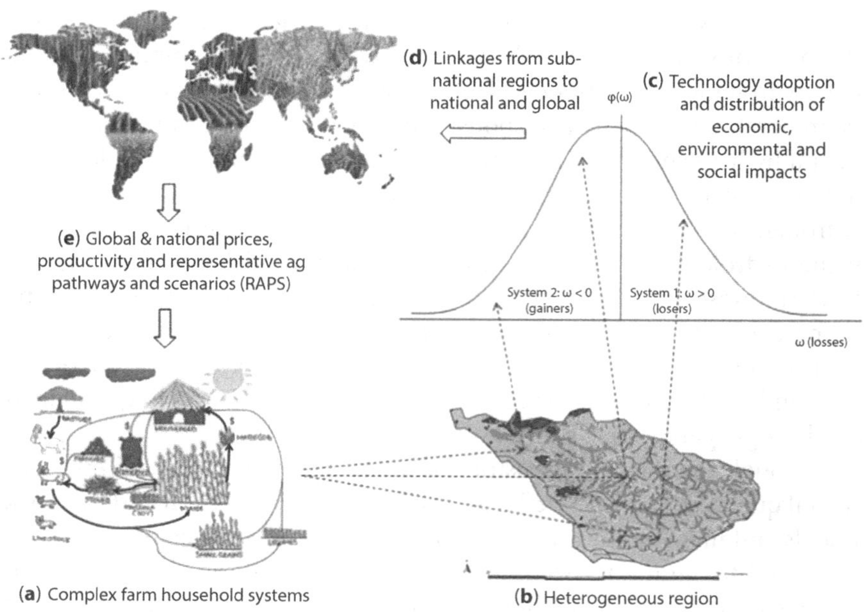

Figure 14.5 The AgMIP Regional Integrated Assessment framework, which places an emphasis on cross-scale connections and analyses distributional consequences in populations of farm families that are diverse.

environment, plants, and soil root-zone continuum, biophysical models (for example, crop or nutrient models) often function at the point/field scales. Estimating surface and subsurface fluxes and biological processes is necessary for scaling up from the point size to the landscape scale.

14.3 Present-Day Farming System Frameworks in Relation to Certain Application Situations

The status of agricultural system science today is then discussed in relation to its potential and limits in supplying data to support the five Use Cases' diverse spectrum of decision makers. In a systems analysis, each use case represents a collection of interactions between people and systems in a specific setting. The Use Cases demonstrate a variety of demands for very diverse applications at various dimensions and degrees of intensification in developing and developed nation contexts. Crop, livestock, and agricultural system models are required for these Use Cases, according to [36]. Here, we explore the issue of whether current models of agricultural systems, data sources, and decision support systems (DSS) are capable of supplying the information required for this Use Cases expanding farming in Africa.

Limitations and capabilities

Additionally, livestock models need to have their numerous shortcomings addressed by NextGen technologies. Even while our knowledge of animal feed requirements is generally reliable, there are still significant mistakes in the estimation of ruminant feed intake, particularly under degraded pasture conditions and with no supplementation. Additionally, few models can forecast the generation of methane, which is becoming increasingly significant as studies of the environmental effects of cattle gain popularity. For validation reasons, additional experimental data must be gathered, particularly in the tropics. Understanding how livestock systems will likely be impacted by climate change is another area that needs additional research. The reactions of animals and herds to rising temperatures, more climatic unpredictability, and harsher feed changes are still impossible to estimate. As a result, our capacity to develop adaptation techniques in cattle systems is limited. Additionally, there are limited predictions of how feed quality and consumption may vary as a consequence of climate change. Better rules for controlling behavioral changes as systems develop are required, as well as a better incorporation of the livestock producers' decision-making process into entire farm models. To accurately assess the influence on the well-being of households that depend on livestock for a significant portion of their income, livestock system models must include both biophysical fluctuations in productivity and expected economic reactions to these fluctuations. A given decrease in the price of meat does not result in a proportionate rise in demand, which is a sign that the demand for animal goods is rather inelastic. Similar to how prices will decrease by a bigger percentage than the amounts given if there is a significant rise in the amount of meat supplied for sale. The demand for and supply of inputs for the production of cattle, particularly those derived from range land, tend to exhibit the same phenomena. The overall result of these quantity-based pricing reactions is to amplify the impact of fluctuations in the availability of livestock products, which may further punish ranchers in terms of income loss and access to funds to rebuild herds if drought or other factors led to loss of flock size.

Additionally, there are drawbacks in the socioeconomic models used to assess the advantages and downsides of various technologies, manage crops and animals, and manage the farm and its resources. The majority of existing economic models, for instance, mimics average responses and use straightforward economic-behavioral assumptions. Uncertainty and financial risks must be considered in this Use Case. Risk behavior may be included into many economic models, but doing so adds a significant amount of complexity and necessitates a lot of data restrictions. The absence of data for constructing, evaluating, and implementing models for this Use Case is a fundamental restriction, even if certain shortcomings are attributable to the models themselves. Although some nations have relevant aggregate agricultural system statistics, such as those found at http://www.nass.usda.gov, Deborah's issue concerning sustainable intensification cannot be answered using just those data. In certain cases, models for addressing intensification techniques are tested and improved using data gathered by researchers who are looking at choices for sustainably intensified production systems.

Tools for decision support (DSS). The development of information systems that enable the computation of sustainability measures for particular agricultural practices has made some progress. Much work is being done in the business sector, but there should be greater public-private cooperation in defining and creating more accurate sustainability criteria. The sorts of decision information systems required to support Deborah's advice to farmers on sustainable intensification alternatives catered to particular locations and agricultural systems haven't been developed much, however, up until now. Janssen *et al.* (this issue) provide further information on the DSS and associated knowledge tool limitations.

14.4 Discussion

According to this assessment of agricultural systems modeling, significant advances have been achieved across a range of disciplines that cover diverse yield frameworks from fields to farms, landscapes, and after. In order to create more thorough system models that take into account biophysical, socioeconomic, and ecological variables to provide a broad variety of framework reactions, it is possible to mix component models from several disciplines. For instance, agricultural, livestock, and economic model combinations was used to investigate agricultural frameworks and to assess the effects of various technologies, policies, and the federal and worldwide implications of climatic alteration. Various methodologies are used in the modeling of farming frameworks and in the use of such frameworks in relation to academic and political issues. The approaches used differ depending on the goals of the developers, the anticipated applications, and the data accessibility.

Theoretical ideas and concepts from a variety of disciplines, such as the farming and ecological sciences, engineering, economics, and statistics, have been effectively used by those who have developed agricultural systems models. Through the efforts of several organizations globally, agricultural system models are still being developed. As seen by recent AgMIP advancements and different CGIAR-led initiatives such as the Global Futures project and the CCAFS project, researchers are becoming more and more interested in making contributions to agricultural systems science. The business sector is becoming more and more interested in utilizing models of agricultural systems.

The Use Cases reviewed in [37] show that in order to create usable models of agricultural systems, a minimal number of component models are required. Particularly, crop frameworks, livestock frameworks, and agricultural system frameworks are among these component models. Crop models take into consideration pressures from weed, insect, and disease stresses to mimic production, resource usage, and outputs of nutrients and chemicals to nearby water, air, and environmental frameworks. They combine data to forecast performance for a variety of inputs and procedures which concern to frameworks utilizing extremely regulated, intense yield methods and contemporary techniques, ranging from subsistence circumstances.

Similar to how humans do, livestock models include factors including breeds, feed supplies, herd management, and climate. Economic models which simulate the nature of field decision takers are integrated with different livestock and agricultural frameworks, inclusive of their linkages, in farming system models. These models are required both at the population level and at the level of the separate functioning so that decision-makers at all levels of government and farmers can assess the economic, ecological, and societal consequences of the bio-physical and socio-ecological heterogeneity of frameworks. We draw the conclusion from this research that various platforms are required for merging frameworks and data for various reasons, and that the development of next-generation frameworks and data must make this requirement for a variety of platforms into consideration.

The use cases examined consist of pertinent instances for a wide range of users, from intensive production systems to small-holder agricultural systems in emerging nations, and from frameworks assisted by farming enterprises to those with limited private sector assistance. There are those that call for the integration of component models to solve socioeconomic, food safety, and ecological challenges at various sizes. Others call for frameworks and related data to assess technology at a farm.

Even though the frameworks that were provided differed in their suitability for each Use Case, the lack of data was a problem that affected all of them. All studies of agricultural systems are built on data. In addition to limiting the data which may be utilized as inputs to address frameworks, this restriction hinders the capacities of current frameworks to consist of features of urgency in most examples, limits analysts' capacities to find frameworks across a huge range of features. Data constraints are more significant than conceptual theory and method gaps. We contend that rather than knowledge gaps, the restrictions of present farming system frameworks and tools are more closely related to insufficient data. This restriction reduces users' trust in models' capacities to provide accurate findings, which in turn reduces their usage for choices or policies.

Particularly acute is this absence of data in lesser developed regions. It holds true for economic frameworks, agricultural and animal production models, and use cases that addressed problems in Sub-Saharan Africa's data-poor regions. However, it is also evident that many data-rich places have a shortage of relevant and accessible data. For instance, thousands of agricultural experts carry out studies every year that contrast management and technology under certain circumstances in spite of the fact that people data are often only accessible to people participating in specific research, despite the fact that they might be very helpful for constructing, testing, and applying models.

The assertion concerning a common restriction across Use Cases may have one exception. The management support for precision agriculture's capabilities and constraints are distinct from those of the others. Although there hasn't been much information about spatial variability at a high resolution in specific farms, it is quickly altering as the private sector develops new sensors and monitoring platforms in response to obvious commercial opportunities. The usage of farming system frameworks to assist agriculturists like Greg (in our example) employ precision agriculture to boost resource usage efficiencies, lower ecological hazards, and hike earnings is anticipated to quickly rise as a result of this interest by the corporate sector. Even while resource-poor agricultural circumstances are not expected to employ the same levels of data gathering and precise input implementation, technological advancements are probably to provide more possibilities there.

Attempts are being made to ameliorate the overall data limitation issue via current national and international efforts. For instance, the Global Open Data for Farming and Nutritious program is supporting international efforts to make data important to agriculture and nutrition open, useful, and attainable for unlimited usage anywhere in the globe. This program has more than 150 partners from countrywide governments, non-government organizations, worldwide organizations, and commercial sector organizations that assist it. It is evident that a better concentrated attempt is required to link the different farming framework modeling, database, data harmonization, and DSS attempts altogether so that the scientific resources that are covered in this inventiveness will supply to a adaptable collection of frameworks, data, and platforms to facilitate worldwide public goods. It is very crucial since it will need more of these instruments to guarantee that agriculture can maintain itself economically and ecologically while producing enough food to feed the world for the next 50 to 100 years. Many organizations are working to remember this scenario. In addition, strategies like private-public partnerships are required to combine the strength of private sector funding with the continuing researches to enhance frameworks and modeling tools, as explained in [38]. This is true for the economic models used in every first three Use Cases which solve problems in Sub-Saharan Africa's data-poor regions, as well as production models for crops and animals. Finally, a plan should involve the proper modification and, in few instances, reprogramming of current component frameworks which mostly incorporate various necessary potentials, depending on the state of the models, data, and knowledge systems at the time. The usage of a variety of techniques, such as mathematical frameworks, minimized form frameworks, extended databases, and modular frameworks that integrate component sub-modules, would make it easier to expand components that react to elements that are not presently taken into account by models. According to [39], these data restrictions might be solved by combining farm-level frameworks and knowledge products with data and frameworks at the landscape scale. AgMIP's recent experiences have shown the benefit of using various models, indicating that it would be pointless to try to create ideal frameworks for crops, animals, and agricultural frameworks. Although there are excellent chances for significant improvements in farming systems data, frameworks, and knowledge systems, these tools have inherent restrictions because of the spatial variability of physical, chemical, genetic, and socioeconomic factors as well as the irreducible uncertainties in model structures. These limits will lead to change dependent on implementations, indicating which future assessments of abilities and restrictions have to be dependent on Use Cases which have been carefully established. According to this review, some modern applications can be served by agricultural systems models as they stand, but major developments are required to produce the next generation of data, frameworks, and knowledge systems, which will be necessary to solve more complicated problems and ensure food safety over the course of the next century.

References

1. V. Veeraiah, D. J. Shiju, J. V. N. Ramesh, Ganesh K. R., S. Pramanik, D. Pandey, A, Gupta, Healthcare Cloud Services in Image Processing, in *Handbook of Research on* Thrust Technologies' Effect on Image Processing, IGI Global, 2023.
2. P. T. Khanh, T. H. Ngọc, and S. Pramanik, Future of Smart Agriculture Techniques and Applications, in *Advanced Technologies and AI-Equipped IoT Applications in High Tech Agriculture*, Ed., A. Khang, IGI Global, 2023.
3. N. Montalvo-Romero, A. Montiel-Rosales, R. Purroy-Vásquez and P. Quechulpa-Pérez, "Agro-Technological Systems in Traditional Agriculture Assistance: A Systematic Review," in *IEEE Access*, vol. 11, pp. 123047-123069, 2023, doi: 10.1109/ACCESS.2023.3329087.
4. S. Mehta, V. Kukreja and A. Gupta, "Next-Generation Wheat Disease Monitoring: Leveraging Federated Convolutional Neural Networks for Severity Estimation," *2023 4th International Conference for Emerging Technology (INCET)*, Belgaum, India, 2023, pp. 1-6, doi: 10.1109/INCET57972.2023.10169991.
5. T. H. Ngọc, P. T. Khanh and S. Pramanik, Smart Agriculture Using a Soil Monitoring System, in *Advanced Technologies and AI-Equipped IoT Applications in High Tech Agriculture*, Ed., A. Khang, IGI Global, 2023.
6. V. Veeraiah, V. Talukdar, Manikandan K., S. B. Talukdar, V. D. Solavande, S. Pramanik, and A. Gupta, Machine Learning Frameworks in Carpooling, in Handbook of Research on AI and Machine Learning Applications in Customer Support and Analytics, Eds., Md S. Hossain, R. C. Ho and G. Trajkovski, IGI Global, 2023.
7. T. Nikaein, P. Lopez-Dekker, S. C. Steele-Dunne, V. Kumar and M. Huber, "Modeling SAR Observables by Combining a Crop-Growth Model with Machine Learning," in *IEEE Journal of Selected Topics in Applied Earth Observations and Remote Sensing*, vol. 16, pp. 7763-7776, 2023, doi: 10.1109/JSTARS.2023.3301124.
8. V. Jain, M. Rastogi, J. V. N. Ramesh, A. Chauhan, P. Agarwal, S. Pramanik, and A. Gupta, FinTech and Artificial Intelligence in Relationship Banking and Computer Technology, in AI, IoT, and Blockchain Breakthroughs in E-Governance, Eds., K. Saini, A. Mummoorthy, R. Chandrika and N. S. Gowri Ganesh, IGI Global, 2023.
9. M. Masood *et al.*, "MaizeNet: A Deep Learning Approach for Effective Recognition of Maize Plant Leaf Diseases," in *IEEE Access*, vol. 11, pp. 52862-52876, 2023, doi: 10.1109/ACCESS.2023.3280260.
10. A. D. Andrushia, T. M. Neebha, A. T. Patricia, K. M. Sagayam, ans S. Pramanik, Capsule Network based Disease Classification for VitisVinifera Leaves, *Neural Computing and Applications*, doi: 10.1007/s00521-023-09058-y, 2023

11. A. Roy and S. Pramanik, A Review of the Hydrogen Fuel Path to Emission Reduction in the Surface Transport Industry, *International Journal of Hydrogen Energy*, https://doi.org/10.1016/j.ijhydene.2023.07.010, 2023
12. S. Pramanik, K. M. Sagayam and O. P. Jena, Machine Learning Frameworks in Cancer Detection, *ICCSRE*, 2021, Morocco, 2021.
13. A. S. Terliksiz and D. T. Altilar, "A Simple and Efficient Deep Learning Architecture for Corn Yield Prediction," *2023 11th International Conference on Agro-Geoinformatics (Agro-Geoinformatics)*, Wuhan, China, 2023, pp. 1-6, doi: 10.1109/Agro-Geoinformatics59224.2023.10233344.
14. N. Mittal and A. Bhanja, "Implementation and Identification of Crop based on Soil Texture using AI," *2023 4th International Conference on Electronics and Sustainable Communication Systems (ICESC)*, Coimbatore, India, 2023, pp. 1467-1471, doi: 10.1109/ICESC57686.2023.10192937.
15. G. Zhipeng, W. Jiang, K. Peng, Y. Deng and X. Wang, "Wetland Mapping and Landscape Analysis for Supporting International Wetland Cities: Case Studies in Nanchang City and Wuhan City," in *IEEE Journal of Selected Topics in Applied Earth Observations and Remote Sensing*, vol. 16, pp. 8858-8870, 2023, doi: 10.1109/JSTARS.2023.3302031.
16. S. Pramanik, M. G. Galety, D. Samanta, N. P. Joseph, Data Mining Approaches for Decision Support Systems, *3rd International Conference on Emerging Technologies in Data Mining and Information Security*, 2022.
17. G. Venkatakotireddy, C. Lakshminatha Reddy, R. R. J. Prabhakaran, T. Nivethitha and M. R., "Development of High-Quality Crops using Optimized Machine Learning in Smart Agriculture Environment," *2023 Third International Conference on Artificial Intelligence and Smart Energy (ICAIS)*, Coimbatore, India, 2023, pp. 532-536, doi: 10.1109/ICAIS56108.2023.10073727.
18. P. C. V. and U. S, "IoT Based Site Specific Nutrient Management System for Soil Health Monitoring," *2022 International Conference on Smart and Sustainable Technologies in Energy and Power Sectors (SSTEPS)*, Mahendragarh, India, 2022, pp. 166-170, doi: 10.1109/SSTEPS57475.2022.00050.
19. G. Roggiolani, M. Sodano, T. Guadagnino, F. Magistri, J. Behley and C. Stachniss, "Hierarchical Approach for Joint Semantic, Plant Instance, and Leaf Instance Segmentation in the Agricultural Domain," *2023 IEEE International Conference on Robotics and Automation (ICRA)*, London, United Kingdom, 2023, pp. 9601-9607, doi: 10.1109/ICRA48891.2023.10160918.
20. S. Praveenkumar, V. Veeraiah, S. Pramanik, S. M. Basha, A. V. Lira Neto, V. H. C. De Albuquerque and A. Gupta, Prediction of Patients' Incurable Diseases Utilizing Deep Learning Approaches, *ICICC*, 2023, Springer, doi.org/10.1007/978-981-99-3315-0_4
21. V. VidyaChellam, V. Veeraiah, A. Khanna, T. H. Sheikh, S. Pramanik and D. Dhabliya, A Machine Vision-based Approach for Tuberculosis Identification in Chest X-Rays Images of Patients, *ICICC*, 2023, Springer, doi.org/10.1007/978-981-99-3315-0_3

22. V. Veeraiah, K. O. Thejaswini, Dilip R., S. K. Jain, A. Sahu, S. Pramanik, and A. Gupta, The Suggested Use of Big Data in Medical Analytics by Fortis Healthcare Hospital, in Adoption and Use of Technology Tools and Services by Economically Disadvantaged Communities: Implications for Growth and Sustainability, IGI Global, 2023.
23. F.-A. Croitoru, V. Hondru, R. T. Ionescu and M. Shah, "Diffusion Models in Vision: A Survey," in *IEEE Transactions on Pattern Analysis and Machine Intelligence*, vol. 45, no. 9, pp. 10850-10869, 1 Sept. 2023, doi: 10.1109/TPAMI.2023.3261988.
24. S. Ahamad, S. Janani, V. Talukdar, T. Sharma, A. Sahu, S. Pramanik, and A. Gupta, Apollo Hospital's Proposed Use of Big Data Healthcare Analytics, in Big Data Analytics Techniques for Market Intelligence, IGI Global, 2023.
25. S. Pramanik, Intelligent Farming Utilizing a Soil Tracking Device, in *Convergence of Cloud Computing, AI and Agricultural Science*, Eds., A. K. Sharma, N. Chanderwal and R. Khan, IGI Global, 2023.
26. J. Hou, Q. Zhai, Y. Zhou and X. Guan, "A Fast Solution Method for Large-scale Unit Commitment Based on Lagrangian Relaxation and Dynamic Programming," in *IEEE Transactions on Power Systems*, doi: 10.1109/TPWRS.2023.3287199.
27. K. Dushyant, G. Muskan, A. Gupta and S. Pramanik, "Utilizing Machine Learning and Deep Learning in Cyber security: An Innovative Approach", in *Cyber Security and Digital Forensics*, M. M. Ghonge, S. Pramanik, R. Mangrulkar, D. N. Le, Eds., Wiley, 2022, https://doi.org/10.1002/9781119795667.ch12
28. R. R. Chandan, S. Soni, A. Raj, V. Veeraiah, D. Dhabliya, S. Pramanik, A. Gupta, Genetic Algorithm and Machine Learning, in *Advanced Bioinspiration Methods for Healthcare Standards, Policies, and Reform*, Hadj Ahmed Bouarara, IGI Global, 2023, doi: 10.4018/978-1-6684-5656-9
29. B. K. Pandey, D. Pandey, V. K. Nassa, A. S. Hameed, A. S. George, P. Dadheech and S. Pramanik, A Review of Various Text Extraction Algorithms for Images, in *The Impact of Thrust Technologies on Image Processing*, Nova Publishers, 2023.
30. L. Zhang, H. Yang, S. Yang and X. Zhang, "A Macro-Micro Population-Based Co-Evolutionary Multi-Objective Algorithm for Community Detection in Complex Networks [Research Frontier]," in *IEEE Computational Intelligence Magazine*, vol. 18, no. 3, pp. 69-86, Aug. 2023, doi: 10.1109/MCI.2023.3277773.
31. R. D. McAllister and J. B. Rawlings, "A Suboptimal Economic Model Predictive Control Algorithm for Large and Infrequent Disturbances," in *IEEE Transactions on Automatic Control*, doi: 10.1109/TAC.2023.3285854.
32. Y. Wang *et al.*, "Can Digital Intelligence and Cyber-Physical-Social Systems Achieve Global Food Security and Sustainability?" in *IEEE/CAA Journal of Automatica Sinica*, vol. 10, no. 11, pp. 2070-2080, November 2023, doi: 10.1109/JAS.2023.123951.

33. A. P. Mazurenko, V. V. Mishin and V. N. Orobinskaya, "On the Issue of Developing an Agent-Based Model of the Hydrolithosphere of a Resort and Recreational Region," *2023 5th International Conference on Control Systems, Mathematical Modeling, Automation and Energy Efficiency (SUMMA)*, Lipetsk, Russian Federation, 2023, pp. 527-530, doi: 10.1109/SUMMA60232.2023.10349611.
34. X.-H. Le, D. H. Nguyen and G. Lee, "Performance Comparison of Bias-Corrected Satellite Precipitation Products by Various Deep Learning Schemes," in *IEEE Transactions on Geoscience and Remote Sensing*, vol. 61, pp. 1-12, 2023, Art no. 4704012, doi: 10.1109/TGRS.2023.3299234.
35. L. Zhu *et al.*, "Coupling SMAP Brightness Temperature into SWAT Hydrological Model for 30-m Resolution Soil Moisture Retrievals," in *IEEE Journal of Selected Topics in Applied Earth Observations and Remote Sensing*, vol. 16, pp. 8319-8333, 2023, doi: 10.1109/JSTARS.2023.3310197.
36. M. Car, B. A. Ferreira, J. Vuletic and M. Orsag, "Structured Ecological Cultivation with Autonomous Robots in Agriculture: Toward a Fully Autonomous Robotic Indoor Farming System," in *IEEE Robotics & Automation Magazine*, vol. 30, no. 4, pp. 77-87, Dec. 2023, doi: 10.1109/MRA.2023.3315934.
37. B. Sisheber, M. Marshall, D. Mengistu and A. Nelson, "Assimilation of Earth Observation Data for Crop Yield Estimation in Smallholder Agricultural Systems," in *IEEE Journal of Selected Topics in Applied Earth Observations and Remote Sensing*, vol. 17, pp. 557-572, 2024, doi: 10.1109/JSTARS.2023.3329237.
38. E. M. Pechlivani, G. Gkogkos, N. Giakoumoglou, I. Hadjigeorgiou and D. Tzovaras, "Towards Sustainable Farming: A Robust Decision Support System's Architecture for Agriculture 4.0," *2023 24th International Conference on Digital Signal Processing (DSP)*, Rhodes (Rodos), Greece, 2023, pp. 1-5, doi: 10.1109/DSP58604.2023.10167922.
39. X. Shen, Q. Wu, H. Zhang and L. Wang, "Optimal Planning for Electrical Collector System of Offshore Wind Farm with Double-Sided Ring Topology," in *IEEE Transactions on Sustainable Energy*, vol. 14, no. 3, pp. 1624-1633, July 2023, doi: 10.1109/TSTE.2023.3241357.

15

Simulating the Effects of Agricultural Land Fragmentation on Farm Efficiency: A Mathematical Modeling Analysis

Diksha Nautiyal[1], Manjunath H. R.[2], Praveen Kumar Singh[3], Umesh Kumar Tripathi[4], Saurabh Raj[5] and Soma Bag[6*]

[1]School of Agriculture, Dev Bhoomi Uttarakhand University, Dehradun, Uttrakhand, India
[2]Department of Physics, Faculty of Engineering and Technology, JAIN (Deemed-to-be University), Karnataka, India
[3]College of Agriculture Sciences, Teerthanker Mahaveer University, Moradabad, Uttar Pradesh, India
[4]Department of Agriculture, Sanskriti University, Mathura, Uttar Pradesh, India
[5]Symbiosis Law School, Nagpur Campus, Symbiosis International (Deemed University), Pune, India
[6]Asadtala Nivedita Kanya Vidyamath, Nandigram, India

Abstract
It is necessary to conduct an integrated review of prior advances at the grassland, farm, and regional levels in order to comprehend the possibilities for sustainable intensification. For this, two theories addressing the state of rice yield in the Philippines were created. We first propose that there are trade-offs between rice productions, labor productivity, gross margin, and N usage effectiveness. We next propose that farming features and socioeconomic circumstances at the farm and local levels influence the administration strategies utilized by agriculturists. Two household surveys that analyzed changes in rice-based agricultural systems through time (2000–2022) and regional variability were used to test these

**Corresponding author*: bagsoma2016@gmail.com
Diksha Nautiyal: ORCID: https://orcid.org/0009-0007-2169-3036
Manjunath H. R.: ORCID: https://orcid.org/0000-0002-1764-0683
Praveen Kumar Singh: ORCID: https://orcid.org/0009-0001-4832-0025
Umesh Kumar Tripathi: ORCID: https://orcid.org/0009-0005-9467-5547

Sabyasachi Pramanik, Niranjanamurthy M., Ankur Gupta and Ahmed J. Obaid (eds.) Mathematical Modeling in Agriculture, (295–320) © 2024 Scrivener Publishing LLC

hypotheses. The production of a dry season rice crop in the area was made possible by an increment in the number of cultivated farms during the previous 50 years and the adoption of superior cultivars. Additionally, direct sowing has taken the position of transplanting, and herbicides have taken the place of manual weeding. As seen by the steady shift from subsistence to commercial agricultural methods and the rising share of hired labor and rice sales, these led to higher rice yields and labor productivity. Our findings suggest that there is a trade-off between rice productions, labor productivity, and nitrogen usage effectiveness since production levels that maximized labor productivity and nitrogen usage efficiency were around 30% and 37% lower in the wet and dry periods than the climatic possible production, respectively. This may be explained at the field level by 1) the usage of transferring as a technique of crop formation that led to better productions but reduced labor yield when contrasted to direct-seeding, and 2) the high N implementation extents that produced enhanced productions but decreased N usage efficiency. In comparison, there was no trade-off between rice productions and profitable success since yield levels that maximized gross margin were about 82% of the climatic capability in both the wet and drier seasons. The second hypothesis's findings weren't always definitive. For instance, the amount of nitrogen applied per hectare was inversely related to field dimension, whereas the time of the starting fertilizer implementation was confidently correlated with house dimension and the quantity of parcels. In the state near the center of the flood framework, better rigorous techniques and improved agricultural performance were seen.

Keywords: Land fragmentation, mathematical model, agriculture

15.1 Introduction

By its emphasis on production capability, soil standards, and precision farming [1], sustainable boosting was recommended as a method to increase yield and resource usage. In developing nations where there are still significant yield disparities, this is especially important. An examination of potential field restrictions is necessary to comprehend the chances for sustainably intensification. Crop management programs need to be tailored to farmers' demands at the field level in terms of higher productions, resource efficiency, labor productivity, and profitability. To ease agriculturists' resource shortages (land, labor and fund), to guarantee fair resource administration, and to lower the biophysical and profitable threat of agriculture, structural adjustments may be required at the farm and regional levels.

In Central Luzon, Philippines, rice-based agricultural systems' yield gaps investigation found that crop establishment technique (transplanting),

Nitrogen (N), potassium (K), seed usage, and direct-seeding) were all significantly correlated with rice yields. Additionally, a portion of the variance in the capability production gap, or the contrast between technically capable and real productions, was explained by the timing of the first and second implementations of fertilizer and insecticides. When comparing farmer yields to input needs to meet the climatic capability in production, technology [2] production gaps which are termed as the contrast between those yields and the top agriculturists' productions were hugely related to insufficient water and nutrition utilization. Through an agricultural standpoint, addressing these restricting variables necessitates the development of' administrative packages' that may attain production levels near to the climatic capable production when assuring higher resource usage capability. Additionally, these management plans must accommodate smallholder farmers' way of life.

The framework for identifying barriers to sustainable intensification might be an aggregated study of farming frameworks that takes drivers at various scales into consideration. This calls for the creation of indicators that capture various aspects of smallholder farming systems as well as the exploration of potential trade-offs and synergies between such indicators. This study aims to identify regional, farm, and field-level barriers to and facilitators of the boosting of rice-dependent agricultural frameworks in the Philippines. We want to know if there are trade-offs between rice productions, labor yield, cumulative margin, and N usage capability for the administrative tools that agriculturists use, and if so, to what extent. We also want to know how farmer natures and socioeconomic states at the farm and local levels influence agriculturists' administrative executions and the preceding indicators. For these reasons, the authors examined two homely surveys: a cross-sectional home scrutiny of 1,800 rice agriculturists assisted in 2015–2022, in the Philippines, and a panel home scrutiny of nearly 150 rice agriculturists conducted between 1918 and 2023.

15.2 Conceptual Foundation

The contrast between climatic capability production (Yp) and real productions (Ya) is the production gap under irrigated settings. Through better crop management, farmers may close production disparities, but this may not be their primary goal. Additionally, it is necessary to assess production gaps and associated management plans in terms of various indicators and explain them in terms of field and local circumstances (Figure 15.1). For this, the authors developed two theories, which are described below.

In accordance with labor yield, cumulative margin, and N usage capability, the first postulate states that administrative techniques linked to smaller production gaps accomplish sub-optimally. The management packages utilized are partly to blame for trade-offs between indicators (step 1 in Figure 15.1). Modern rice cultivars offer more Ya (and Yp) than current types, but they also need more labor during threshing, harvesting, and crop management. When compared to manual weeding, the use of herbicides may increase both labor and output since they allow for more timely management of weeds because they take less labor. [3] found that transplanted rice had significantly greater yields than direct-seeded rice in terms of crop establishment, when direct-seeding required less labor in crop growth but better labor force for weed management. These instances show how interactions between field-level management techniques might impact farm performance.

Closing yield gaps is anticipated to result in increased earnings, expenses, labor demands, and fertilizer implementation levels (step 2 in Figure 15.1).

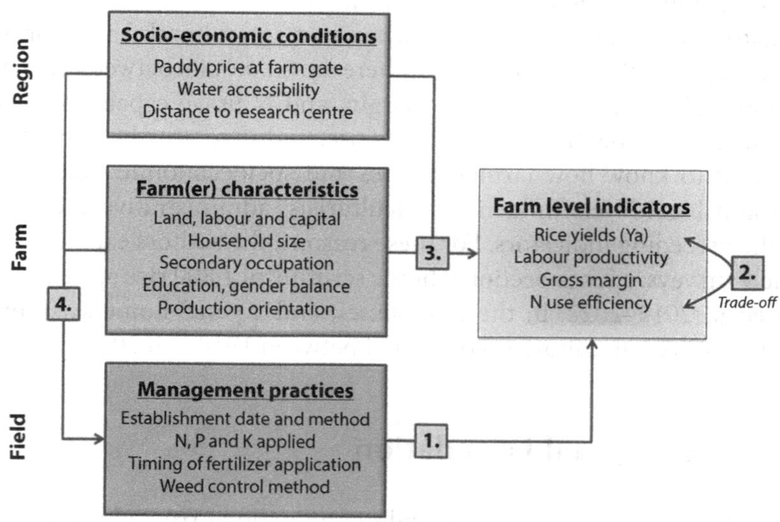

Figure 15.1 An integrated evaluation of the management techniques and rice production disparities in the Philippines. [4] go into depth on how management strategies and rice yields relate to one another. Although not clearly shown in this figure, temporal dynamics were considered throughout the investigation. In this study, we evaluate the performance of various administrative packages in respect of various indicators (step 1); the presence of deals between maximizing the goals of other farmers and reducing yield gaps (step 2); and the effects of field level determinants and local natures on rice production gaps, farming performance, and administrative steps (steps 3 and 4).

If output costs are lower and input costs are higher, there are likely to be trade-offs between increasing cumulative margin and reducing production gaps. Closing yield gaps also necessitates more labor for crop management tasks like fertilization, weeding, pest control, and reaping, but the use of labor-reducing technologies (such as tractors, herbicides, and mechanical threshers) that are capital intensive can help to balance the capable trade-offs between higher yield and higher labor yield. The law of minimizing returns causes sub-optimal N usage efficiency, which raises the threat of crop loss because of dwelling, and stopping production gaps in grassland irrigated rice frameworks needs considerably higher quantities of N (160-210 kg N ha1,).

The accessibility of farming resources (land, labor, and fund), as well as agriculturist characteristics and regional conditions, can all be attributed to the variability seen in farming level indicators, according to the second hypothesis. This is seen in Figure 15.1 steps 3 and 4. For instance, [5] found that the utilization of financing, the cost of trans-planting labor, and the supervision of hired labor were some of the key determinants of the crop growth technique chosen by rice farmers in Nueva Ecija. Due to the increasing significance of capital and non-rice revenue in the area, farmers in the Philippines embraced direct-seeding and hiked mechanization. Since farmers are more likely to invest in fertilizers if they have access to greater funding, we anticipate non-rice income to have a positive correlation with fertilizer usage. Furthermore, we believe that smaller farms would have more labor available, which will result in earlier crop planting and fertilizer application dates. Regional circumstances may also influence the timing of crop establishment and land preparation because they control the distribution and availability of water. Due to variations in water accessibility and the execution of land reform initiatives, they may also have an impact on farm level metrics. As farming and local level factors are probably to have an impact on the management strategies employed by farmers and their performance, knowing the restrictions to intensify crop administration needs an aggregated examination of drivers at several sizes. Temporal trends should also be taken into account since restrictions are not constant.

15.3 Resources and Techniques Household Polls

The Worldwide Rice Research Organization performed two household surveys in Central Luzon, and the data from both surveys were utilized in this research (Figure 15.2). A wet season crop is grown between May and June and October and November, while a dry season crop is grown between January and April. Double rice cultivation is popular in the area. The dry season crop can only be grown with irrigation and the wet season crop can be grown either way.

In agricultural systems based on rice, there have been changes in family characteristics and crop management throughout time. The survey area for the rice fields was narrowed down to a certain kilometer along the major route. 105 and 60 households, respectively, are interviewing rice farmers, in the wet season and dry season, respectively. Because of water restrictions and the growth of various crops, fewer families farmed rice in the DS than in the WS, resulting in a smaller sample size. Figure 15.3 shows crop administrative trends, performance metrics, and farm level circumstances.

While the quantity of municipalities was chosen in each state according to the province's rice area, 15 municipalities totaling more than 2,000 hectares of total rice region were chosen at random. Four communities in each municipality were chosen at random. A systematic sampling method was used to identify 20 farmers in each hamlet, yielding a sample of 1,800 homes that grow rice a semi-structured survey was used. The economic

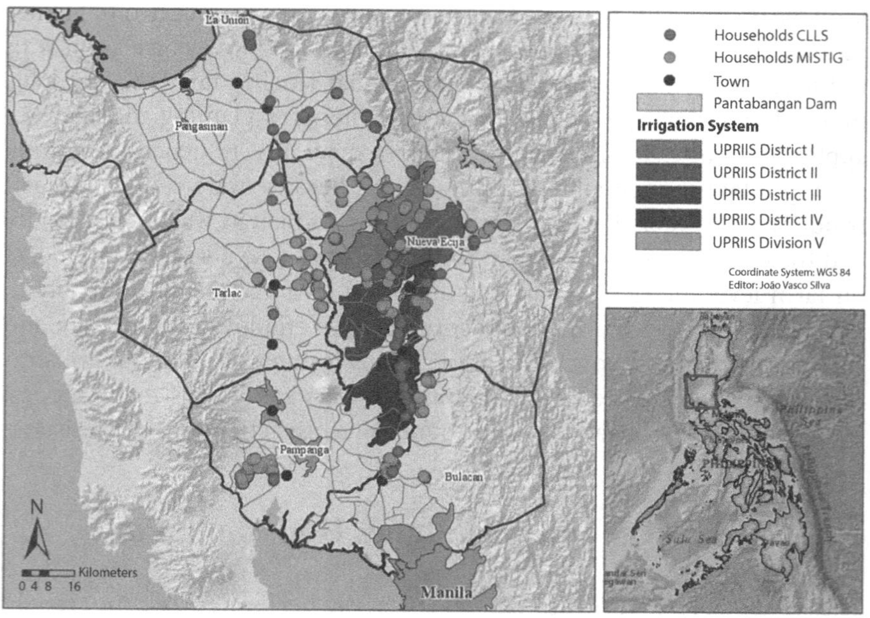

Figure 15.2 A map of the research area that shows where the houses were scrutinized and the Pampanga River Aggregated Irrigation Framework are located. Despite being physically located in the center of Luzon they do not belong to Administrative Region 3.

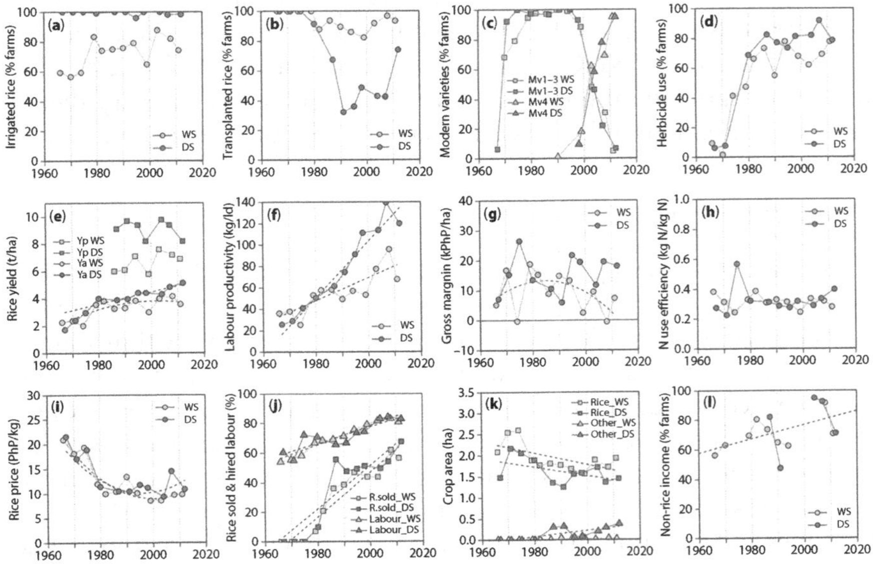

Figure 15.3 Temporal alterations in rice-dependent agricultural frameworks in the Philippines, during the last 50 years in terms of crop management methods (a–d), performance factors (e–h), farm(er) natures, and local circumstances (i–l). [6], which provides further information regarding the implementation of Yp in (E), might be consulted. Paddy costs were normalized to 2006 values utilizing the client cost index, and dashed lines represent mathematically major linear and quadratic regressions. In the Supplementary Material (Fig. A8), variations in household characteristics over time are shown. Wet season is abbreviated as WS, while dry season is abbreviated as DS. Central Luzon Loop Survey, the data source.

feasibility of rice agriculture as an origin of revenue for rural people relies on the yields it produces. A lower input-output cost ratio, for instance, is likely to encourage the closure of production gaps in agriculturists' farms while a higher input-output cost ratio will impede the usage of inputs, showing an economic optimum production well less than Yp. This is assuming that farmers behave in a way that maximizes gross margins. As a result, the following estimate of the cumulative margin (PhP ha1) from rice farming was utilized as an economic indicator:

Rice Revenue (PhP per hectare) = Rice Kept (kg per hectare) + Paddy Price (PhP per kilogram) were delivered via tablets to collect data on farm level circumstances and the biggest rice parcel's crop management in relation to 2013. The indicators are calculated slightly differently based on the homely scrutiny as the CLLS consists of data for all of the plots that each farming household cultivates (80% of households cultivate two fields or more per season, out of which 60% cultivate more than one field per season), whereas the MISTIG survey only includes data for the highest parcel that every agricultural home cultivates. The significance of every theory and how the various indicators are calculated are described below.

Production of rice

In addition to achieving larger economic rewards, increasing rice production is crucial to ensuring family rice self-sufficiency. Rice yield (Ya, t ha1) is a measure of rice production. When data for many farms within a single field are present, as in the case of the CLLS, Ya are computed as the ratio between rice output at 15% moisture available and the total area of every rice farm farmed by every lone farm. Ya are the production level recorded by every agriculturist for his or her biggest parcel for the MISTIG survey, and this was normalized at 14% moisture content. Rice agriculturists' sowing times and variations in rice types were taken into consideration while simulating climatic potential yields (Yp) using ORYZA v3 for the years 2001 to 2022. The labor productivity metric (kg rice labour-day1, ld) exemplifies the hardship of farming. Adoption of labor-saving technology has contributed to an increase in labor productivity in Central Luzon.

The quantity of rice utilized to pay in kind for the reaping and threshing activities per unit area is deducted from the total amount of rice harvested to get the "Rice Kept" figure. The "Paddy Price" is the unitary rice cost that is paid to each family. The consumer price index was used to normalize past costs described in the CLLS to 2019 costs (Central Bank of the Philippines, accessed on October 19, 2014), however historical prices recorded in the MISTIG survey are reported as 2014 values. The authors presume that agriculturists have a non-economic incentive to produce rice (e.g., rice self-sufficiency at the household level), so our predictions of cumulative margin relate to a "best case scenario" as 1) the grades of rice merchandized by every home is typically less than the amount preserved after reaping because homes can keep rice for seed or can deposit it for homely use, and 2) family labor was not included as a price in the analysis.

Environmental sustainability

As rice agriculture in the lower land regions of SE Asia is an intense action with two to three rice harvests cultivated in a lone year, nitrogen usage capacity (NUE) is a pertinent measure to evaluate the ecological sustainability of rice agriculture. Additionally, the usage of mineral fertilizers, notably N, is crucial for its long-term production and sustainability. A low NUE indicates significant environmental losses for a given N intake, and a higher NUE mitigates this threat.

Performance of management packages in Step 1

The CLLS was used to describe the administrative executions utilized by rice agriculturists, and linear mixed models were used to evaluate their success in matters of production, labor productivity, cumulative margin, and N usage effectiveness (step 1 in Figure 15.1).

Step 1: Accounting for management packages

Through successive subsets of the CLLS, the management programs that farmers have used over the last 50 years have been identified. To account for interactions between various management practices, we examined management packages rather than individual practices. In order to identify differences in water management, irrigated fields and rainfed fields were first divided. Second, rice types grown in traditional (Tv) or contemporary (Mv) fields were distinguished. Due to the little time overlap between Mv3 and Mv4 category of variations (Figure 15.3c) for a definition of [7] further separation within Mv was not made. Third, direct sowing and transplanting were distinguished using the crop establishment technique. Fourth, farms in which hand-weeding was done or pesticides were sprayed were divided using weed management. Lastly, fields with farms with "Lower N executed" (i.e., less than the entire average N implementation) were distinguished from those having "Higher N executed" (i.e., having N implementation similar to or more than the cumulative average N implementation level).

Mixed linear model

A linear mixed framework, taking into account administrative package and state as constant outcomes and year in home as random effects, was used to assess the performance of administrative packages (having a minimum of 10 observations) related to the various performance metrics. To account for variations in access to markets, inventions, and irrigation (e.g., interval to a research organization), province effects were added. To obtain year-based environmental features and changes in technology usage for every family, year within homes was added. By fitting the similar framework to the WS and DS data individually to account for biophysical variations (such as solar radiation and precipitation), season effects were controlled for. The R program predicaments was used to determine the predicted averages and standard flaws for every administrative package, and the R package nlme was used to fit the linear mixed models. Differences were evaluated at a 5% level of significance.

Step 2: Comparison of trade-offs between indicators

As shown in step 2 of Figure 15.1, trade-offs between indicators were first evaluated by contrasting the excellent-performing fields in terms of every indicator, and then every field was evaluated utilizing boundary line analysis.

A comparison of the most effective farms

We utilized the top 15% of Ya, labor productivity, cumulative margin, and NUE to spot several "farm types," that the authors showed as the fields pursuing a certain goal. As an example, the top farms that had Ya over the 90th percentile of this criterion, or "forming farms" in terms of rice output, were categorized as "production" farms. In the same way, the terms "labor productivity maximizing farms," "gross margin maximizing farms," and "NUE maximizing farms" were defined. Mathematically notable average differences in field level indicators between the four agriculturists' aims taken into consideration were evaluated using Tukey's test. Utilizing combined data over the years, this survey was conducted using the CLLS for the WS and DS individually.

Analyzing boundaries

Boundary line analysis was used to examine the association between Ya and the other variables. The fundamental justification for using this benchmarking technique was that, although having equal yields (Y), fields or farms can have various labor yield, cumulative margins, or NUE owing to variations in, respectively, mechanization levels, input-output pricing, and N administration. As a result, the approach was used to: 1) determine the level of an indicator that is best for every supplied Ya, 2) predict the Ya at which each performance indicator is at its highest level, and 3) evaluate how otherwise fields with comparable Ya accomplished related to the various indicators. By collecting data from specific farms, we may get quantitative insights into the synergies and trade-offs between raising Ya and optimizing different variables. In three independent processes that are detailed in the Supplementary Material (Figure 15.1a), the boundary lines were determined for the WS (irrigated and rain-fed) and the DS (irrigated) data independently. Because the MISTIG survey is the currently, indicative, and thorough home analysis obtainable for the manufacturing framework under consideration, we employed it in our study to bypass the surprised results of technical development with time. R software was used for model fitting and data processing.

Steps 3 and 4: Indicator drivers and crop management

In order to 1) understand how farm performance in one season (for example, DS) can result in farm performance in another season (for example, WS), and 2) determine if the "best-performing farms" were compatible with time in either WS or DS, steadiness in field performance through every seasons and for many years, was analyzed. To deal with this in the first aspect, the CLLS was used to estimate multiple pair-wise regressions using field performance in season i 1 as the independent variable and rice production in season i as the dependent variable. In order to find how often every field was designated as the greatest-produced field in the WS or in the DS, a frequency analysis was done with regard to the latter attribute. Similar evaluations were done for gross margin, NUE, and labor productivity.

The drivers of crop administrative methods (step 4 in Figure 15.1) and farm level indicators (step 3) were examined utilizing multiple linear regressions [8] registered on the CLLS having pooled data from various years. The ordinary least squares (OLS) method was used to estimate the intercept and coefficients of the various frameworks. Ya (t ha), labor yield (kg ld ha), cumulative margin (PhP ha), net energy consumption (kg N kg 1 N), nitrogen usage (kg N ha), initial fertilizer implementation timing and crop yield date were all taken into consideration as dependent factors. The mean across many fields was used to aggregate the three administrative variables at the field level. The aforementioned management techniques were chosen because, according to [9], they clearly explain resource or efficiency yield disparities. Following the examination of correlations between independent variables, model estimation was carried out independently for the WS and DS utilizing the lm function of the R [10] program. In order to interpret parameter estimates as elasticity, all variables were ln-transformed, i.e., the dependent variable's percentage changes in the event of a 2% change in the uncontrolled variable. Data on non-rice income, which is defined as revenue from sources other than farms, were not gathered for the T-tests [11] were used to contrast the average of field and local variables across categories of fields having different administrative techniques in the occasion of categorical administrative characteristics (such as crop growth approach). Prior to analysis, field and local variables were log-transformed to homogenize variations amongst the various groups. To further understand the resource yield difference, additional group comparisons were made between farms utilizing herbicide or manual weeding, and every macronutrients (N, P, and K) or select macronutrients (N, P, or K). Due to the little time overlap between these two kinds of variations (Figure 15.3c), no group contrasts between Tv and Mv were made.

15.4 Results

Changes in rice farming throughout time

Figure 15.3 summarizes the notable temporal alterations in administrative techniques, performance metrics, agriculturist features, and local states during the last 50 years. It is important to note that the CLLS time period begins in the year when the semi-dwarf type IR8, which was the catalyst for the Green Revolution in Asia, was released. As a result, the region's rice-based agricultural systems underwent a number of structural changes, mainly in the Philippines.

The Pantabangan Dam's inauguration in 1978 increased the number of fields utilizing irrigation in the WS and made it possible to grow a second rice crop in the dry season (Figure 15.3b). Without irrigation and the availability of short-cycle modern varieties, this would not have been possible. Beginning in the 1980s, direct seeding progressively replaced the laborious transplanting of rice that farmers had previously employed to establish the crop (Figure 15.3b). This change was most noticeable in the DS and hardly noticeable in the WS. However, the percentage of farmers transplanting rice increased significantly once again between the 2007 DS and the 2012 DS. Modern rice types were adopted regularly, without significant disruption. In both seasons, rice yields grew with time: Ya was ca. 3.6 t ha1 in 2011 WS, 5.2 t ha1 in 2012 DS, and 2 t ha1 in 1966 WS and 1967 DS (Figure 15.3e). In contrast to the WS, where yield growth was only shown up until 1990 before stagnating at about, mathematically significant production advancement was seen in the DS (3.8 t ha−1). The average production gaps in the WS and DS were 3.0 and 4.6 t ha1, respectively. In the WS and at a rate of 1.1 kg ld1 every year overall, labor productivity grew.

Due to production improvements and the implementation of labor-saving technology, 2.6 kg ld1 per year in the DS (Figure 15.3f). After 2000, the observed average labor productivity was about 120 kg ld1 in the DS and 70 kg ld1 in the WS, respectively. The WS's average gross margin for growing rice grew to 13.5 kPhP ha1 (about 250 € ha1) in 1985 and rapidly dropped afterwards, but the gross margin in the DS did not exhibit a statistically significant rise (Figure 15.3g). For NUE, which stayed around, no mathematically major alterations over time were seen in both seasons was 0.3 kg N kg 1 N (Figure 15.3h).

Significant alterations also took place in socioeconomic situations and agriculturist characteristics. Prices for paddy dropped by around half in the years 1966–1967, 20 PPH per kilogram to about 10 PhP kg1 starting in 1998 to 1999, and the temporal patterns were almost the same for both seasons (Figure 15.3i). The amount of rice sold increased significantly in time in both seasons: in the years 1979–1980, farmers sold around 5% of their whole crop; in 2011–2019, output went up to over 60% (Figure 15.3j). While there was a little rise in the domain of different crops (like maize and vegetables) throughout the DS, the area of rice cultivation decreased by around 0.5 ha between 1966 and 2012 (partly as a result of land being converted to residential areas) (Figure 15.3k). In both seasons, the proportion of hired labor to total labor climbed linearly from around 60% to 80%, indicating that hired labor has become more significant over time as compared to family labor (Figure 15.3j). As a result, rice farmers in Central Luzon changed from producing for subsistence to producing for the market during the previous 50 years. Finally, owing to work possibilities in other sectors of the economy and remittances, the fraction of fields having accessibility to non-rice income tends to rise in time (Figure 15.3l).

Performance of management packages in Step 1

Figure 15.3 provides a summary of the most popular management programs during the last 50 years. Irrigation + Modern Variety + Transplanting + Herbicide + N, P was, in essence, the package that was most often utilized in the WS. Starting in the late 1980s, seeding was used as a crop establishment technique rather than transplanting. The differences across management programs are mostly caused by the amount of N used and the usage of transplanting, which is consistent with [12] findings. Crop establishment technique was the key factor in determining labor productivity, especially in the DS where direct seeding outperformed transplanting significantly in terms of labor output. Due to the high variability of this metric, no major differences in cumulative margin were found for the various administrative packages. The rate of N application connected with every administrative package was the primary factor in determining NUE in both seasons, low N implementation (i.e., N application rate lesser than the final mean N implementation rate) produced a higher NUE (> 0.4 kg

N kg1 N) than higher N implementation. Finally, to attain this goal, newer categories and higher N rates are needed. High Ya and these did not negatively impact labor productivity, but they significantly reduced NUE and may have increased WS risk.

Direct seeding and pesticide usage, as opposed to transplanting and manual weeding, were related with levels of labor productivity. Comparing direct sowing to transplanting, Ya was somewhat harmed; however, there were no differences between manual weeding and pesticide application. Additionally, there were no major differences in economic performance amongst management packages. The findings imply that, regardless of the weed control strategy, escalation is hugely accomplished with administrative packages utilizing contemporary types and higher N levels. To balance production and drudgery goals, the choice of crop planting technique is essential.

Step 2: Comparisons of indications

In the WS and DS, respectively, the production gap between Yp and the yield of production-maximizing fields was on average 1.2 and 2.4 t ha1 (or 16% and 25% of Yp in relative notations) (Figures 15.4a and b). The production difference between Yp and the production which maximized cumulative margin, labor yield, and NUE was almost twice as large as what was observed for farms whose output was maximized: on average, 2.3 t ha1 (34%) in the WS and 40 %) in the DS, or 3.7 t ha1 (Figures 15.4a and b). Yield-maximizing fields recorded labor productivity that was hugely lesser than that of labor yield-maximizing fields in both seasons (Figures 15.4c and d) and solely in the WS (Figures 15.4e and f). Yield-maximizing fields also recorded hugely lesser cumulative margins than gross margin-maximizing fields. Additionally, farms that focused on increasing output had a NUE that was much lower than farms that prioritized increasing NUE, but significantly higher than farms that prioritized increasing labor productivity (Figures 15.4g and h).

The MISTIG survey was used in a similar study (2013 DS - 2014 WS) in order to cross-validate the earlier outcomes of Figure 15.4. Greatest labor yield was recorded in the WS at 186 kg per day and in the DS at 243 kg per day, with yields of 4.9 and 4.9, respectively.

While less than Yp (7.2 and 8.8 t ha1 in the WS and DS), the yields that maximized labor yield are comparable to the mean Ya (4.6 and 5.6 t ha1 in the WS and DS). NUE was maximized by a similar Ya and was about 1 N = 0.76 kg N kg. The greatest profit from rice growing was about 66 and 92 kPhP ha-1 in the WS and DS, respectively, were attained with productions of 6.4

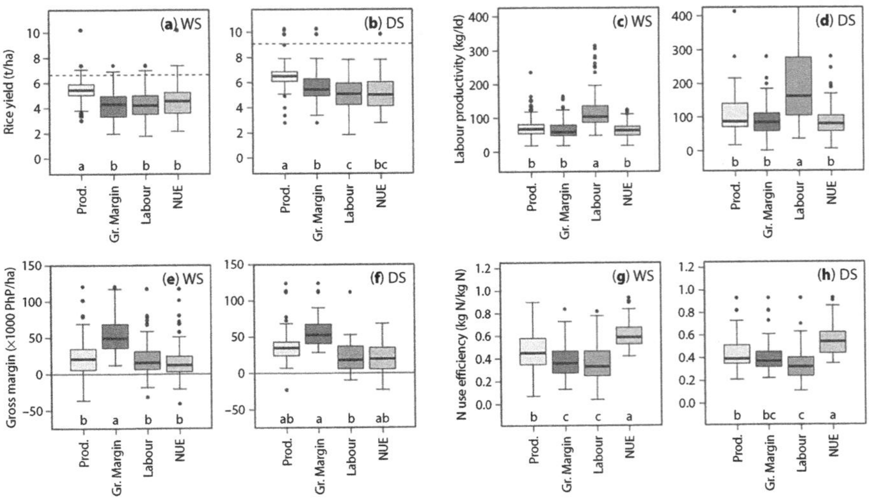

Figure 15.4 The variation in farm level indicators for several farmer goals (highest output, highest cumulative margin, highest labor yield, and highest NUE) in the Philippines. WS data are shown in (a), (c), (e), and (g), whereas DS data are shown in (b), (d), (f), and (h). At the bottom of each box plot, different letters show mean differences between farmers' aims that are significant at the 5% level. Data for the WS and DS were aggregated across 12 years and 11 observation years, respectively. The climatic capability yield (Yp), as predicted, is shown as horizontal dashed lines in the figures (a) and (b).

and 7.6 t ha-1. The yield-maximizing gross margin and Yp differed by 0.8 (11%), and 16%) in the WS and DS, 1.4 t ha-1, respectively. These findings suggest that although yields that maximize labor productivity also maximize net unit economics (NUE), they underperform in terms of cumulative margin (i.e., closing the production gap has a huge impact on cumulative margin). The research for the current year (Figure 15.5) that is not biased by cost fluctuations reveals that agriculturists with higher productions may have higher cumulative margins, contrary to what the past analysis (Figure 15.4) implies for farms that want to maximize their gross margins. This held true even when using profit as the economic indicator rather than gross margin, which is represented as the difference between returns and prices of material inputs, own farm, hired labor, and own power.

Multiple pair-wise regressions' findings show that rice productions in a season, like WS was positively associated to rice production.

Yields throughout the course of a season (for example, DS), but only three out of ten years showed this association to be statistically significant at the 5% level. However, because to the low adjusted-R2 of these regressions (21%) that was moreover shown for different rice bowls in SE Asia,

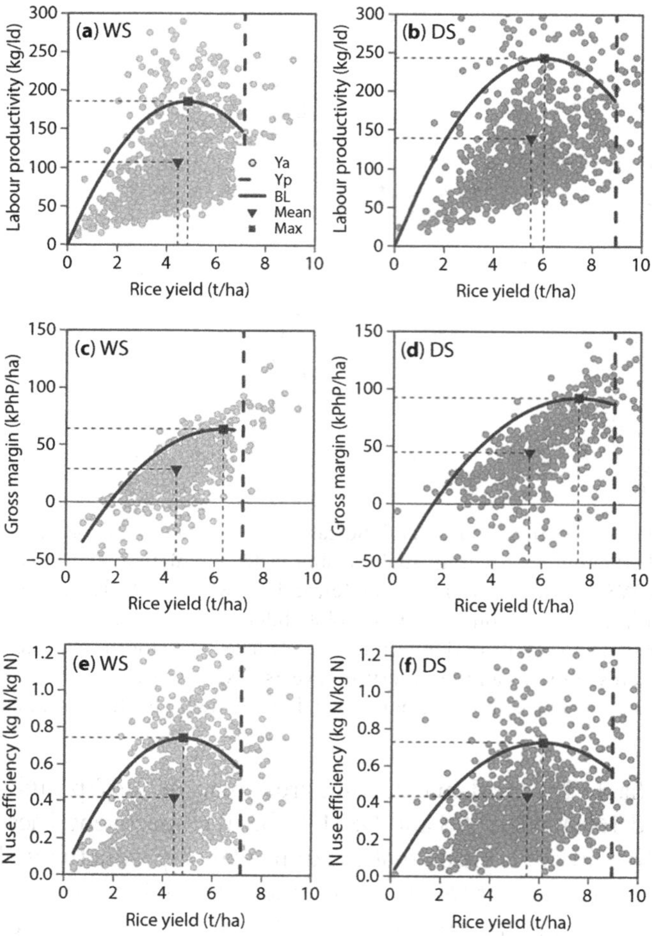

Figure 15.5 The correlation between rice productions and labor yield, cumulative margin, and N usage efficiency for rice-dependent agricultural frameworks in the Philippines. While DS data solely contain irrigated observations, WS data consists of both rain-fed and irrigated monitoring. Background information is included in Figure 15.6a. Legend: Ya stands for actual farmer productions, Yp for climatic potential production, BL for fitted boundary-line, Mean for average across fields, and Max for greatest anticipated by the boundary line.

rice yields in one season was not a strong indicator of rice productions in a subsequent season? While field performance was evaluated relating to labor yield, cumulative margin, and NUE in general, there were similarly weak connections between seasons (data not shown).

In conclusion, it was discovered that rice fields in Central Luzon performed more continually throughout years of the similar season than over the course of the two successive seasons (WS and DS). It supports the independent analysis of both seasons as well as the relationship between farm performance and farm(er) characteristics, principal factors influencing farm performance.

The province, farm size, the percentage of engaged labor (+), field power (+), off-farm earnings, paddy cost (+), and Ya were the most significant variables in the WS (i.e., elasticity 0.15%). Farm size increased by 1%, and Ya decreased by 0.25% as a result. The percentage of hired labor (0.16%) and the quantity of parcels (0.10%) showed a significantly less pronounced beneficial association. Additionally, there were -ve correlations between Ya and off-farm earnings as well as between Ya and the degree of the home head. Greater Ya was also seen on leased as opposed to owned property.

Farms utilizing machineries for farm construction had lesser Ya than fields utilizing animal draught power, but both variables were positively linked with paddy cost (+) and the percentage of engaged labor (+) at the 10% level of significance. In spite of the season, ya were always greater in Nueva (the state in the center of the irrigation framework).

In both seasons, there was a similar correlation between labor productivity and the percentage of hired labor (+) as well as paddy price (+). The latter had a rather significant impact: a 1% rise in paddy price was linked to an increase in labor productivity of around 0.70%. Off-farm income had a negative correlation with labor productivity during the WS, but this measure had a positive correlation with labor-saving technology (such as equipment and animal draught power). During the DS, farm size (+) and other revenue () were also related to labor productivity. Pampanga and Pangasinan had lower labor productivity in both seasons than Nueva Ecija.

In a previous research, the authors discovered that crop establishment technique and date, as well as timing of the initial fertilizer implementation, were the administrative executions citing the efficiency production discrepancy. Here, we discovered a relationship between hired labor, farm size, and crop planting technique (Figures 15.6a–d). Direct seeding farms have bigger farm sizes than transplanting farms for the WS periods between 1986 and 1999. This outcome was anticipated given that direct sowing involves less labor and hence is better suited for bigger farms.

Additionally, the majority of the time there were significant differences between the two practices in terms of hired labor, showing that transplanting depends more on hired labor.

Following our previous study, fertilizer (N and K) implementation and weed management were the management strategies that explained the resource yield discrepancy. The findings were only notable for 2009 WS

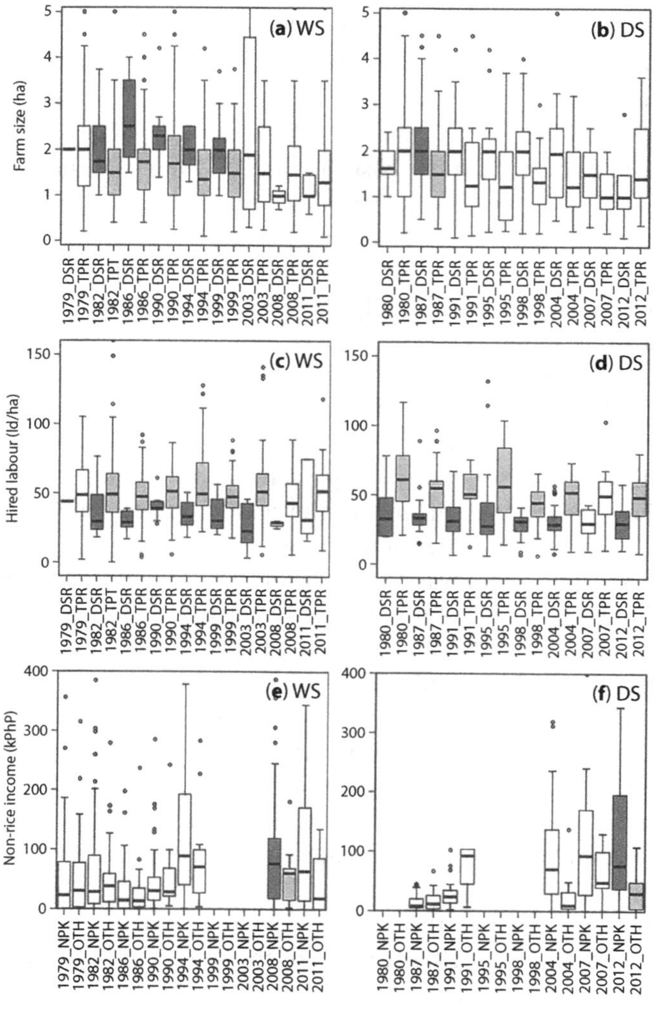

Figure 15.6 The differences in farm size, hired labor, and non-rice revenue across management techniques that cause rice yield discrepancies in the Philippines. Keep in mind that since data gathering methods vary from year to year, income figures shouldn't be compared. When different management techniques (such as direct sowing (DSR) vs. transplanting (TPR) and the use of N, P, and K in a certain year) are compared, light and dark grey colors show major differences; white indicates no significant differences.

and 2019 DS; however, non-rice revenue seemed to be higher on fields utilizing N, P, and K than on fields utilizing either N, P, or K in the most current times (Figures 15.6e and f). Farm size was the primary factor in N usage in both seasons, indicating that lesser N rates were used on bigger fields. Additionally, there were correlations between N usage and the percentage of engaged labor (+), farm power (+), off-farm income (), gender balance (+), and, to a lesser degree, education (+) during the WS. Finally, in both seasons, much less N was administered in Bulacan than in Nueva. For the majority of the factors examined in weed control, there were no discernible differences between utilizing a pesticide and manual weeding.

15.5 Discussion

At the farm, and local levels, we carried out an aggregated study of rice production gaps and connected administrative tools employed by rice agriculturists in the Philippines. Based on two premises, we sought to identify barriers to or drivers of intensification of rice farming: (1) there exists a trade-off between reducing production gaps and improving performance of different indicators (labor yield, cumulative margin, and NUE); and (2) strengthening of administrative operations employed by agriculturists (finding field performance) may be described by field and local situations. We go through our key conclusions in respect to these assumptions, the consequences of field structural alteration, and engaged labor at various levels in the sections that follow.

Closure of the yield gap and farm performance

For the purposes of yield gap analysis, a benchmark of 80% Yp for exploitable yield potential has been recommended. This limit was set because producing more than it would be neither commercially feasible nor ecologically or socially beneficial for farmers. According to our data, the relative yield gap closing for the most recent period (year 2014) was about 68% of Yp for farms that maximized labor productivity or NUE, and approximately 86% of Yp for farms that prioritized gross margin (Figure 15.6). These findings support our initial theory in part and suggest that an exploitable production capacity of 82% Yp would be a relatively higher bar to prevent trade-offs between yield and social, environmental, or economic performance while yet being sufficient to do so. This is consistent with findings for Dutch arable farms.

The indicator whose accomplishment improved the greatest over time (Figure 15.3f) was labor productivity, particularly when direct seeding was used as a crop establishment technique in place of transplanting (Figure 15.3b). Similarity to the researches of [13], hired labor (+) and paddy price (+) were the main variables linked to labor productivity. In spite of

N application rates rose with time, particularly in the DS, there were no discernible trends for NUE (Figure 15.3h), and the wide range of this indicator (0.2-1.3 kg N kg1 N) was described by the quantity of N utilized in various administrative packages. NUE was greater in Nueva than in different other regions in the WS, probably as a result of the nearness to PhilRice, which is supporting site-based nutrient management. In general, lesser levels of N registered lead to greater NUE.

In the current time period of data (Figure 15.5), a synergy between higher Ya and higher cumulative margin was seen, but not when data from a period of 50 years were aggregated (Figure 15.4). While the latter can be skewed by different paddy costs pooled over several years (Figure 15.3i), the former can be described by the higher paddy costs earned by Filipino agriculturists relative to different Asian locations. To stop the importation of cheap rice from elsewhere, the government of the Philippines has intervened to enforce high paddy prices.

Since the gross margin from rice is only around 17.3 kPhP ha (1 season) (about 300 € ha 1), this rice pricing strategy has a negative impact on consumers and a marginally beneficial one on farmers. A significant increase in gross margin does not seem to be a result of improved crop management for either restrictions or intensification-inducing factors. Between farms adopting direct seeding or transplanting as a crop growth technique, there were obvious differences in field dimension and engaged labor (Figure 15.6). Compared to direct sowing, which requires less hired labor and is more often utilized on larger farms, transplanting is labor-intensive. In addition, despite the fact that direct-seeding has been proven experimentally to need more labor for weeding than transplanting, we observed no differences in labor consumption for weed management between the two establishing procedures (data not shown). This might account for why direct-seeding farms have higher labor productivity but somewhat lower Ya than transplanting fields, although there aren't any significant yield differences between them.

In Nueva Ecija, crop planting took place sooner than in different provinces. It might be attributed to greater water stress in states not taken by UPRIIS (Figure 15.2) and to Nueva Ecija's superior water accessibility and nearness to research institutes, which may favor earlier crop establishment. As an unequal distribution of water was seen within UPRIIS, institutional variables at the local level can also influence water obtain ability and foundation dates, even if it reduces regional rice output and rice cultivated region. Additionally, the WS saw a positive correlation between planting rice earlier in the season and higher productions and cumulative margins, but the DS saw the inverse correlation, albeit the evidence in both cases was not particularly strong. Our data showed that this link was only present in the WS, contrary to our expectations that past crop planting will enable agriculturists to harvest the crop sooner when the cost is still higher.

Management of nutrients

While families with off-farm income applied smaller quantities of N in the WS, there was a -ve correlation between field dimension and N implementation rate in both seasons. Lesser N implementation rates might be a tactic to reduce crop lodging caused by typhoons, but it's unclear how this would affect off-farm revenue. Low N application rates have the potential to result in lower Ya but higher NUE in the near term, but can cause a loss in soil fertility over time owing to the depletion of soil nitrogen reserves. Even while non-rice revenue show to be larger for the earlier category in the current periods, the authors moreover discovered that there were no major differences in non-rice earnings between fields utilizing N, P, and K and farms using N, P or K (Figures 15.6e and f). This shows that the availability of revenue other than rice throughout the growing season is not strongly related to the usage of fertilizers.

Numerous farms may still improve their N usage efficiency (Figure 15.5). Farmers can achieve this by minimizing the amount of nitrogen applied prior to or during sowing, using one or two split implementations at panicle initiation and at flag leaf extrusion, and applying a cumulative quantity of nitrogen consistent with the anticipated production levels and the quantity of naturally occurring nitrogen content in the soil. Because of its huge variable character in flooded soils, timing adjustments and the use of split implementations are easier to apply than predicting indigenous soil N content. However, [14] pointed out that agriculturists often restrict the quantity of implementations to only once or twice in places with higher pay rates, or greater opportunity costs for labor (like the Philippines), and this is likely to have significant implications for fertilizer scheduling. According to our study, fields with bigger families are likely to fertilize later than fields with small households (although we are unable to determine the reason for this), and farmers preferred using their own fields over those that were leased out or shared, although the evidence for both trends is not very strong. Additionally, farmers who applied their first fertilizer at later dates in DS had low productions, cumulative margins, and N usage efficiency than agriculturists who registered their fertilizer at before time dates; this finding is in line with the present technical guidance.

In Central Luzon, there has been a progressive shift over the last 50 years from subsistence to commercial rice-dependent agricultural frameworks (Figure 15.3). For instance, because of greater accessibility to urban labor markets and the growing significance of non-rice earning for rural sustenance, family income structurally shifted away from land-saving and toward labor-saving technology. Furthermore, land reform initiatives failed to guarantee land ownership and reduce farm size. According to anthropological studies on daily politics in Central Luzon, the social interactions between landowners and landless families were the driving forces behind these developments.

The effects of (hired) labor at various levels are analyzed in light of the aforementioned dynamics. Due to the advent of labor-saving technology, there was a decrease in the utilization of labor on the ground level. Saving labor at the farm level enhanced labor productivity [15], often at the price of real yields (Figure 15.4 and Figure 15.5), and enabled family members to participate in actions away from the farm. These contributed to an increase in the value of engaged labor over family labor (Figure 15.3j), for example, in crop planting activities (Figure 15.6), together with favorable labor markets. N usage in the WS was likewise strongly correlated with the percentage of hired labor. Because they increased the job opportunities for landless families, these improvements in crop and field administration had an impact on the area. These labor dynamics suggest that, despite their beneficial effects for the rural economy, rice agriculture has largely become a secondary action for landowner households in Philippines that in addition to the likelihood of crop failure in the WS, cannot be conducive to more intensification and closing the production gap.

15.6 Conclusions

Over the last 50 years, there have been significant changes in Central Luzon's rice-based agricultural practices. Improved cultivars have been widely adopted, and the percentage of irrigated fields has increased in terms of crop management. Additionally, direct seeding replaced transplanting as a technique of crop establishment, and N application rates rose. Herbicides predominated over manual weeding, especially in the DS. The steady shift from subsistence to commercial yield frameworks shown by these shifts is further confirmed by the rising significance of hired labor and rice sales. As a consequence, although there were no discernible changes in cumulative margin and N usage efficiency in time, there was an improvement in rice productions and, particularly, in labor yield. Our findings partly support the study's initial premise since there was a trade-off between maximizing labor productivity or N usage efficiency and eliminating yield gaps, but not as much between the two. The best real yields were attained in the field while using high N levels with transplanting as a crop establishment technique. However, transplanting produced less labor productivity than direct sowing and heavy N application led to sub-optimal N usage efficiency. Contrarily, there were no significant differences in cumulative margin among administrative packages, suggesting that under the current pricing ratios in the study area, there was no compromise between production and economic goals. These findings are also true at the field level, where real productions that maximized N usage efficiency or labor yield were around 30% less than climatic potential yields, whereas actual yields that maximized gross margin were only about 18% less. Understanding the role that input mix efficiency plays in the observed trade-offs (i.e., the yield losses linked with a poor mixture of inputs) should be the main goal of future study. Results relating the second hypothesis were not particularly concluding, despite the fact that there was less correlation between parameters at the field level and management methods or field performance measures. Smaller farms used greater N application rates for nitrogen management, whereas bigger families and farms with more parcels applied their first fertilizer later in the wet and drier seasons, respectively. In comparison, bigger farms used direct-seeding as a crop establishment technique which required less hired labor than transplanting. Contrary to different variables at the field or local levels, the difference in sowing dates was mainly connected to various states in the WS (a proxy for example, water accessibility). In conclusion, farm and regional factors seem to influence, but not restrict, the management strategies utilized by agriculturists in the Philippines.

References

1. P. T. Khanh, T. H. Ngọc, and S. Pramanik, Future of Smart Agriculture Techniques and Applications, in *Advanced Technologies and AI-Equipped IoT Applications in High Tech Agriculture*, Ed., A. Khang, IGI Global, 2023.
2. T. H. Ngọc, P. T. Khanh and S. Pramanik, Smart Agriculture Using a Soil Monitoring System, in *Advanced Technologies and AI-Equipped IoT Applications in High Tech Agriculture*, Ed., A. Khang, IGI Global, 2023.
3. S. Dey, A. Marino and A. Bhattacharya, "Target Characterization Using the Polarimetric Scattering Trace Correlation," in *IEEE Geoscience and Remote Sensing Letters*, doi: 10.1109/LGRS.2023.3334734.

4. S. Pramanik, Intelligent Farming Utilizing a Soil Tracking Device, in *Convergence of Cloud Computing, AI and Agricultural Science*, Eds., A. K. Sharma, N. Chanderwal and R. Khan, IGI Global, 2023.
5. W. Zhang et al., "Online Recognition of Small Vegetable Seed Sowing Based on Machine Vision," in *IEEE Access*, vol. 11, pp. 134331-134339, 2023, doi: 10.1109/ACCESS.2023.3336944.
6. Z. Yu et al., "RiceMapEngine: A Google Earth Engine-Based Web Application for Fast Paddy Rice Mapping," in *IEEE Journal of Selected Topics in Applied Earth Observations and Remote Sensing*, vol. 16, pp. 7264-7275, 2023, doi: 10.1109/JSTARS.2023.3290677.
7. P. Pal, R. P. Sharma, S. Tripathi, C. Kumar and D. Ramesh, "NSGA-III Based Heterogeneous Transmission Range Selection for Node Deployment in IEEE 802.15.4 Infrastructure for Sugarcane and Rice Crop Monitoring in a Humid Sub-Tropical Region," in *IEEE Transactions on Wireless Communications*, vol. 22, no. 6, pp. 3643-3656, June 2023, doi: 10.1109/TWC.2022.3220146.
8. S. Pramanik, K. M. Sagayam and O. P. Jena, Machine Learning Frameworks in Cancer Detection, *ICCSRE*, 2021, Morocco, 2021.
9. G. Singh and J. Singh, "A Fog Computing based Agriculture-IoT Framework for Detection of Alert Conditions and Effective Crop Protection," *2023 5th International Conference on Smart Systems and Inventive Technology (ICSSIT)*, Tirunelveli, India, 2023, pp. 537-543, doi: 10.1109/ICSSIT55814.2023.10060995.
10. S. Pramanik, M. G. Galety, D. Samanta, and N. P. Joseph, Data Mining Approaches for Decision Support Systems, *3rd International Conference on Emerging Technologies in Data Mining and Information Security*, 2022.
11. R. Bansal, A. J. Obaid, A. Gupta, R. Singh, and S. Pramanik, "Impact of Big Data on Digital Transformation in 5G Era", *2nd International Conference on Physics and Applied Sciences (ICPAS 2021)*, doi:10.1088/1742-6596/1963/1/012170, 2021.
12. M. Mukelabai, C. Derks, J. Krüger and T. Berger, "To Share, or Not to Share: Exploring Test-Case Reusability in Fork Ecosystems," *2023 38th IEEE/ACM International Conference on Automated Software Engineering (ASE)*, Luxembourg, Luxembourg, 2023, pp. 837-849, doi: 10.1109/ASE56229.2023.00191.
13. O. Hosam, R. Abousamra, A. Ghonim and K. Shaalan, "Utilizing Machine Learning to Develop Cloud-Based Apprenticeship Programs Aligned with Labor Market Demands," *2023 IEEE 10th International Conference on Cyber Security and Cloud Computing (CSCloud)/2023 IEEE 9th International Conference on Edge Computing and Scalable Cloud (EdgeCom)*, Xiangtan, Hunan,China,2023,pp.446-451,doi:10.1109/CSCloud-EdgeCom58631.2023.00082.
14. C.-T. Chiang, "A Fruit Citric Acid Concentration Detector for Fruit Farm IoT Applications," in *IEEE Sensors Journal*, vol. 23, no. 20, pp. 25340-25345, 15 Oct.15, 2023, doi: 10.1109/JSEN.2023.3312703.

15. S. Praveenkumar, V. Veeraiah, S. Pramanik, S. M. Basha, A. V. Lira Neto, V. H. C. De Albuquerque and A. Gupta, Prediction of Patients' Incurable Diseases Utilizing Deep Learning Approaches, *ICICC*, 2023, Springer, doi.org/10.1007/978-981-99-3315-0_4

16

Simulating the Effects of Land Use Policies on Agricultural Productivity: A Mathematical Modeling Perspective

Vinaya Kumar Yadav[1], Sushila Arya[2], Asha Rajiv R.[3], Devendra Pal Singh[4], Siddharth Ranka[5] and Dharmesh Dhabliya[6]*

[1]*Department of Agriculture, Sanskriti University, Mathura, Uttar Pradesh, India*
[2]*School of Agriculture, Dev Bhoomi Uttarakhand University, Dehradun, Uttrakhand, India*
[3]*Department of Physics, School of Sciences, JAIN (Deemed-to-be University), Bengaluru, India*
[4]*College of Agriculture Sciences, Teerthanker Mahaveer University, Moradabad, Uttar Pradesh, India*
[5]*Symbiosis Law School, Nagpur Campus, Symbiosis International (Deemed University), Pune, India*
[6]*Department of Information Technology, Vishwakarma Institute of Information Technology, Pune, Maharashtra, India*

Abstract

Implementation fragmentation has long been a problem in agricultural modeling. Numerous models have been developed and there is a great deal of duplication. Models are sometimes improperly connected, reusing model components is uncommon, and it is frequently challenging to utilize frameworks to supply feasible remedies in the farming industry. We contend that in order to change this, it is necessary for a community that is open, self-reliant, and dedicated to build agricultural models, together with the related data and tools, as a shared resource. Recent advancements in information and communications technology may help

**Corresponding author*: dharmesh.dhabliya@viit.ac.in
Vinaya Kumar Yadav: ORCID: https://orcid.org/0009-0005-6050-7756
Sushila Arya: ORCID: https://orcid.org/0000-0001-7492-4757
Asha Rajiv R.: ORCID: https://orcid.org/0000-0002-4058-2178
Devendra Pal Singh: ORCID: https://orcid.org/0009-0000-7479-8631

Sabyasachi Pramanik, Niranjanamurthy M., Ankur Gupta and Ahmed J. Obaid (eds.) Mathematical Modeling in Agriculture, (321–344) © 2024 Scrivener Publishing LLC

such a community. We look at how these advancements may be used to design and put into practice the newest data, models, and decision-making tools for farming yielding frameworks. The authors' goal is to evaluate pertinent technologies in their growth, maturity, and capability to assist the farming modeling fraternity. The technologies under consideration include techniques for cross-disciplinary collaboration in development and for engaging users and stakeholders. Our qualitative analysis indicates that, as a general evaluation ultimatum, the compatibility of data origins, modular granular open frameworks, testimonial data-sets for implementations, and particular user need analysis techniques requirement to be directed to enable farming modeling to move to big data age. It will make it possible to incorporate new data sources and have considerably stronger analytical capabilities. In order to effectively address the demands of beneficiaries and those who create implementations for their frameworks, farming frameworks modeling must rapidly adapt and find latest data and Information and Communication Technologies (ICT). This acceptance necessitates the extensive acceptance of a collection of good practices called standard operating procedures (SOPs).

Keywords: Land use policy, agriculture, mathematical model

16.1 Introduction

The field of ICT [1] is evolving quickly. People can connect at lightning-fast rates at any moment to digital [2] technology. Through telephone and internet [3] service providers, even people in rural, developing countries are increasingly able to connect online. In-season crop yield and growth, soil moisture, and different dynamic variables may all be remotely monitored in real time [4] using satellite [5] and drone capabilities. With the help of high-performance computing [6], enormous volumes of organized and unstructured [7] data (also known as "big data") may be processed quickly utilizing cutting-edge sensing technologies [8], as well as to scale and verify models in novel ways. These capabilities may now be made accessible to many end users with previously unthinkable ease and affordability thanks to web and cloud technology. Society now awaits far better-grade data to be accessible to help everyday decision-making as a consequence of these and other developments.

The knowledge that our frameworks and decision support systems did not keep up with technology must limit our excitement for these new agricultural science innovations. In fact, various systems utilized in these frameworks were created in the 1990s and 2000s, long before the present day's sophisticated data gathering, computation, storage, access, and processing technologies, programming languages, and coding standards were available. As a result, we identify two separate potential means for using contemporary ICT to simulate agricultural systems. In the first place, technological developments like big data [9], crowd-sourcing (i.e., collecting data and information from dispersed networks of interviewee) and remote sensing [10].

The science of modeling agricultural [11] systems may be advanced by performance computing. Second, the profession and use of agricultural systems modeling [12] may be transformed by new technologies by accomplishing it in a far more accessible, dispersed, adaptable and collaborative manner. The logic of farming frameworks modeling is advancing gradually and incorporating many newer Information Communication Technologies to move forward the research on an example-by-example base, as is seen in current special issues of various journals. A paucity of applications employing agricultural systems models is the result of the operation and use of farming frameworks modeling not developing as quickly. For cropping systems models, [13] have addressed the necessity to build an essential feedback from application to research. Therefore, the emphasis of this analysis is not on ICTs that are pertinent for modeling scientists engaged in universities or research institutions, but rather on strategies for facilitating the participation of players outside of the academy. A "next generation" modeling fraternity (NextGen) will be created as a result of achieving this level of participation, as suggested in [14]. NextGen will consist of modelers and framework designers employed across regimen, spatial degrees, and temporal degrees to utilize newer data sources, produce and address newer frameworks, as well as software coders who will create the NextGen modeling systems, data processing implementations, and visualization tools.

Here, the authors take an ICT-centric perspective to the visualized NextGen of farming frameworks and the assisting modeling fraternity. Our goal is to evaluate pertinent technologies in terms of their development potential, maturity, and ability to serve the farming modeling fraternity. The technologies under consideration include techniques for cross-disciplinary collaboration in development as well as user and stakeholder involvement.

Five use cases which were developed to assist the sight for a NextGen modeling fraternity are used to evaluate current ICT advancements. For more information, see the introduction overview by [15]. Making educated choices may be aided by a NextGen of implementations built on farming frameworks modeling for businesses, governments, and agriculturists in the food chain. Figures 16.1 and 16.2 show the knowledge

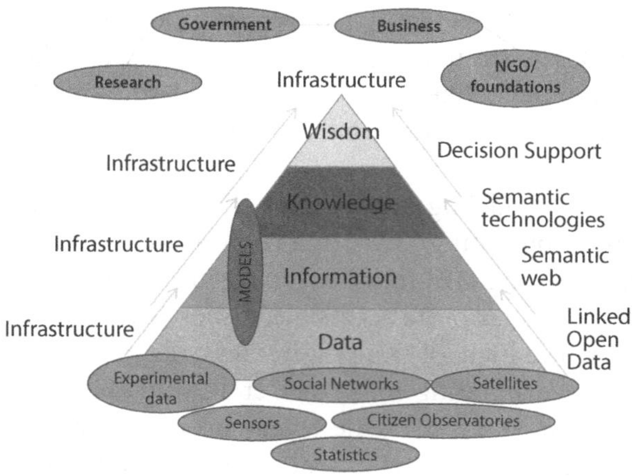

Figure 16.1 Knowledge pyramid which connects data to information to knowledge and wisdom uses data as the basis in applications that address decision-taking via wisdom in study, government, industry and foundations.

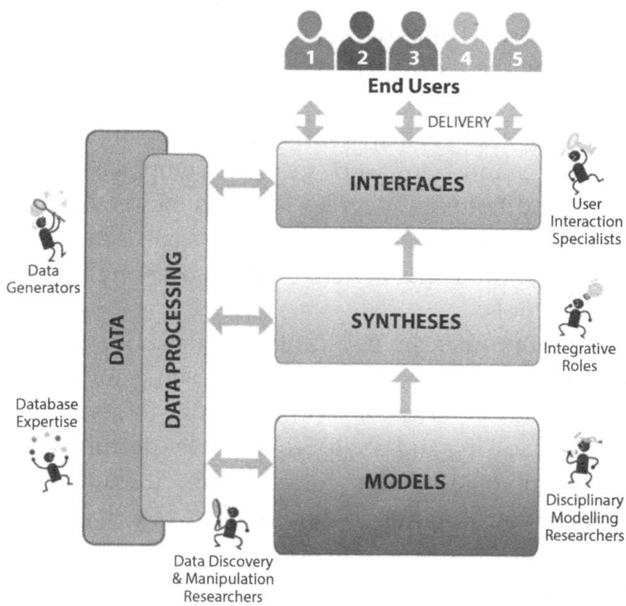

Figure 16.2 Application chains are a way of describing how data and information flow via levels of modeling, syntheses, and interfaces with end users while taking into account the roles of various players along the information chain.

chain and application chain ideas, which provide contrasting viewpoints on the importance and location of modeling in the larger context of ICT and decision-taking are employed to slackly arrange the paper's content. A knowledge chain is a series of interconnected stages that transform data into knowledge, information, and ultimately wisdom for use in decision-making. According to this viewpoint, data serve as the raw material which when coupled with attributes related to explanation and quality, yields information. Knowledge [16] may be created by connecting information to other sources of information and organizing it into causal chains. In the end, knowledge acts as an input for choices made depending on wisdom that can only be found in the brain of a decision-taker and cannot be digitized. Application chains are the subject of a different viewpoint. The infrastructure required for agricultural models to operate must include both software (such as in levels of data accessibility, processing, evaluation, and conception) and hardware (such as servers, computation power, and storage), as shown in Figure 16.2. Applications geared toward end users provide information and expertise depending on the data in the architecture, such as a production projection to a supply chain management, consequences on agricultural revenue of a regulation alteration, and estimations of disease-based crop harm to an agriculturist. Implementation chains can be straightforward or confusing, and they can consist of operations like data accessibility, extraction, transformation and aggregation. They can also include a single or more frameworks, the aggregation of output from various frameworks, and steps to transform, analyze, and visualize model output. The whole range of users of the NextGen ICT architecture, including primary data collectors, database experts, software coders, modelers, and end-users of knowledge and data, must be taken into account while designing the application chains. This document is arranged in accordance with the many levels of the focuses on the NextGen of farming frameworks model implementations, knowledge chain, and the encompassing agents and aspects of the implementation chains. The consumers and use cases where NextGen frameworks may have a significant responsibility are presented in Section 16.2. The players involved in the application chain are described in Section 16.3. Newer technologies which may help with the demonstration of modeling assessments to users are covered in Sections 16.4 and 16.5, along with growths in data as the foundation of every modeling endeavor.

16.2 Upcoming Applications of NextGen Farming Frameworks

Exemplary usage cases

The shareholders who determine the outputs and features of NextGen farming modeling systems may be represented by a variety of use cases. The most applicable ICT technologies to address in more depth in this chapter were chosen using five of these use cases. These are the usage cases:

1. A farmer in Africa who employs a decision-support tool depending on modeling and analysis outputs.
2. A scientist working in a global agricultural research station who is interested in evaluating various technological alternatives and developing new technologies for sustainable escalation.
3. A project manager in a donor organization who evaluates several project ideas to determine which ones have the best chance of succeeding.
4. A precision agriculture farm adviser who helps farmers uses cutting-edge data streams via farming administrative software.
5. An adviser for food supply chain businesses who assesses the environmental impact of goods and value chains using online technologies.

A broad variety of agricultural frameworks, inheritor, and needs for data and modeling elements have all been represented by the use cases that have been selected. In use cases 1, 2, and 3, smallholder farming systems are highlighted because meeting the demands of these frameworks is seen to be crucial for ensuring food security in evolving areas, since smallholder fields provide the majority of the world's food. The use cases purposefully included a variety of consumers in order to highlight the possibility for future real-world model implementations. ICT for farming modeling has often only been thought of from the perspective of one user type, namely the researcher or scientist.

The five use cases were examined from an ICT viewpoint in this research, and the shortcomings of the present farming modeling frameworks for dealing with every use case was noted. Then, a final combination of these five ICT remedies is utilized to inspire the remainder of the study. The authors have suggested a prototype implementation chain for every use case which will address these weaknesses.

African agriculture expansion is the first use case

Issue statement

In a region of South Africa where various fields are tiny, wages are poor, and agriculturists often raise maize and beans as basic crops for their home's sustenance and to sell for money, Jan is employed as a field adjunct officer. Some homes could raise animals or cultivate their own food. The adjunct service's mission is to assist agriculturists in producing maize at better and more consistent yields while also offering nutrition advice to help them get enough protein and other essential elements for their families' well-being. Jan learns about the new maize and bean kinds that local farmers may now purchase. The bean types are better resilient to a typical foliar illness, and these new kinds are more tolerant of heat and drought. Using tiny amounts of inorganic fertilizer, animal manure, and agricultural leftovers, Jan moreover provides knowledge on how to optimize nutrition administration of the crops. He is also knowledgeable with a novel CGIAR-developed method for partly harvesting rainwater to boost the amount of water available to farm and vegetable crops.

Present shortcomings

Jan isn't a modeler, yet he may nevertheless gain from the results of models that simulate farming productivity and farming-scale economies. Here, Jan's NextGen tool must have the capacity to join or be coupled with current data regarding local variables (soils, weather, genetics, home economics, local markets) with field-scale frameworks in order to forecast the feasibility of utilizing the newer types. If these data ever exist, they are mainly hard to obtain. Weather information is often regarded as confidential and therefore not readily available. For crop production modeling, good quality, localized soil data are either nonexistent or only accessible at a scale which is impractical at the farm level. Except in circumstances when a current research study has been conducted in the region, information regarding household demographics and economics is seldom accessible. In any event, it's possible that these data include private information that shouldn't be made public until the data have been anonymized. It is necessary to use pre-configured models that are suitable for smallholder systems in this area, including elements for mixed live-stock/cropping systems that are directed by data related to administrative operations that are widespread in the area (for example, planting date, cropping density, variations used, etc.).

The architecture as it is will prevent Jan from getting a straightforward response. Current frameworks will replicate these frameworks, but gathering the necessary data and configuring the framework will take significant work and close coordination between modelers and primary data gatherers. Researches like this one would normally be conducted utilizing current technologies to provide general recommendations for management for an area, with findings not tailored to specific farms.

NextGen application chain in draft

Jan utilizes the NextGen tools to assist provide recommendations to every field family which is realistic, and is implemented, and provides the greatest results in accordance with better steady output, greater revenue, and better nutrition. Farms differ in dimension, labor accessibility, soils, and different features. Jan asks the farmer for information that he can enter in his smartphone that contains NextGen applications made specifically for the agricultural systems in his area. These apps allow Jan to figure out what combinations of system elements could work best for different farm

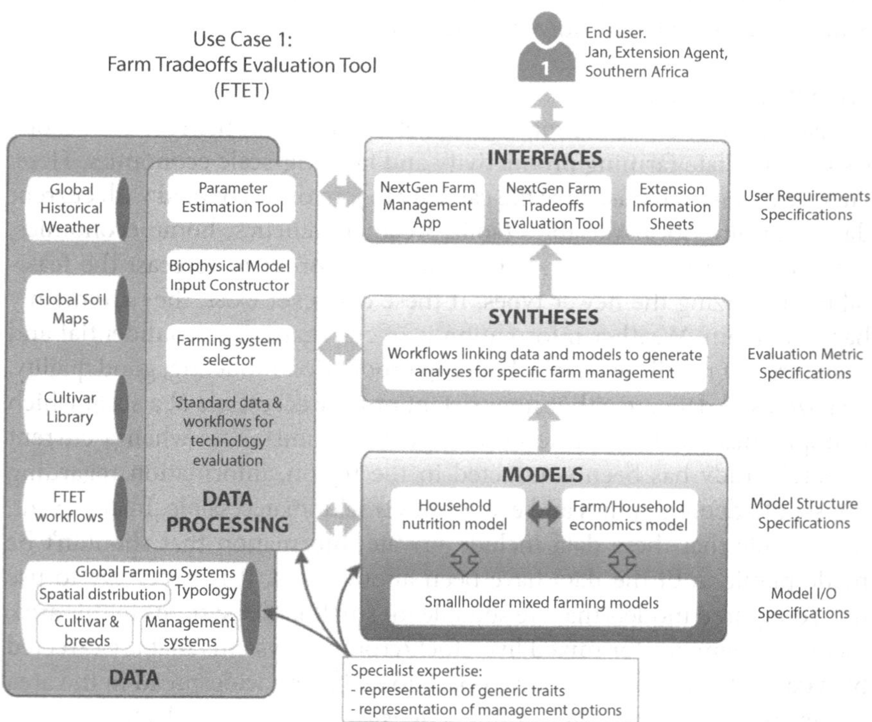

Figure 16.3 The parts of the infrastructure for data, modeling, and dispatch based on implementation chains to provide use case 1.

circumstances. Additionally, this software offers instruction files for added information sheets that highlight the aspects of crop and agricultural methods which are most probably to be successful with the field member in the local language or languages. Figure 16.3 depicts the design-time story in visual form.

The authors' evaluation of the five use cases led us to the conclusion that they call for a combination of current technological and data breakthroughs. This comprehensive study must be qualitative and is based on the characteristics of the five use cases and the preliminary implementation chains, both of which may have been built differently. The authors see a strong focus on data aggregation in the widest aspect, having a requirement to assign to data from many domains and sources (both private and public).

Jan is beforehand acquainted with the user interfaces and accessible choices since he has used the NextGen applications in the past to assess changes to cropping system management. To assess the effectiveness of the new kinds, he makes use of the NextGen Farm Tradeoffs Evaluation Tool (FTET). He will be able to identify combinations of system elements that could match certain farm settings the best with the aid of the app, and he will be able to enter these administrative frameworks into the Global Farming Systems Typology Database [17]. For usage with the FTET, the NextGen database already contains soil characteristic and weather information unique to the farms in Jan's area. A workflow called the FTET was developed for analyzing synergies and tradeoffs between administrative choices and total farm gain and nourishment. The tool's elements include a nourishment analysis depending on inputs and outputs from the field, farm productivity using biophysical models, and family income forecasting for each scenario. Using the FTET user interface, Jan's input data from every home and the suggested enhanced variety may be incorporated to the workflow. Jan writes, distributes, and holds discussions about extension information sheets in the regional language which outline the elements of crop and agricultural frameworks which are better to be successful with the field household. These sheets are based on the results from the FTET.

Just one of the five case studies, number three, depends on the synthesis of enormous quantities of better-organized sensor data. The remaining four scenarios will mostly need the use of "big data" approaches to synthesis multiple-purpose datasets like agricultural frameworks typologies, from various origins of largely unstructured data. The majority of use cases rely on the availableness of sufficient input data for utilization in modeling implementations that is of high quality, accessible, and ideally open.

These data needs seem to be any modeling system's easy pickings, but in the reality of modeling today, they are sometimes unexpectedly challenging to get. For instance, it is necessary to have complete, relevant, and adequate soil data that are appropriate for use in crop models. As it is finished, the GlobalSoilMap program will assist in releasing this restriction, but local data would be required to rely on these generic data sets in order to record the heterogeneity.

Although linked data and the semantic web [18] may facilitate easier use case realization and data exchange, they don't seem to be absolutely necessary. On the other hand, standardized data protocols [19] are crucial since they would enable data to be shared, found, aggregated with data from other origins, and utilized in many implementations and analyses in diverse modeling areas. It is probable that a NextGen framework will ultimately be largely based on the application of Linked Open Data principles. The consumers of our prototype implementation chains are formerly pretty distinctly defined as a consequence of the use-case methodology; the major demand for NextGen is to more thoroughly explain what these users really need using cutting-edge requirements analysis methods. Such visuals cannot be efficiently designed or produced using current techniques. It's interesting to note that in certain situations, using mobile and online channels to deliver knowledge has advantages over using desktop-based solutions.

Finally, we find a clear requirement to transition from monolithic frameworks to stable, resilient, and well-defined elements in matters related to modeling methodologies. In the latter method, models are driven flexibly by outside programs rather than being long, big containers of analytical procedures. Since applications may easily show on a library of aids made accessible via online protocols for data accessibility, processing, and computing, models must develop into latest algorithms which may be invoked successfully in a service-based framework. In terms of infrastructure, implementing the service-oriented architectures supporting the implementations depends on the availableness of utilities for data and frameworks.

Except for Use Case 2, where model linking and modeling frameworks are more important than an adaptable ecology in which a user may inspect the possibilities of frameworks. Users must be able to execute the resultant configuration continuously in a reliable atmosphere once a workflow has been built. Virtual machines [20] and containers, which are pre-configured sets of an OS with loaded programs which may be located easily on H/W, would probably perform a significant part in recording these settings. The NextGen modeling infrastructure will probably include a system which enables academics to create and share these workflows, consisting of links to several data sources.

16.3 Envisioned Consumers of the Application Chain Beneficiaries

By enabling possessors to easily comprehend both the organic systems which they control via their agricultural methods and every other's ways of operation, NextGen apps may not only increase comprehension but also transform the power dynamic. Massive possibilities are now available for those who have been underserved due to the fast-growing digitalization of society and the increased accessibility of internet and mobile technologies in rural areas. For instance, agriculturists in Africa may now get text messages about where to buy seeds, fertilizer, and crop insurance, which enables them to have choices depending on the most recent data. The polarity between scientists and end users has to be broken down for NextGen applications, and a wide range of shareholders, consisting of companies, agriculturists, people, the government, NGOs and research organizations, must be consulted. As stated in the initial paper, the development of NextGen models begins with a comprehension of the data needed by different shareholders, and proceeds backward from there to identify the frameworks and data required to convey that data in the format desired by users. Information Communication Technology provides a variety of ways for analyzing user needs, from conventional approaches like user requirements evaluation to cutting-edge methods like user-centered technique, where software is created quickly in close collaboration with the end user. The latter method guides and modifies development in every iteration based on user needs and requirements.

To our understanding, no use of user-centered framework approaches for agricultural models has been documented in the scientific literature. Contrary to other decision-support interfaces to current models, agricultural and cultural models were mostly built from a push viewpoint to reveal the functionality and the extensive data demands of the framework, not to enhance end-user usability. These results differ from the way models have traditionally been created, where the scientist has always been the main user. For the NextGen of apps, this is most likely not the case.

The emergence of mobile communication, social media, text messaging, radio, TV, and applications made for mobile phones has also increased the number of options to connect with end customers. But access to SMS and voice message services is expanding quickly, and it is anticipated that end users in village regions of developed nations will leave out the phase of using PCs and have direct utilization of mobile phones and also tablets. Currently, there still exist geographic regions where smallholder agriculturists need the mobile networks for data sharing. This pattern indicates that there is a significant amount of untapped capability to increase the quantity of data offered to agriculturists and processors throughout the supply chain, both as their demands aren't yet well defined and as assistances tailored to their usage aren't yet established. Because of smaller displays and varied handling requirements, mobile technology delivers a unique user experience. Therefore, if mobile graphics are to persuade people to continue using them, they must be simple but effective. They often only deal with a small number of data points, which are conveyed to the end user in a powerful visual style with a focus on color and readability.

According to [21] and other researchers, a longer-term plan for the NextGen modeling framework will require to focus on educating newer framework creators and different knowledge framework experts from the implementation fraternity, particularly in poor nations. Users and developers from developing economies may provide distinctive viewpoints that can direct the development and implementation process of the model to incorporate major applicable elements essential to their user groups. For instance, a model developer in Asia can stress the significance of soil P in production restrictions of that locality—a modeling element missing in various current farming yield frameworks created in places where P is utilized at considerably greater input levels. Model users who have in-depth understanding of the decision-taking needs of actors in developing economies and the Information Communication Technologies that are accessible to them (such as cell phones) may provide to elements which cater to these actors' specific demands.

The distribution of the finished goods to consumers and access to NextGen data products both need software development. Both non-proprietary items created in the public sector and proprietary products created by private enterprise are likely to be included in the top layer shown in Figure 16.2.

Farming information standard data sources

We identify three primary conventional techniques for gathering information relevant to modeling agricultural systems:

1. Data collection by governments is done for administrative, information management, and monitoring objectives. These data are often standard in arrangement and are typically gathered on a regular foundation until they are necessary for policy. These data consist of countrywide statistical data, meteorological data, tracking data for allowances and taxes, and data to track environmental performance.

2. To fulfill particular project requirements, research projects gather data (e.g., farm and home surveys, multi-dimensional panel data, soil sampling). These data are mainly unstructured (i.e., not consistent in format) and incidental (i.e., gathered on an uneven foundation). These statistics are often useless since they are not shared.
3. Businesses gather data for their own activities, consisting of agriculturists and business-to-business service providers. They often don't exchange data because of rivalry or privacy issues. In comparison to the developed world, the availableness of data at the level of agricultural families and fraternities is lower in the developed world. Many pieces of potentially useful data are now accessible for study thanks to these sources. These data, however, are often locked, accessible exclusively for certain reasons, or poorly maintained for future accessibility.

Mobile communication, crowdsourcing, and remote sensing are three new data technologies that are becoming mature enough to be used in agricultural systems modeling. The development of crowdsourcing, also known as citizen science, civic science or volunteer GIS, is considerably aided by the rise of mobile technology. Cell phones, GPS [22], and tablets serve as sensors or tools which upload data straight to the internet with precise time and position details. These methods are sometimes seen as a chance for "near sensing": the use of field instruments with sensors to record observations, such as temperature readings based on information from a mobile smartphone. Additionally, there are specialized technologies that collect more location-specific data, like leaf area index sensors and UAV. These crowdsourcing [23] tools provide us the chance to collect more data more cheaply. In these situations, residents contribute to the data collection process via volunteer actions like biodiversity measurements, mapping, and early warning. As a result, individuals, groups, and smallholders may manage and contribute to public records. This opens up a wide range of opportunities (particularly given how young the technology is), including some cutting-edge applications like IIASA's Geowiki. Occasionally, crowdsourcing is set up as a public event, such as simultaneous air quality calculations across France on similar days and and times.

These processes are carried out semi-automatically and on an as-needed basis in every modeling project in the majority of contemporary integrated modeling systems. Several organizations are working to find solutions to this problem, including the Consortium of Universities for the Advancement of Hydrologic Science, Data for Environmental Modeling

(D4EM) [24], and the GeoSciences Network (GEON) [25]. The necessity to assure interoperability by coordinating syntax (formats) and semantics (definitions) grows as data availability rises. The creation of new products and all forms of analysis are given new chances by improved data interoperability. But the required standardization has not yet taken place.

Issues relating to transparency, confidence, privacy, and intellectual property

The goal of NextGen farming frameworks modeling is to make previously inaccessible data and tools available in a manner that will encourage discovery, composition, and implementation by a broad range of academics and majors. Based on [26] research, Figure 16.4 shows correlations between farms, climatic data, and soil data for Europe. Confidentiality [27], privacy [28], and intellectual property problems, as well as data and open source software, continue to be significant. There are connections between openness and confidentiality (as well as related subjects); for more information on cooperative models.

One crucial element is that license agreements should facilitate accessibility. In our opinion, open-access agreements that are straightforward and clearly define the rights and responsibilities of data users should be supported for use in agricultural modeling. Scientists from various fields have had to deal with elaborate licenses that end serves mainly to enforce the data owners' preferences and that impede others who want to reuse the information. Data (and software) licenses must be (i) as straightforward as possible; (ii) compatible with one another (like the Creative Commons); and (iii) non-invasive, meaning that licensees shouldn't be required to pass all license conditions on to their users. These characteristics that support openness will need to be combined with various degrees of permit control on access to data and software (such as the need of attribution). Different communities of practice collect various types of data that are essential to NextGen (for example, weather reports) on behalf of many users. In these circumstances, NextGen will need to influence data-licensing agreements rather than taking the initiative. However, information about management techniques, yields, or cultivars that is predominantly acquired throughout the course of agricultural production does not fall into this category. The NextGen community must take active steps to create a sharing and reuse-friendly licensing environment for this data.

Issues with confidentiality and privacy may occur in several contexts. Agronomic activities, geographic location, economic data, and nominal records, for instance, might expose a category's name, position, or enterprising

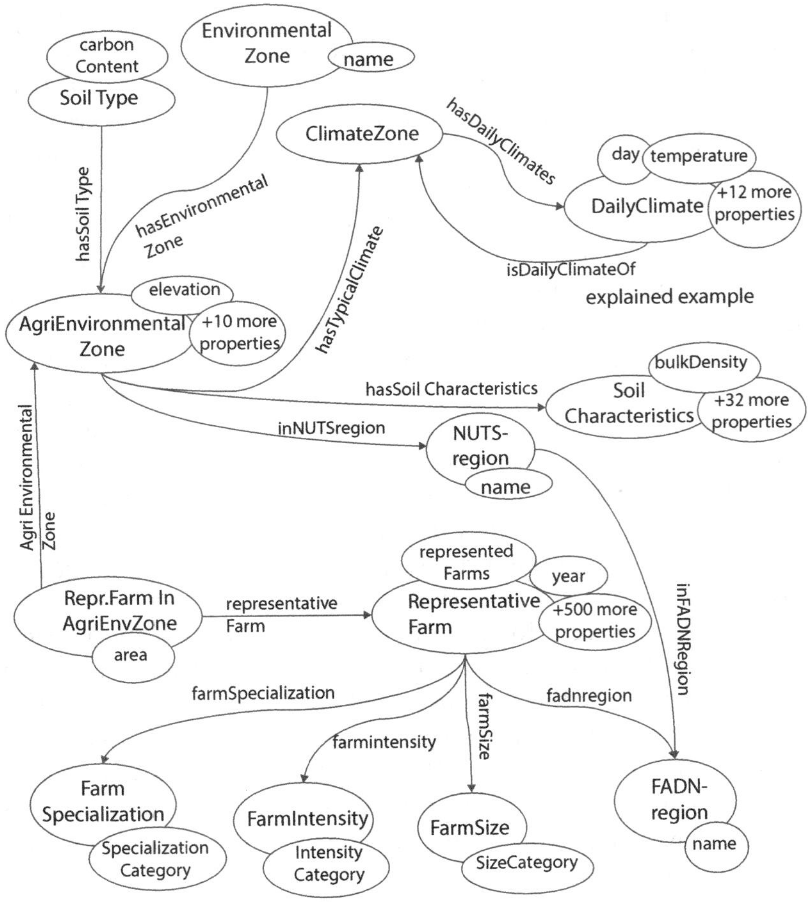

Figure 16.4 A schematic diagram illustrating the interconnections between farms, climate, and soil data specifically pertaining to Europe.

know-how. Agricultural data may also include sensitive or secret information. Data obfuscation procedures are seldom required to allow revelation and safeguarding privacy, even when such information might be secured by not sharing it. Additionally, there exist methods for assuring privacy preservation when doing calculations. The NextGen community must make an investment in standards that safeguard confidentiality and privacy.

Interpreting and visualizing data and model outputs

The inclusion of data visualization tools that facilitate everyday study and engagement with large data is a significant issue for NextGen. When analyzing enormous datasets, the conventional step of loading a file, processing it, and calculating certain scientific parameters may not be effective. In order to uncover findings that are relevant, the analyst may need to attempt a number of procedures and techniques. Problems arise when visualization and analysis are loosely coupled with large data because data transport time might be longer than the time required for data processing. In the field of computer science known as visual analysis, analytical approaches are combined with data administration and visualization. In order to find the anticipated and unearth the unpredicted, to facilitate timely, defendable, and comprehensible evaluations, as well as to effectively communicate those evaluations, visual analytics tools and techniques are used. Algorithms, representations, and large data management are all components of visual analytics. Modern data management and analytics tools that are now being employed doesn't yet fully satisfy the criteria for big data visual analysis, since they are unable to handle enough data sufficiently quickly or in a manner that will aid in understanding. By the utilization of several technologies, such as grid computing, cloud environment, distributed computing, data management and analysis tools have started to intersect. However, these new structures do not sufficiently take into consideration visualization. To integrate visual analytic tools into these architectures, further improvements in data management and analytical calculation will be necessary. Big data analytics have been sped up by high performance computing (HPC) [29]; however, its value for data exploration may be limited by data throughput. Data pre-processing [30] techniques have included principal component analysis and hierarchical clustering implementations.

Model development principles and techniques, creation, composition, and reuse of models

Different perspectives on the interaction between models and software, as well as the designers' institutional contexts and scientific worldviews, all have an impact on how agricultural systems are modeled. In each of these areas, competing viewpoints have existed since the discipline's inception and continue to do so. These various viewpoints lead to various decisions about the level of detail used to describe biophysical processes. Even when adopting a similar scientific viewpoint, a researcher who builds a model independently will adhere to a different model specification and implementation process than a large team working on a formally managed project. The implementation of models by researchers for whom they are engineering artifacts often takes the form of stand-alone programs or modeling frameworks. When the quantity of parts has grown, it becomes easy to put them all together. If it may appear natural and simple to build huge models this way, it is not the case. If a framework is encoded in a coding language, new constraints are added; yet, this knowledge is seldom reflected in the framework's design or application. Software-implemented models often do not state their dependencies or presumptions, leaving it up to modelers to handle integration. In order to help researchers with the technical connecting of frameworks to establish mathematical processes, various attempts have been made to automate aggregation by offering modeling frameworks. A modeling system is a collection of software libraries, classes, and elements that a software developer may put together to create a variety of implementations which employ statistical frameworks to carry out challenging analysis and prediction jobs. Although many modeling systems declare to be domain-independent, a certain discipline is the driving force behind many of their needs. The Common Modeling Protocol, BIOMA, and the Galaxy-dependent FACE-IT are just a few of the frameworks that have been created and utilized in agro-ecosystem modeling by various research groups. There is currently no agreement on how to apply component-level modularity in agro-based frameworks.

Experience with the AgMIP model and SEAMLESS

According to inter-comparisons, and pSIMS, a significant quantity of software for transforming and translating data between various units, formats, and resolutions requires to be written even when a precise modeling system is utilized. In certain circumstances, translation tools are needed to aggregate models and data sources which don't follow generally accepted standards. In other instances, the need for translation tools arises from the varying standards followed by various groups.

For the same model implementation, Lloyd *et al.* (2011) examined four modeling frameworks [31]. A fivefold variance in model dimension and a threefold difference in the use of system-based functions relative to the overall amount of code were found when they looked at the invasiveness of modeling frameworks (i.e., the degree of alteration in model code needed to house a system). These results suggest that framework-specific constraints have a significant influence on framework application and that lightweight systems do actually have a lesser carbon footprint. Models created by other academics are seldom reused by the organizations that initially established modeling frameworks, despite the benefits they were meant to provide in terms of facilitating software development. Similarly, additional problems like model granularity make it difficult to reuse modeling software.

Model fineness

Software for integrated modeling aims to increase model reuse and guarantee the accuracy of outputs. This may be done by striking the correct balance between the projected benefit of component reuse and the invasiveness of the modeling system, as determined by the quantity of code modification required to incorporate a model element in a framework. The granularity of the model components, or the degree to which a bigger entity must be divided into smaller bits, is a crucial consideration in this balancing. Setting the border between one model or sub-model and the next entails choosing the module granularity, which may be a subjective and complicated process.

Most modelers, whether creating their own components or utilizing those created by others, consider the overhead of doing so. The majority of presently deployed agricultural models often include closely connected sub-components (perhaps for better performance), making component replacement a time-consuming procedure that requires extensive code disaggregation and reorganization. Further calibration can also be required.

Finding sub-models that are compatible with both the software interfaces and the scientific criteria of a model becomes more difficult with highly disaggregated components. With greater granularity, there are more linkages. Larger, more sophisticated sub-modules, on the other hand, decrease the likelihood that a reuser may identify an appropriate component because of the more complicated interface. Any modeling system should include integrated model calibration and validation in addition to unit testing of individual components, but this importance increases as more components are shared and may be drawn from a broader selection. There aren't many instances of component reuse across modeling frameworks that are well documented, suggesting that this hasn't occurred much yet. Following a time of trial and error when integrating components across modeling frameworks, best practices for granularity for component design may be created.

Model development process

The open-source movement has established most of the collaborative development approaches. However, using open-source techniques is not a must for collaborative growth; various closed-source elements use comparable project management techniques. The Cathedral and the Bazaar are two prominent project governance models that continue to rule software development in different ways thanks to Raymond's groundbreaking work from 1999. Code is exclusively shared inside the development team according to the Cathedral model, and decisions are made according to a rigid hierarchy of duties. In comparison, the Bazaar approach enables a huge number of developers to donate over the internet having upgrades, revisions, and additions. Almost exclusively Cathedral models of yield, with a single custodian overseeing all the code, have been employed in the creation of agricultural models.

When a solitary entity or a smaller collection of people undertakes control for defining the design of a modeling framework, the Cathedral approach is the most straightforward kind of collaborative development. The approach's key advantage is that a precise description of what defines a certain model at a particular moment is provided. There will just be too many peer players for Cathedral approaches to cooperation to be practical for various components of a NextGen farming modeling framework.

Interactions and infrastructure

Consumer-facing interfaces

Today, most consumer and corporate software is supplied via cloud-hosted services accessible online using a web browser, mainly running on a mobile device or tablet. The method is known as software as a service. User manuals are mostly a thing of the past thanks to Web 2.0 interfaces that are intuitive. As a consequence of these advancements, operational costs are significantly decreased, usability is often improved, and most importantly, the workforce required to build, deploy, and maintain these systems has dramatically increased. In a short period of time, technology enabling information delivery through smartphones and other devices has advanced quickly. Other components of a distributed modeling framework are already in widespread usage, including cloud and web-dependent computation, massive data transfer through the web, and software as a service (SaaS) [31]. For instance, SaaS is being utilized to provide researchers skills for managing research data, publishing data, modeling agriculture, analyzing genomes, and modeling plants. Commercial cloud products from companies like Amazon Web Services, Google, and Microsoft Azure provide strong platform services in addition to on-demand computation and storage. For various areas of cloud computing, standards have been created (such as the Open Cloud Computing Interface and Cloud Data Management Interface).

16.4 Conclusion and Research Plan

For the demands of assignee and on assisting those who create implementations of their frameworks, farming framework modeling must fast acquire and take in cutting-edge data and ICT technologies. This adoption necessitates the broad acceptance of a collection of best operations called standard operating procedures (SOPs). If agricultural models are to serve society to the fullest extent possible, a single focus on one technology will not be enough; concurrent advances in execution would be required via the whole spectrum of ICTs and approaches discussed in this chapter. In order to build efficient computational chains, model developers must first adhere to good coding practices that have long been recognized in software engineering. For instance, with regard to the modeling it, distinctly isolating code which applies model equations from their user interfaces would facilitate chain construction. Simpler model coding is required at the same time. We may allow NextGen framework applications which are substantially lower, in terms of lines of framework-based code, than current frameworks by offering suitable domain-based architectures and functions as libraries. Such advancements will have a significant beneficial effect since the cost of growth and maintenance tends to scale with the number of lines of code. Second, it's imperative that interoperability between data and models become the norm. Intuitive Web 2.0 interfaces should connect every component of the system, and REST (Representational State Transfer) APIs should be connected for programming accessibility. This strategy would enable self-documenting interfaces and applications, and it will make it easier to navigate between and integrate the many modeling and data components. This design is not widely used in agricultural models at the moment. To provide access by any authorized user without the requirement to install local software, software and data should be hosted in the cloud or delivered through services. Third, there is a pressing need to enhance software and data access. To motivate an active group of implementation developers to offer value-added products to end users, the majority of data and modeling elements should theoretically be free in both senses of the word, with lower to no price and fewer to no limitations on usage. The modeling community must facilitate the simple upload and publishing of new data, software, and workflows. To promote software and data exchange, the "publish then filter" strategy should be adhered to. Instead of imposing onerous curation procedures that would unavoidably restrict data and code sharing, feedback mechanisms (like ratings and post-publication reviews) might be offered to discover what is excellent. For providing contributors constructive and quantifiable feedback, mechanisms (such as digital object IDs) must be implemented for citing contributions of data and software and for monitoring accessibility to contributions. Fourth, there has to be widespread advertising of the data integration difficulty. The effectiveness of analytical solutions, such as modeling, is being slowed down by the slicing of data in several domain-based repositories and the relatively limited interoperability between domains and scales. To enable data exchange across disciplines, standard vocabularies and ontologies need to be created as a first step. The crucial byproduct of vocabulary development is that it forces the disciplines to collaborate, shattering the barriers that now prevent integrated modeling from progressing. Data from several dispersed sources may be interpreted by using linked data protocols. Fifth, reference data with quality requirements, must be used to apply NextGen agricultural models. As newer implementations may be quickly evaluated and benchmarked, reference data sets for implementations (at various sizes, areas, and objectives) must be established and released online. Sixth and lastly, employing cutting-edge software design methodologies, user needs of beneficiaries who would profit from implementations of farming framework models need to be studied. This might make it possible to develop a collection of output-presentation elements which are in the public domain well-suited for solving farming issues, and usable in a variety of applications.

In conclusion, to enable farming modeling to embrace the big data age, it is necessary to address the interoperability of data sources, modular granular frameworks, reference data sets for implementations, and particular user needs analysis approaches. This will make it possible to integrate new data sources and have considerably stronger analytical capabilities. We think that a community for NextGen farming frameworks which adheres to these best practices and fulfills the research program is liable to grow significantly and encourage more cooperation both within and across communities. We anticipate it will provide much improved capabilities to assist in ensuring sustainable food production in light of both the current and future climate change.

References

1. V. Veeraiah, K. O. Thejaswini, Dilip R, S. K. Jain, A. Sahu, S. Pramanik, and A. Gupta, The Suggested Use of Big Data in Medical Analytics by Fortis Healthcare Hospital, in Adoption and Use of Technology Tools and Services

by Economically Disadvantaged Communities: Implications for Growth and Sustainability, IGI Global, 2023.

2. S. Ahamad, S. Janani, V. Talukdar, T. Sharma, A. Sahu, S. Pramanik, and A. Gupta, Apollo Hospital's Proposed Use of Big Data Healthcare Analytics, in Big Data Analytics Techniques for Market Intelligence, IGI Global, 2023.
3. Reepu, S. Kumar, M. G. Chaudhary, K. G. Gupta, S. Pramanik and A. Gupta, Information Security and Privacy in IoT, in *Handbook of Research in Advancements in AI and IoT Convergence Technologies*, Eds., J. Zhao, V. V. Kumar, R. Natarajan and T. R. Mahesh, IGI Global, 2023.
4. V. Veeraiah, V. Talukdar, Manikandan K., S. B. Talukdar, V. D. Solavande, S. Pramanik, and A. Gupta, Machine Learning Frameworks in Carpooling, in Handbook of Research on AI and Machine Learning Applications in Customer Support and Analytics, Eds., Md S. Hossain, R. C. Ho and G. Trajkovski, IGI Global, 2023.
5. S. Pramanik, "An Effective Secured Privacy-Protecting Data Aggregation Method in IoT", in *Achieving Full Realization and Mitigating the Challenges of the Internet of Things*, Eds., M. O. Odhiambo and W. Mwashita, IGI Global, 2022, doi: 10.4018/978-1-7998-9312-7.ch008
6. S. Praveenkumar, V. Veeraiah, S. Pramanik, S. M. Basha, A. V. Lira Neto, V. H. C. De Albuquerque and A. Gupta, Prediction of Patients' Incurable Diseases Utilizing Deep Learning Approaches, *ICICC*, 2023, Springer, doi.org/10.1007/978-981-99-3315-0_4
7. R. Bansal, B. Jenipher, V. Nisha, Jain, Makhan R., Dilip, Kumbhkar, S. Pramanik, S. Roy and A. Gupta, "Big Data Architecture for Network Security", in *Cyber Security and Network Security*, Wiley, 2022, https://doi.org/10.1002/9781119812555.ch11
8. D. Pradhan, P. K. Sahu, N. S. Goje, H. Myo Tun, M. M. Ghonge, R. Rajeswari and S. Pramanik, "Security, Privacy, Risk, and Safety Toward 5G Green Network (5G-GN)", in *Cyber Security and Network Security*, Wiley, 2022, doi: 10.1002/9781119812555.ch9
9. A. Mandal, S. Dutta, S. Pramanik, "Machine Intelligence of Pi from Geometrical Figures with Variable Parameters using SCILab", in *Methodologies and Applications of Computational Statistics for Machine Learning*, D. Samanta, R. R. Althar, S. Pramanik and S. Dutta, Eds., IGI Global, 2021, pp. 38-63, doi: 10.4018/978-1-7998-7701-1.ch003
10. Y. Meslie, W. Enbeyle, B. K. Pandey, S. Pramanik, D. Pandey, P. Dadeech, A. Belay, A. Saini, "Machine Intelligence-based Trend Analysis of COVID-19 for Total Daily Confirmed Cases in Asia and Africa", in *Methodologies and Applications of Computational Statistics for Machine Learning*, D. Samanta, R. R. Althar, S. Pramanik and S. Dutta, Eds., IGI Global, 2021, pp. 164-185, doi: 10.4018/978-1-7998-7701-1.ch009
11. P. T. Khanh, T. H. Ngọc, and S. Pramanik, Future of Smart Agriculture Techniques and Applications, in *Advanced Technologies and AI-Equipped IoT Applications in High Tech Agriculture*, Ed., A. Khang, IGI Global, 2023.

12. V. VidyaChellam, V. Veeraiah, A. Khanna, T. H. Sheikh, S. Pramanik and D. Dhabliya, A Machine Vision-based Approach for Tuberculosis Identification in Chest X-Rays Images of Patients, ICICC, 2023, Springer, doi.org/10.1007/978-981-99-3315-0_3
13. T. H. Ngọc, P. T. Khanh and S. Pramanik, Smart Agriculture Using a Soil Monitoring System, in *Advanced Technologies and AI-Equipped IoT Applications in High Tech Agriculture*, Eds., A. Khang, IGI Global, 2023.
14. S. Mehta, V. Kukreja and A. Gupta, "Next-Generation Wheat Disease Monitoring: Leveraging Federated Convolutional Neural Networks for Severity Estimation," *2023 4th International Conference for Emerging Technology (INCET)*, Belgaum, India, 2023, pp. 1-6, doi: 10.1109/INCET57972.2023.10169991.
15. Y. Cai, J. Llorca, A. M. Tulino and A. F. Molisch, "Joint Compute-Caching-Communication Control for Online Data-Intensive Service Delivery," in *IEEE Transactions on Mobile Computing*, doi: 10.1109/TMC.2023.3297598.
16. K. Dushyant, G. Muskan, A. Gupta and S. Pramanik, "Utilizing Machine Learning and Deep Learning in Cyber Security: An Innovative Approach", in *Cyber Security and Digital Forensics*, M. M. Ghonge, S. Pramanik, R. Mangrulkar, D. N. Le, Eds., Wiley, 2022, https://doi.org/10.1002/9781119795667.ch12
17. G. Taviti Naidu, K. V. B. Ganesh, V. VidyaChellam, S. Praveenkumar, D. Dhabliya, S. Pramanik, and A. Gupta, Technological Innovation Driven by Big Data, in *Advanced Bioinspiration Methods for Healthcare Standards, Policies, and Reform*, Hadj Ahmed Bouarara, IGI Global, 2023, doi: 10.4018/978-1-6684-5656-9
18. R. R. Chandan, S. Soni, A. Raj, V. Veeraiah, D. Dhabliya, S. Pramanik, and A. Gupta, Genetic Algorithm and Machine Learning, in *Advanced Bioinspiration Methods for Healthcare Standards, Policies, and Reform*, Hadj Ahmed Bouarara, IGI Global, 2023, doi: 10.4018/978-1-6684-5656-9
19. D. Mondal, A. Ratnaparkhi, A. Deshpande, V. Deshpande, A. P. Kshirsagar and S. Pramanik, Applications, Modern Trends and Challenges of Multiscale Modelling in Smart Cities, in *Data-Driven Mathematical Modeling in Smart Cities*, IGI Global, 2023, doi: 10.4018/978-1-6684-6408-3.ch001
20. B. K. Pandey, D. Pandey, V. K. Nassa, A. S. Hameed, A. S. George, P. Dadheech and S. Pramanik, A Review of Various Text Extraction Algorithms for Images, in *The Impact of Thrust Technologies on Image Processing*, Nova Publishers, 2023.
21. X. Lin, L. Kundu, C. Dick and S. Velayutham, "Embracing Ai in 5g-advanced Toward 6g: A Joint 3GPP and O-RAN Perspective," in *IEEE Communications Standards Magazine*, vol. 7, no. 4, pp. 76-83, December 2023, doi: 10.1109/MCOMSTD.0005.2200070.
22. D. Kaushik, M. Garg, Annu, A. Gupta and S. Pramanik, Application of Machine Learning and Deep Learning in Cyber Security: An Innovative Approach, in *Cybersecurity and Digital Forensics: Challenges and Future*

Trends, M. Ghonge, S. Pramanik, R. Mangrulkar and D. N. Le, Eds., Wiley, 2022, doi: 10.1002/9781119795667.ch12
23. B. K. Pandey, D. Pandey, V. K. Nassa, A. S. George, S. Pramanik and P. Dadheech, Applications for the Text Extraction Method of Complex Degraded Images, in *The Impact of Thrust Technologies on Image Processing*, Nova Publishers, 2023.
24. X. Xue *et al.*, "Computational Experiments for Complex Social Systems—Part III: The Docking of Domain Models," in *IEEE Transactions on Computational Social Systems*, doi: 10.1109/TCSS.2023.3243894.
25. V. Veeraiah, D. J. Shiju, J. V. N. Ramesh, Ganesh Kumar R. S. Pramanik, A. Gupta, and D. Pandey, Healthcare Cloud Services in Image Processing, in *Handbook of Research on Thrust Technologies' Effect on Image Processing*, IGI Global, 2023.
26. X. Liu, Y. Zhang, Z. Zhen, F. Xu, F. Wang and Z. Mi, "Spatio-Temporal Graph Neural Network and Pattern Prediction Based Ultra-Short-Term Power Forecasting of Wind Farm Cluster," in *IEEE Transactions on Industry Applications*, doi: 10.1109/TIA.2023.3321267.
27. S. Pramanik, A New Method for Locating Data Hiding in Image Steganography, *Multimedia Tools and Applications*, doi: 10.1007/s11042-023-16762-3, 2023.
28. S. Pramanik, An Adaptive Image Steganography Approach Depending on Integer Wavelet Transform and Genetic Algorithm, in *Multimedia Tools and Applications*, 2023, https://doi.org/10.1007/s11042-023-14505-y
29. A. D. Andrushia, T. M. Neebha, A. T. Patricia, K. M. Sagayam, and S. Pramanik, Capsule Network based Disease Classification for VitisVinifera Leaves, *Neural Computing and Applications*, doi: 10.1007/s00521-023-09058-y, 2023.
30. V. Veeraiah, D. J. Shiju, J. V. N. Ramesh, Ganesh K. R., S. Pramanik, D. Pandey, and A. Gupta, Healthcare Cloud Services in Image Processing, *Handbook of Research on Thrust Technologies' Effect on Image Processing*, IGI Global, 2023.
31. W. Lloyd, O. David, J. C. Ascough II, K. W. Rojas, J. R. Carlson, G. H. Leavesley, P. Krause, T. R. Green, and L.R. Ahuja, Environmental modeling framework invasiveness: Analysis and implications, *Environmental Modelling & Software*, Vol. 26, Issue 10, October 2011, pp. 1240-1250.

17

Quantifying the Economic Benefits of Agricultural Extension Services: A Mathematical Modeling Analysis

Rajeev Kumar[1], Satendra Kumar[2], Pradeepa P.[3], Akanchha Singh[4], Karun Sanjaya[5] and Ankur Gupta[6*]

[1]*Department of Agriculture, Sanskriti University, Mathura, Uttar Pradesh, India*
[2]*School of Agriculture, Dev Bhoomi Uttarakhand University, Dehradun, Uttrakhand, India*
[3]*Department of Electrical and Electronics Engineering, Faculty of Engineering and Technology, JAIN (Deemed-to-be University), Karnataka, India*
[4]*College of Agriculture Sciences, Teerthanker Mahaveer University, Moradabad, Uttar Pradesh, India*
[5]*Symbiosis Law School, Nagpur Campus, Symbiosis International (Deemed University), Pune, India*
[6]*Department of Computer Science and Engineering, Vaish College of Engineering, Rohtak, Haryana, India*

Abstract
This study proposes concepts for a newer era of farming system frameworks which might satisfy the requirements of an expanding end-user community as demonstrated by a collection of Use Cases. To get sustainable local, regional, and worldwide food security, the authors foresee newer data, models, and cutting-edge technologies that might speed up the innovation process. We list desired attributes for frameworks and discuss a few possible improvements we anticipate for the individual framework elements and their aggregation. We provide an application mechanism which will connect a "pre-competitive" space for framework creation to a "competitive space" for the production of knowledge products and

**Corresponding author*: ankurdujana@gmail.com
Rajeev Kumar: ORCID: https://orcid.org/0009-0009-4143-5776
Pradeepa P.: ORCID: https://orcid.org/0000-0003-3869-1364
Akanchha Singh: ORCID: https://orcid.org/0009-0008-3360-9951

Sabyasachi Pramanik, Niranjanamurthy M., Ankur Gupta and Ahmed J. Obaid (eds.) Mathematical Modeling in Agriculture, (345–372) © 2024 Scrivener Publishing LLC

via public-private associations for newer data architecture. Additional testing and assessment of current frameworks, the creation and testing of modular framework elements and aggregation and connections between framework aggregation plans and newer data administration and realization tools would all provide the foundation for specific model upgrades.

***Keywords*:** Mathematical model, agriculture, data model, economy

17.1 Introduction

A convergence of multiple strong factors has led to the notion of developing a newer era of farming framework data, models, and knowledge elements (NextGen). First, there is growing agreement that a sustainable and better productive farming is necessary to address the local and global problems to food security in the twenty-first century. According to this agreement, it would be beneficial to develop new and experimental instruments that might be utilized to evaluate the sustainability of existing and potential frameworks, create better sustainable frameworks, and control sustainable frameworks. In turn, these various but connected problems generate a need for improvements in data and analytical skills. Second, a new generation of models may be built on an extensive and expanding body of information about the techniques that govern farming frameworks.

The ability to tackle this realization in newer and potent ways to create more yielding and sustainable farming frameworks is made possible by data management, modeling, processing capacity, and information technology. In order to solve the difficulties of global food security in a sustainable manner, we envision accelerating development with the next generation of agricultural systems models. But based on our analysis, the group of farming framework modelers cannot go on as normal in order to be a helpful component of this technique of farming creation. In this chapter, we use the Use Cases proposed in [1], as well as our combined involvements with farming frameworks, data, and modeling, to express the features we believe the newer era of models, data, and knowledge elements requires to carry out this approach. The next generation of models that we envision includes a number of novel features, including links to a variety of knowledge goods that would make it possible to utilize the models and other information products. We refer to these newer frameworks and knowledge elements as "second generation" or "NextGen" [2] since they would mark a significant shift from the present generation of agricultural models [3].

17.2 Creating New Models for the Future: A Demand-Driven, Prospective Strategy

Recognizing that the majority of approach was done by researchers in research or academic organizations and has therefore been more driven by research and academic concerns than user demands is an initial phase to knowing the promise of farming systems frameworks. The process of creating a newer era of frameworks which are intended to satisfy user demands must thus be turned "on its head" by beginning with user requirements and moving backward to the frameworks and data required to find important model outputs.

The NextGen Use Cases described by [4] demonstrate that entire users want whole-farm frameworks, and smallholder farms in developed countries in particular require models that account for interactions among various crops and often animals. The capacity to replicate intercropping or crop-livestock interactions is, however, restricted in many agricultural systems models, which only represent a single crop. Why? One argument is that a lot of models were created in more developed nations where important commodity crops are grown. A different reason is that single crop frameworks are simpler to build, use less computer power, and rely on a smaller data set than crop rotation, inter-crop, or crop-livestock systems models. Scientists are also reacting to sponsor sources which are better inclined to encourage discipline science and scientific organizations that recognize scientific advancements. In a lab or academic context, component techniques within a lone crop or a single profitable consequence are easier to study and may provide more publishable results. In many academic contexts, creating meaningful decision-making tools for agriculturists or governmental decision-takers is at best a minor issue.

For example, to analyze the deals found in efforts to increase the sustainability of farming frameworks, more aggregated agricultural-system frameworks were authorized as being necessary for many years by researchers. To connect econometrically derived financial simulation frameworks with biophysical crop implementation and ecological process frameworks, for instance [5] suggested approaches. [6] examine the many researches which have combined bio-physical and financial frameworks of different sorts for farming-level or landscape-scale analysis. [7] outline a complicated bio-physical agricultural system framework technique. Currently, AgMIP has created software tools that allow the construction of crop and livestock implementation models at the landscape scale, allowing for the connection of these models to economic models and data from farm surveys. Although these examples demonstrate advancements in more thorough, integrated methods to farming framework modeling, these modeling techniques are much complicated and need a lot of data, which poses new issues for both models and users. Modularization is one technique that may allow for increased framework complexity while maintaining reasonably user-friendly and understandable models, as we will describe below. There are other approaches as well, such as adjusting the complexity of the model to the temporal and geographical scales.

Even if it is obvious that framework creation has to be more correlated with user requirements, it is equally critical to acknowledge that science educates shareholders about what can be significant and feasible. Who could have predicted even a long time ago that farming decision support tools would make utilization of information gathered by UAVs connected to simulation models of farming frameworks? So although the creation of models and data must be motivated by user-defined requirements, it should also be forward-looking, utilizing the greatest research and the inventiveness of creative scientists. A model-development open pre-competitive environment connected to a competition for knowledge product growth.

Current frameworks developed from academic agricultural research, as [8] explained in their work on the past evolution of farming systems frameworks. The creation of the models started with research tools that were largely driven by the need to better grasp fundamental techniques and their implications on system performance, despite the fact that there was a notion which "decision support" was crucial. Framework development will continue to be only tangentially related to user demands as long as it is driven largely by academic and research outputs. Therefore, a newer group of institutional structures and incentives is required to reorient model development towards user demands. A possible organizational structure for these novel configurations is shown in Figure 17.1. The diagram illustrates the connections between the "competitive space" of knowledge product growth and the "pre-competitive space" of fundamental research and model development. The healthcare [9] industry's attempts to work together on fundamental research while engaging in product growth gave rise to the idea of "pre-competitive space". The arrows that connect these two "spaces" point in both directions to symbolize the necessary and inevitable exchange of ideas.

The fierce space element depicted in Figure 17.1 is essentially absent from the model creation methodology currently in use. If this competitive space does survive, it is in the private sector where proprietary administrative assistance is offered, and links in Figure 17.1 from the production

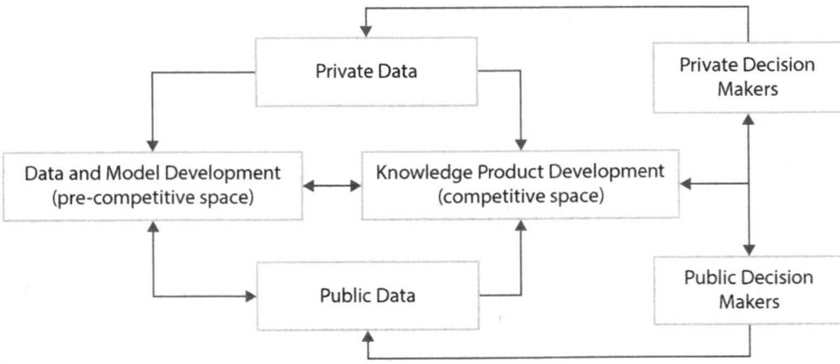

Figure 17.1 Links between the competitive space of knowledge product growth and the pre-competitive space of model and data creation.

of competitive knowledge products back to the generation of data and models are mostly absent. By tying on-farming decision support tools [10] to databases which may be utilized for framework building and analysis, we demonstrate in Figure 17.2 how this connection from private decision makers to frameworks and public data can be formed.

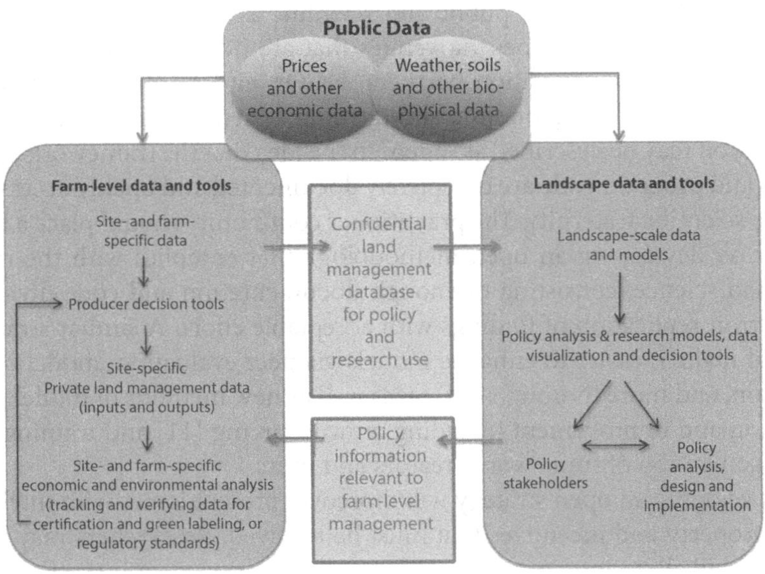

Figure 17.2 Connections between data and decision-making resources at the field and landscape sizes.

The way research institutions do their business will probably need to change and it could be necessary to form public-private partnerships (PPPs) with well-defined responsibilities and bounds for the development of certain NextGen goods. PPPs are one method for research and business to work together to create newer applied knowledge which may support the development of newer products and services. In PPPs, it is typical for both public and commercial partners to contribute financing and collaboratively develop the research issues which will later be addressed by research organizations and academic institutions. PPP structure presents a variety of issues. For instance, PPPs have been controlled in the European Union to prevent unfair competition. According to EU regulations, there must always be multiple private partners involved, and the research partner owns the intellectual property rights to the knowledge evolved (such as tools, frameworks, articles, and approaches), which it may then license to private partners for commercial use.

Openness is crucial for a NextGen community of practice

Here, being open implies encouraging and enlisting others to participate; it also means being prepared to collaboratively determine priorities with a larger stakeholder group (such as research programs, private industry, etc.).

Thirdly, being open to public and scientific assessment of procedures, instruments, and outcomes via venues that are not only scientific and it includes partners, regulation partners, NGOs, and third parties. Only a small number of the economic and farming systems frameworks now in utilization may be described as "open" in the idea that the framework equations and program code are completely documented and openly accessible to the scientific fraternity. The practitioner community should place a high focus on developing an open methodology that complies with the rules of good science, consisting of enough documentation and code dividing to permit replication of findings with acceptable effort. A similar strategy would make it easier to enhance models via peer evaluation, model comparison, and more thorough testing, as well as new methods of model production and improvement including crowdsourcing [11] and training the next generation of framework creators and users.

Developing an open strategy will moreover present issues with intellectual property and incentives that must be resolved. Researchers may now be more likely to join, according to current experience with the Farming Model Inter-contrast and Development Project, a new scientific community committed to an open approach, but it has also highlighted several barriers to participation.

A presiding "super-model" can in time appear, but the only method to understand that this framework is destined to permit a multi-model ecology to prosper. Another benefit of an open technique is that it would promote the exposure of competitive frameworks and modeling techniques, rather than a lone "super-model." As modelers address difficult issues like depiction of heterogeneity and dynamics and links across scales, we anticipate the emergence of alternative methodologies. Trade-offs must be made for models to be tractable, and an open strategy is required to make it easier to try out different solutions.

Recent initiatives to promote a more open technique to farming model evolution include noteworthy examples. Following completion of the primary project-related growth, the bio-economic farming framework FSSIM was released with a license which permitted continued usage and expansion. It is noteworthy that the model's open sourcing was integrated with training sessions; however, this did not result in a large-scale model adoption by the community or the spontaneous collection of the necessary data. With the help of more than 20 engineers, the DSSAT crop modeling fraternity is making an attempt to provide its technology open-source. The model is widely used around the world and has received substantial documentation from the Global Trading and Analysis Project, which also enables users to modify its normal framework. Other researchers may use the freely accessible documentation and the IMPACT framework established by the Worldwide Food Regulation Research Organization. A user-friendly, lucid instrument for effect evaluation was required by prospective users, according to experience, which led to the development of the TOA-MD framework for technology acceptance and sustainability evaluation of farming frameworks. Users have access to testimonial and a self-tutored education curriculum for the TOA-MD model, and there is a developing user community.

Demand-driven model development's objective will be accomplished by the pre-competitive space of framework creation and the competitive space of knowledge product evolution must be strengthened. The present situation seems to be that the modeling fraternity is rigorous in analytical capacity but weaker in its ability to connect to user command, and that the designers of user-based farming-level elements (such as those that provide data from mobile devices) are weaker in analysis. The potential for "gains from trade" via promoting more engagement between the two populations therefore seems to exist. Finding the vital research that might allow improved service delivery to knowledge-product consumers must be a crucial component of this engagement. Further, as shown in the NextGen examples, expanding a larger fraternity which may supply data and analytics for public expenditure and regulation decision-taking provides public good value.

New methods for gathering, managing, and using data

Frameworks modeling at the field, regional, and landscape scales now has new prospects because of the growth in the availability of various types of data as well as the capacity to organize and utilize it. Examples of potential private and public data types are shown in Figure 17.2, which may be created and utilized for both landscape-scale expenditure and regulation research examples and farm-level management Use Cases. In order to support investment decisions and evidence-based policymaking, a few of these data will be collected and utilized at the field level, while others will be produced and utilized for landscape-scale evaluation. While field-level decision-taking [12] and landscape-scale evaluation serve various causes, they altogether rely on two types of data: private data, such as site- and field-based features of the farm effect, as well as site- and field-based administrative agreements, and public data, like costs and different openly obtainable financial data.

Several farm-level data and decision-making tools from both public and commercial sources

There are sources being used right now, and they're changing quickly. The general organization of these tools, the data they utilize as inputs and the results they produce are shown in Figure 17.2. The overall organization of the data and frameworks required for landscape-scale research and policy evaluation is shown in Figure 17.2. The utilization of public data for pricing, weather forecasts, and regulation data, private site and field-based input usage data, and outcome-dependent data which may be used for all field-level administrative options and landscape-scale policy decisions are fundamental characteristics of landscape-scale models. Landscape-scale data fall into three basic categories: publically accessible bio-physical data, like scaled-back climate and soils data, the private site- and field-based data derived from creator- and organization-produced databases [13], and the publicly accessible financial data, including pricing and policy information. In order to offer accurate information about financial and ecological consequences and trade-offs. Most places on the globe often do not have access to such data. Because of this, these models' implementation depends on publicly accessible data on farm management amassed regularly via special-purpose surveys. One significant consequence of the structure shown in Figure 17.2 is the integration of knowledge product planning, farming system frameworks, and data collecting at the farm level. In Section 17.4, we revisit this subject.

A systems perspective

The NextGen examples make it abundantly evident that whole-farm system techniques are required. Agro-ecosystems, also known as managed ecosystems, include biological, physical, and human constituents that operate at a range of sizes (such as the cell, organism and farm). Fields are part of broader ecological and human frameworks that operate at continental, national, and global sizes as well as regional ones (such as watershed and population). However, the need for system-level knowledge [14] should not be interpreted as negating the requirement for component-level tools. In fact, there will be a rising need for instruments to enhance control of soil fertility, pests and diseases, and various aspects of on-farm control, especially in the better specific, industrial frameworks. However, these elements won't be able to meet objectives of sustainable management unless they are included in a larger systems approach. Without a systems approach, for instance, fertilizer and pesticide usage cannot be handled successfully to take into account possible off-farm consequences on water quality.

Multiple significant implications of the systems approach exist for second-generation models. A collection of interdependent subsystems interact at each system level. This shows that models of modular subsystems may be combined to create models of big, complicated systems. Software engineering is impacted by the ongoing important topic of the extent of modularization that may be achievable. For instance, many crops are now modeled separately from animals, as mentioned in [15]. For a variety of Use Cases, systems with many interacting livestock, crops, and crop-livestock interactions are required. This highlights the requirement for these relating elements to be included in a modular "plug and play" framework. These biophysical production framework elements also link with environmental, financial-behavioral, and behavioral system constituents. The links between the subsystems demonstrate the necessity for standardized methods to connect the inputs and outputs of the subsystems. As we said above, a number of increasingly complicated systems models have been created; however, as of right now, each modeling system employs a distinct method for integrating models, making it difficult for model components from various developers to simply interact with one another. The appropriate amount of complexity for Use Cases is a key problem presented by the systems approach, and is further examined. Environmental modeling [16] research shows that the benefits of complexity generally fade with time. The benefit of "minimum data" or "parsimonious" techniques has also been shown in ecological modeling. In order for model makers to move away from creating application-specific models, both modularity and parsimony are necessary. This is especially true for complicated farming systems frameworks and financial frameworks.

[17] also serve to highlight the requirement of a systems approach. Crop rotations have a crucial responsibility in attaining better sustainable control of input-based frameworks since they are crucial for managing soil fertility and soil pests. It's also possible that switching to more varied frameworks which utilize crop rotations and combine crops having livestock will be essential to increase the sustainability of larger-scale frameworks. The commercial-crop examples also show the necessity of conducting evaluations of landscape-scale effects, such as those on water quality (due to chemical and soil runoff and leaching), biodiversity (due to effects on fish and other wildlife), and greenhouse gas discharges (due, for example, to soil management and fertilizer use). Designing and evaluating frameworks which achieve the objectives of "sustainable intensification" and "climate-smart agriculture" need similar sorts of assessment.

Credibility, ambiguity, and improved model

The utility of frameworks for decision-taking is clearly constrained by model credibility, according to both present and prospective model users. By making it simpler for prospective users to comprehend the many applications of frameworks and framework outputs if they are relying on straightly engaging with frameworks and framework output, enhanced knowledge goods might play a significant role in this area. As opposed to this, many end users of framework outputs, as mentioned by [18], are just interested in the information that models can give and have no intention to engage entirely with the frameworks themselves. For instance, a lot of prospective users mistakenly see them as predictive tools and are unaware of their true purpose as exploratory tools used in combination with future situations. Users who employ suitable knowledge products may avoid misusing models and make acceptable interpretations.

Clear messaging and recording of model performance

Improvements in knowledge products that relate data and frameworks to users, as shown in Figure 17.1, might make it easier for modelers to play a significant role in building model credibility. The requirements for a "useful" model vary depending on the application, which might range from research to on-farm control to regulation of decision-taking. For certain scientific applications, high levels of precision can be required, yet for regulation evaluation, the production of data can be more crucial for its timeliness than for its accuracy or precision. Therefore, proper performance criteria are required for model assessment, including criteria for improving modular component performance as well as final framework performance in producing outputs requested by end users.

To evaluate the performance of complex systems models under present and potential future situations, many kinds of formal model assessment methodologies have been developed. While evaluation out of sample also uses future possibilities, which adds extra uncertainty, [19] note that assessment out of sample also uses graphical and qualitative methodologies. [20] present a useful contemporary example utilizing an agent-dependent framework which blends bio-physical and financial components.

By the creation of the Coupled Framework Inter-contrast Project, framework inter-comparison—different strategy for framework enhancement—was used in a variety of scientific domains, consisting of the field of climate frameworks. In order to compare model outcomes, discover significant discrepancies in model outputs, and eventually enhance frameworks and their performance in relation to the aforementioned measurement, procedures for the usage of "reference scenarios" may be established. Some academics believe that the usage of model "ensembles" may be used to describe model uncertainty, albeit this interpretation is debatable. The usage of model ensembles as a technique to enhance the performance of farming systems frameworks has been proven to act superiorly in certain ways than individual frameworks. The ensemble approach's need for a large number of different, independently created models is one of its drawbacks. There are often insufficient diverse models, for instance in economic modeling, to have framework inter-comparison or ensemble techniques helpful. Inter-comparison is difficult in economic models as well because of diverse approaches to measuring model inputs and outcomes. Once again, well-designed information products might make it easier to utilize and comprehend model ensembles, enhancing the proper application and legitimacy of farming systems frameworks.

17.3 Potential Improvements to Model Elements

As stated in the contrasting study by [21], several component changes would be required for NextGen systems frameworks to satisfy critical Use

Cases. Here, we go through relevant cross-cutting concerns and certain discipline-based element advancements which NextGen farming systems frameworks will need for the many primary implementations, including the five Use Cases and others.

Overlapping problems

Data collection and availability is an overarching subject that is covered by [22]. A significant barrier to model development and use is a shortage of adequate data or accessibility to available data. Here, we concentrate on universal issues in model creation. Humans manage agricultural systems for humans. Model design must take into account the goals of the users of the data produced by frameworks and the actions of decision takers whose actions are modeled. The capacity of the majority of models now in use to capture decision-makers' economic or other behavioral incentives is limited. This is a general modeling subject since the farmers' management choices are influenced by the productivity of their crops and livestock, financial costs and rewards, and ecological and social effects. Behavior has to be included into NextGen models in a number of different ways.

If the authors want to design frameworks which offer data on reliability for specific Use Cases or which provides data to field officers to aid decision taking, we must first have a better grasp of the goals of decision takers. Decision-makers would want various types of data if, for instance, yield risks management are a top priority as opposed to if it is not. Notably, the decision-takers' conduct does not require to be represented if the objective is to educate decision-taking rather than to anticipate actions.

Second, the nature of the decision takers must be represented for frameworks which are intended to forecast or project feasible results or repercussions of choices taken by agriculturists. In addition to the complexity that may be modeled for bio-physical manufacturing processes, this necessity adds a significant amount of complexity. A crucial problem that has to be solved in the construction of NextGen models is determining which behavioral models are best applicable to the examples (such as gain expansion, risk control, obtaining social status, or different social or ecological goals).

The social aspect of agricultural decision-taking, especially how social interactions and conventions affect decision-taking, has to be well grasped and reflected in frameworks, according to the third point. Technology adoption, for instance, could be influenced by social learning. Economic research is actively modeling social interactions, but data requirements are large and there aren't any empirical generalizations which can be utilized to frame models yet. While different social researchers moreover investigate social interactions, their approaches are often qualitative and difficult to incorporate into quantitative models.

Illustrating diversity

The enormous amount of biological, physical, financial, and social variability of farming frameworks, in both geography and time, is a significant finding from the growing availableness of field-level data. This point is illustrated by the farms that make up the Use Cases. For example, smallholder maize-dependent fields in Africa have coefficients of variants of key features like field dimension that are in the range of 100% or higher. In the United States, commercial crop farms have coefficients of variation that are moreover substantial, varying from 55% to 155%.

For example, heterogeneity has significant effects on how the authors model farming frameworks. For typologies like small and big farms in an area, "representative farms" have been used in many modeling studies. For certain uses, these typologies may facilitate appropriate approximations of heterogeneity. But data illustrating the variance in site-based circumstances throughout the landscape in the formation of a probability distribution may be necessary for accurate portrayal of bio-physical techniques (e.g., crop development, chemical leaching and erosion). The relevance of variation among farms for the study of erosion and chemical leaching in sharply sloping mountain-side farming was shown by [23] as an extreme example. This also holds accurate for examination of financial distributional concerns like alterations in poverty price, which calls for a framework that can estimate the earning distribution.

Illustrating dynamics

Crop production may be influenced by soil moisture and nourishment, crop rotations, and different system dynamics. Crop development takes place throughout time during the growing season. In order to accurately capture geographical heterogeneity and temporal features, the majority of process-dependent bio-physical implementation frameworks of farming systems are dynamic. The challenge of accurately simulating human conduct in a dynamic environment, however, significantly increases the complexity of systems models. Economic activity is influenced by anticipation for choices are taken in order to produce future results, and knowledge is gathered when realizations are seen and choices are made. Advances in the dynamic are particularly needed in one area.

Although discussing the difficulties in modeling system dynamics would go beyond the scope of this work, it is evident that development in this area is both necessary and difficult. Basic research on next-generation models should focus on how to make this development in a manageable and practical fashion, but it should be driven by the sorts of applications which shareholders anticipate for frameworks and associated knowledge products.

Creating a route and a scenario

Everything that affects an agricultural system cannot be modeled. Since certain components (also known as "drivers" or "exogenous" variables) in models keep on estimates denoted by the modeler or the framework user, models are based on a logical framework. A crucial, little-examined part of modeling is how these drivers are established or altered to reflect the circumstances in which the research is being conducted. The problem is currently getting greater attention in climate research, but the larger modeling development community has to pay it more attention as well. In particular, the user must comprehend the environment in which the evaluation or "simulation experiment" is undertaken and the kinds of impacts which are being assessed if frameworks are to be connected to end users via knowledge products.

Based on the collaborating modeling techniques which were evolved, for example, a collaborative process of route and scenario generation is one approach to address this problem. Stakeholders may contribute to the paths and scenarios of the future, and their involvement also makes it easier for them to apply the results of the models. "Apps" which assist users of climatic consequence evaluations in good comprehending climate assessment scenarios and better interpreting framework findings are an instance of a knowledge product. For this reason, AgMIP is creating an "impacts explorer" as a form of knowledge product. Presently, new methods for scenario building are being investigated that make use of fresh data and cutting-edge computer science techniques like machine learning.

Crop management

To reduce the uncertainty around how crops would react to altering climatic circumstances, including alterations in mean variables and alterations in extremes, a number of agricultural processes need significant advancements in knowledge and modeling capabilities. In order to quickly and effectively translate fresh information from the field to crop frameworks, experimentalists and modelers must collaborate from the start to make sure that the correct research queries are asked as implementations are arranged, crucial field data are aggregated at the proper times, and process-dependent comprehension is noted.

Under the majority of climate change scenarios, extreme weather and climatic occurrences are portrayed to be more recurring and severe, with considerable financial and social implications in farming areas across the globe. Crop models must accurately depict the applicable effects of weather and climatic extremities. This entails a better knowledge of the criteria that define a weather or climatic occurrence as "extreme" for various crops in various places, as well as the modeling methods that need improvement to depict agricultural reactions and their variations to these severe occurrences. Successive durations of production-minimizing climatic controls, such as two or more consecutive years of dry weather, which are often seen in sub-Saharan Africa, may be very harmful. Flooding during a crop season or lengthy periods of record-high temperatures are examples of other severe occurrences.

Genetics

The biological complexity linked with genetic supremacy, ecological and control influences, and links between plant development techniques make it difficult to establish predictive ability which scales from genotype to phenotype. To integrate complex features at the gene network, and entire plant levels, crop model advancements are required. Through connections with model coefficients, phenotypes are connected to alterations in genomic areas. In this issue, [24] talk over how genetic data might be included into models for future crops.

In agricultural locations all around the globe, crops are by then enduring increased levels of CO2 and temperature. It is becoming clearer how increased rates of CO2 and temperature increase will link to influence crop production, but crop models still need to be updated to reflect this new knowledge. Beyond the present emphasis only on nitrogen and water interactions, water related to soils and crops are further significant, and optimizing C and nutrient cycling, and various nourishment interactivities, plays a critical role in sustainable intensification. It is essential to significantly enhance the simulation of each of these processes, as well as how they interact and are managed, particularly when (i) stress is present, and (ii) soil biology-mediated processes are taken into account.

In most models, weeds, diseases, and insect pests are not appropriately or not at all included as production-minimizing variables related to food yield and financial effect. The section in question addresses this significant issue.

Ozone

Although there is evidence that ozone harm to agricultural yield is significant, crop modeling studies seldom take this into account. The effects of ozone on agricultural yields are known, but damage mechanisms and functions still need to be studied. The incorporation of O_3 response functions and contrasts of response functions with process-dependent methods like leaf conduction, and entire canopy conductance parameterizations are examples of model advancements with relation to ozone impacts on crops. To discover a great deal more concerning the AgMIP procedures for sensitive site implemented framework should take into account the various reactions of various crop species and types.

Crops and farming practices

A lone crop-by-crop technique served as the foundation for the field of crop modeling. These newer agricultural framework simulation tools include the complexity which advances with numerous interacting biophysical and socioeconomic elements, signaling the emergence of a new paradigm that advances beyond "crops" to "farming systems". To minimize uncertainty and mathematical dispersion, these field system frameworks have a tendency to emphasize field-scale interactions between system elements while simulating cropping systems with less detail or using more frugal or "minimum-data" approaches. Now that NextGen models have access to more computing power and data, such conflicts between framework complexity and modeling skills may be more effectively resolved. Ex ante testing of transformations, uncertainty propagation linked to framework parameters and structure, and the quantity and varieties of implemented crops (like vegetables and fruits) are all required from the standpoint of crop modeling.

The capacity to implement multi-year crop rotations is now included into many models. Multiple crops may be grown on the same field according to certain models. For instance, [25] created a pea-barley intercropping model depending on STICS which permits an inversion of height dominance across species throughout the crop cycle and a tropic relationship between crop development rate and the possibility of N_2 fixation. The advancement of technology will allow modelers to quickly include multiple crops in farms and numerous crops throughout time as the norm in relation to rotations and intercrops. Using the improved modeling settings, the response of these more complex cropping approaches may then be examined through various sustainable escalation control options. In the same way, inversion studies may be carried out to identify the best cropping systems and management approaches for certain intended results.

Livestock rising

Other enhancements can involve the creation of broad, standardized feed libraries connected to a GEO-WIKI to enhance feed mapping worldwide. These libraries might potentially be used to derive feed quality functions for various agro-ecological situations. Expanding current home data collection techniques to incorporate pertinent livestock data would be one approach to achieve this. More comprehensive typologies of agricultural and animal production systems might be helpful as a first step in tackling heterogeneity. These typologies might be created using current farm household, agro-ecology, farm, rangeland, population, market, and other geographical data. Indicators of intensification, gender dimensions of family labor and control over assets and income, and operation size are required for the mapping of NextGen production systems.

Better geographical data are required, including productivity metrics and standardized feeds that are spatially explicit. These data should ideally be connected to crowdsourcing and massive data rescue projects. It is also necessary to improve farm and regional statistics on production costs for various livestock technologies. Rarely accessible, this data is essential for both regional and worldwide research. With the use of these data, bio-spatial analysis of livestock production gaps might be performed in order to direct investments and find chances to employ livestock as a means of advancing agriculture, reducing poverty, and protecting the environment. Future-looking analyses, such as assessments of the climate impact, need livestock components of future scenarios. In all cases, the cattle industry needs better and more consistent storylines. These plotlines might be created as a part of the global and regional "representative agricultural pathways" that AgMIP and other research teams are currently creating. Currently, only the worldwide "shared socio-economic pathways" used in climate impact assessments include such narratives; for further information.

Rangelands and pastures

All animal production methods depend on pastures and rangelands, which are often tightly incorporated with systems for growing crops (such as pasture in rotation). Although these systems' biophysical components and the data needed to describe them are basically comparable to crop production systems', next-generation models must take into account a number of their characteristics. A more open and data-rich environment would make it easier to examine different long-term dynamical representational strategies, which might help with a number of crucial management and assessment problems for grasslands. One of the major objectives of grassland management is to manage grass swards (as well as desired forb and species) to preserve desirable plants; however, modeling tools have only been somewhat helpful in this regard. In order to understand longer-term changes in species that may affect productivity capacity, vulnerability to degradation, and carbon dynamics (especially woody invasion), models that describe vegetation dynamics are also useful. Awareness of the possible benefits and risks of depending on grassland fodder supplies requires an awareness of the unpredictability from year to year. The development of new models that improve our capacity to predict this danger would be a significant and significant step forward. Improved communication between the groups involved in agricultural modeling and the ecologists researching long-term vegetation dynamics is required.

Grazing animals are the main consumers of fodder supplies, yet most

Only tangential relationships exist between grazers (livestock or animals) and grassland models. The potential of models to contribute to significant developing concerns would be improved with better inclusion of grazing impacts on grasslands, grazing distributions across landscapes, fodder demand/consumption, live-stock/wildlife migration, etc. For instance, it may be challenging to assess holistic grazing management using present modeling frameworks since various management components change in response to a range of distinct indications from the land and assumptions about future circumstances. New data-management feedback loops inside models may be implemented via a system that incorporates user demand into the model development process. Such interactions between information producers and consumers might guide data gathering (for instance, via drone or remote sensing) to simplify model application. Understanding how well cattle consume fodder resources, what is required to support wildlife populations, and how much grassland production may be available for alternative purposes (such as biofuels) all depend on models that better depict grazer-grassland interaction.

Animal illness

Animal illnesses that are highly infectious pose a serious danger to agriculture in both industrialized and developing nations. Diseases may spread purposefully by humans as a kind of bioterrorism, develop from reservoirs of wildlife, or be chronic in populations of cattle. Models are necessary to understand how a disease will spread and to assist decision-makers in creating the best therapies. These models must account for the disease's epidemiology as well as how agricultural operations, particularly the movement of cattle by farmers, impact those behaviors. Significant advancements in this field have been made recently, often drawing on research on human disorders. For instance, it is now feasible to parameterize an epidemiological model using data on cattle migration. In human epidemiology, there are signs of a model comparison movement; cattle disease epidemiology might potentially benefit from this strategy.

New techniques for genetic control

There is a lot of ongoing research being done on innovative genetic insect control techniques. The majority of this research is now focused on the insect vectors of human illnesses like malaria, but the same methodology may also be used to study insect pests that damage crops and, of course, the insect vectors of diseases that affect cattle. The main benefit of these strategies is that they all entail self-sustaining treatments that propagate organically across a pest population, albeit the regulatory concerns related to them are complicated since virtually all of them are categorized as genetically modified. Joint population and genetic dynamic models, many of which are explicitly spatial, are used in this field's cutting-edge modeling work. Over the next several decades, this subject is probably going to be one of the most significant and fascinating modeling applications in agriculture.

Information technology, data management, and precision management

In the next years, integrated agricultural technologies—which are described as the combination of enhanced genetics, agronomic input, information technology, sensors, and intelligent machinery—will be crucial to agriculture. Economic factors, the need to produce more food with less land and water for a growing population, and the desire to save resources to lessen the environmental effect of food production will all be driving forces behind these developments. While many of these technologies have the possibility to be modified to circumstances in other regions of the globe, these developments are now taking place in the commercial-scale industrialized agricultures of the world. Nowadays, practically everywhere on the globe, farmers in rural regions have access to low-cost information regarding pricing thanks to mobile phones. Unmanned aerial vehicles are also anticipated to quickly adapt to local circumstances and be utilized to perform tasks like monitoring crop growth and insect prevalence and enhancing management choices. These technologies are quickly resulting in the automation of numerous production operations, notably the operation of equipment and choices about input application rates, in large-scale, capital-intensive agricultural systems.

Beginning in the middle of the 1990s, agriculture became automated, leading to farmers and agricultural firms having access to a lot of data. Unmanned aerial vehicle (UAV) imaging and yield monitoring may now be combined to create a prescription map that the farmer can follow. Should access be made more available, this private data might potentially be of tremendous use to the scientific community. In certain parts of the globe, farmers now have extremely high resolution historical agricultural production data for their land. Despite these quick improvements in the complexity and automation of farming equipment, a critical component of the puzzle—the analysis of the enormous quantity of newly accessible data to provide the farmer a roadmap of what action to take where and when—remains missing. Farmers now handle the majority of variable rate applications using empirical and rule-of-thumb methods rather than a systems approach that takes into consideration the interplay of the soil, crop, management, and weather. To meet these informational demands, it will be necessary to integrate genetics, agronomic management (production input), weather, soil, information technology, and machines in a systemic manner. It would also affect the larger food system, opening up new data and analytical issues for agricultural system modeling.

The social and economic aspects

It is also necessary to generalize behavioral presumptions and look at how they affect investment and policy analyses. The majority of economic models simply assume profit maximization. On risk modeling, there is a wealth of material that might be used. Models for agricultural systems might take use of recent advances in the literature on expectation development and behavioral economics.

A demand-driven strategy that starts with user-identified outcomes and indicators is required, as stated in Section 17.2. In the context of sustainability, several outcomes are of interest, and users often want to understand trade-offs and synergies across economic, environmental, and social factors. In order to meet a variety of Use Cases, modeling methodologies need to take a few crucial outcomes into account poverty and the distribution of income. The majority of economic models estimate various aspects of farm family income; nevertheless, in order to assess income distribution and poverty in farm household populations, a comprehensive definition of revenue sources is required. In addition to means or averages, population-level outcomes are required, as mentioned in the discussion of heterogeneity in security in nutrition and food. Existing models simulate food production, but none of them fully captures the availability, access, stability, and usage of food at the household and regional levels. Data on food intake at the household and individual levels throughout time is a significant restriction. Mobile data collection techniques are currently being developed. In order to examine concerns with nutritional variety and sustainable diets, it is also required to represent this data in different nutrient currencies than k calories. Health impact modules might be created using earlier research on the health effects of pesticide usage on agricultural workers and other occupational dangers. Big data may be utilized, as it is everywhere, to better understand effects (in this example, data from medical records or insurance claims. Figure 17.3 shows the components of the value chain for the agriculture industry. Landscape-scale

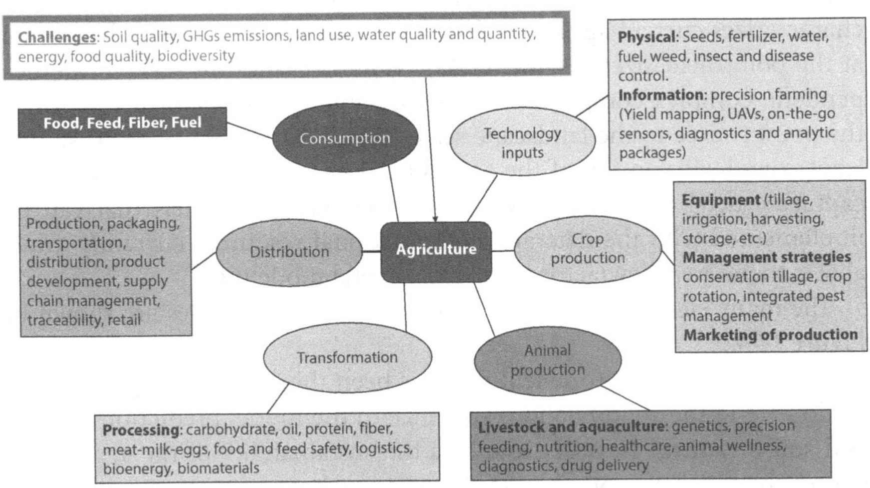

Figure 17.3 Components of the value chain for the agriculture industry.

and global-scale models may be connected to specialized health outcomes models, according to [26]. In terms of relevant measurements, research on many elements of gender effects and outcomes has evolved. With improved data, current farm household models and impact assessment models might be updated to include examination of gender implications related to new technology. The examination of effects by age and health status is in a similar position.

Equity and vulnerability

Different agricultural improvement techniques have clear winners, but they can have unwanted "casual ties" and unfavorable incentives. Understanding these dynamics is crucial from the perspective of development in order to guarantee that the right policies are created to uphold equitable opportunities for all societal segments. For instance, wealthy farmers are often the ones that embrace innovations first. This element may alter power dynamics in marketplaces, which would have an impact on less wealthy farmers. In order to avoid expanding the gap and making the poorer farmers more vulnerable, it is crucial to provide alternative choices and safety nets for them. Moving from single farm models to multi-farm and regional models should help us better understand these dynamics. The capacity to represent distributional effects and susceptibility is being enhanced by methods that use population-based data complexity of the environment and system

A focus on vertical fluxes of energy, water, carbon, nitrogen, and nutrients between the atmosphere, plants, and the soil root zone continuum characterizes current agricultural system models, which generally operate at the point/field sizes (Figure 17.4a). The integration of several interdependent, complex components is necessary for a holistic upscaling from the point source to the landscape scale (Figure 17.4b), which adds significant complexity on top of the agricultural system itself. The best way to capture crucial interactions while also maintaining models that can be implemented with the data and computational resources currently available is therefore a key factor in environmental modeling.

The many elements connecting point to landscape scales are shown in Figure 17.4.

Using a model on crop yield that has been thoroughly compared, "one could extend the validity and confidence in the model prediction of AET and soil moisture". In order to assess the sustainability of the aggregated approach, consisting of impacts on water and soil resources quality and amount as well as ecological estimate, next-generation frameworks must take into account the lateral linkages via the landscape and regional scales.

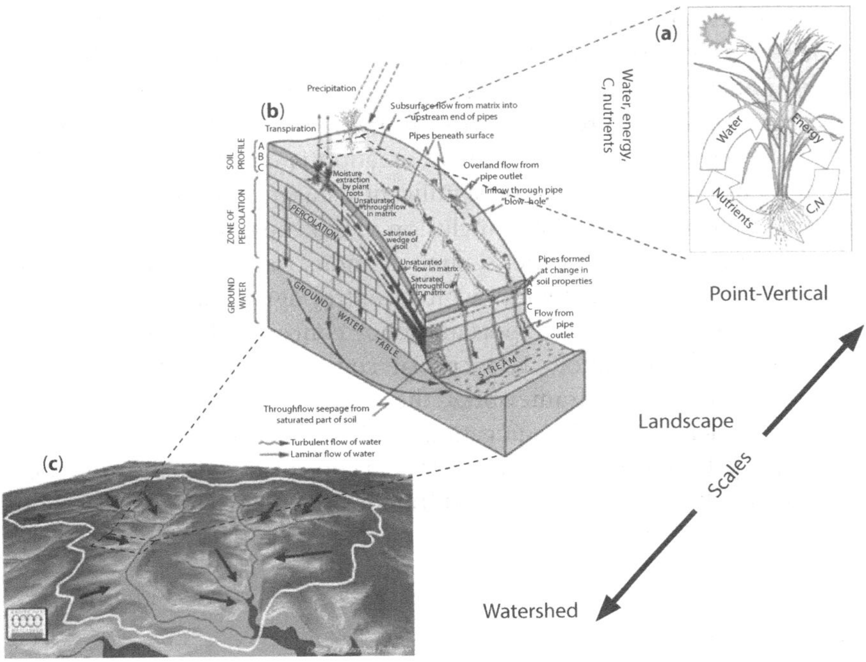

Figure 17.4 Lateral linkages between different scales and environmental elements.

The focus of efforts to develop the next generation of modeling will be on creative ways to make model results more understandable so they can be used in decision-making. The development of model simplification strategies may potentially make use of new techniques for assessing uncertainty. For instance, the use of meta-models and the identification of processes that don't affect certain situations might allow for simplification without compromising outcomes.

17.4 Conclusions

In order to attain sustainable regional and worldwide food security, the authors foresee newer frameworks and knowledge components which might speed up the invention technique. We suggest a user-directed technique to farming system framework growth which will link a cooperative "pre-competitive space" for framework growth and development to a "competitive space" for knowledge product evolution. This approach is based on the evaluation of a group of applications and interactivities with shareholders. Additionally, the authors pinpoint desired features for models as well as some of the future improvements we see for the individual model parts and their interaction. An implementation technique that might connect a pre-competitive framework growth space to a competitive space for knowledge element creation is explored in the closing chapter. This approach calls for significant data improvements as well as multiscale model development. Inclusion of stakeholders via research pertinent to important applications, such as industrialized agricultural systems, smaller-scale frameworks which control agriculture in developed countries, and analysis for important policy concerns like climate change mitigation and adaptation, is a crucial component. Additional testing and analysis of current frameworks, the creation and testing of modular framework elements and aggregation, and connections between framework aggregation platforms and newer data control and visualization tools would all be the foundation for fixed framework enhancements, like those covered in this chapter.

References

1. P. T. Khanh, T. H. Ngọc, and S. Pramanik, Future of Smart Agriculture Techniques and Applications, in *Advanced Technologies and AI-Equipped IoT Applications in High Tech Agriculture*, Ed., A. Khang, IGI Global, 2023.
2. V. Jain, M. Rastogi, J. V. N. Ramesh, A. Chauhan, P. Agarwal, S. Pramanik, and A. Gupta, FinTech and Artificial Intelligence in Relationship Banking and Computer Technology, in *AI, IoT, and Blockchain Breakthroughs in E-Governance*, Eds., K. Saini, A. Mummoorthy, R. Chandrika and N.S. Gowri Ganesh, IGI Global, 2023.
3. V. Veeraiah, V. Talukdar, Manikandan K., S. B. Talukdar, V. D. Solavande, S. Pramanik, A. Gupta, Machine Learning Frameworks in Carpooling, in *Handbook of Research on AI and Machine Learning Applications in Customer Support and Analytics*, Eds., Md S. Hossain, R. C. Ho and G. Trajkovski, IGI Global, 2023.
4. S. Mehta, V. Kukreja and A. Gupta, "Next-Generation Wheat Disease Monitoring: Leveraging Federated Convolutional Neural Networks for Severity Estimation," *2023 4th International Conference for Emerging Technology (INCET)*, Belgaum, India, 2023, pp. 1-6, doi: 10.1109/INCET57972.2023.10169991.
5. A. Makhloufi and A. Kallel, "Inversion of a New Designed ANN-Based 3-D-RTM Emulator by Continuous MCMC Technique to Monitor Crop Biophysical Properties Using Sentinel-2 Images," in *IEEE Transactions on*

Geoscience and Remote Sensing, vol. 61, pp. 1-14, 2023, Art no. 4406714, doi: 10.1109/TGRS.2023.3297363.

6. T. Nikaein, P. Lopez-Dekker, S. C. Steele-Dunne, V. Kumar and M. Huber, “Modeling SAR Observables by Combining a Crop-Growth Model with Machine Learning,” in *IEEE Journal of Selected Topics in Applied Earth Observations and Remote Sensing*, vol. 16, pp. 7763-7776, 2023, doi: 10.1109/JSTARS.2023.3301124.
7. A. B. Potgieter, *et al.*, “Advances in the Study of Biochemical, Morphological and Physiological Traits of Wheat and Sorghum Crops in Australia Using Hyperspectral Data and Machine Learning,” *IGARSS 2023 - 2023 IEEE International Geoscience and Remote Sensing Symposium*, Pasadena, CA, USA, 2023, pp. 1952-1955, doi: 10.1109/IGARSS52108.2023.10282230.
8. T. H. Ngọc, P. T. Khanh and S. Pramanik, Smart Agriculture using a Soil Monitoring System, in *Advanced Technologies and AI-Equipped IoT Applications in High Tech Agriculture*, Ed., A .Khang, IGI Global, 2023.
9. S. Praveenkumar, V. Veeraiah, S. Pramanik, S. M. Basha, A. V. Lira Neto, V. H. C. De Albuquerque and A. Gupta, Prediction of Patients' Incurable Diseases Utilizing Deep Learning Approaches, *ICICC*, 2023, Springer, doi.org/10.1007/978-981-99-3315-0_4
10. V. VidyaChellam, V. Veeraiah, A. Khanna, T. H. Sheikh, S. Pramanik and D. Dhabliya, A Machine Vision-based Approach for Tuberculosis Identification in Chest X-Rays Images of Patients, *ICICC*, 2023, Springer, doi.org/10.1007/978-981-99-3315-0_3
11. B. Bansal, V. Nisha Jenipher, R. Jain, R. Dilip, M. Kumbhkar, S. Pramanik, S. Roy and A. Gupta, “Big Data Architecture for Network Security”, in *Cyber Security and Network Security*, Wiley, 2022, https://doi.org/10.1002/9781119812555.ch11
12. A. Mandal, S. Dutta, and S. Pramanik, “Machine Intelligence of Pi from Geometrical Figures with Variable Parameters using SCILab”, in *Methodologies and Applications of Computational Statistics for Machine Learning*, D. Samanta, R. R. Althar, S. Pramanik and S. Dutta, Eds., IGI Global, 2021, pp. 38-63, doi: 10.4018/978-1-7998-7701-1.ch003
13. K. Dushyant, G. Muskan, A. Gupta and S. Pramanik, “Utilizing Machine Learning and Deep Learning in Cyber Security: An Innovative Approach”, in *Cyber Security and Digital Forensics*, M. M. Ghonge, S. Pramanik, R. Mangrulkar, D. N. Le, Eds., Wiley, 2022, https://doi.org/10.1002/9781119795667.ch12
14. Y. Meslie, W.Enbeyle, B. K.Pandey, S. Pramanik, D.Pandey, P.Dadeech, A. Belay, A. Saini, “Machine Intelligence-based Trend Analysis of COVID-19 for Total Daily Confirmed Cases in Asia and Africa”, in *Methodologies and Applications of Computational Statistics for Machine Learning*, D. Samanta, R. R. Althar, S. Pramanik and S. Dutta, Eds., IGI Global, 2021, pp. 164-185, doi: 10.4018/978-1-7998-7701-1.ch009

15. G. Weikmann *et al.*, "Multiyear Mapping of Water Demand at Crop Level: An End-to-End Workflow Based on High-Resolution Crop Type Maps and Meteorological Data," in *IEEE Journal of Selected Topics in Applied Earth Observations and Remote Sensing*, vol. 16, pp. 6758-6775, 2023, doi: 10.1109/JSTARS.2023.3294107.
16. G. Taviti Naidu, KVB Ganesh, V. VidyaChellam, S Praveenkumar, D. Dhabliya, S. Pramanik, and A. Gupta, Technological Innovation Driven by Big Data, in *Advanced Bioinspiration Methods for Healthcare Standards, Policies, and Reform*, Hadj Ahmed Bouarara, IGI Global, 2023, DOI: 10.4018/978-1-6684-5656-9
17. Ransinghe R.A.D. M, Samarakoon S.M.A. D, Waththaladeniya N. M., Ihalagedara I.H.U. B, Udara Srimath S. Samaratunge Arachchilllage and N. Kodagoda, "Smart System to Optimize Organic Crop Rotation Using Precision Agriculture Data," *2023 7th International Symposium on Multidisciplinary Studies and Innovative Technologies (ISMSIT)*, Ankara, Turkey, 2023, pp. 1-6.
18. G. Karerangabo, E. Armando, O. Gatera and E. Mugisha, "Internet of Things Based Quality and Environmental Monitoring System for Cassava Value Chain," *2023 IEEE AFRICON*, Nairobi, Kenya, 2023, pp. 1-6, doi: 10.1109/AFRICON55910.2023.10293506.
19. W. Zhang *et al.*, "Underwater Image Enhancement via Weighted Wavelet Visual Perception Fusion," in *IEEE Transactions on Circuits and Systems for Video Technology*, doi: 10.1109/TCSVT.2023.3299314.
20. A. Muzirafuti, S. Lanza, D. Paltrinieri, G. Faina, M. El Hafyani and G. Randazzo, "Geomorphology in the Digital Age: A Case Study Project of Quantitative Geomorphology from Images," *2023 IEEE International Workshop on Metrology for the Sea; Learning to Measure Sea Health Parameters (MetroSea)*, La Valletta, Malta, 2023, pp. 513-518, doi: 10.1109/MetroSea58055.2023.10317332.
21. H. Rahmani *et al.*, "Next-Generation IoT Devices: Sustainable Eco-Friendly Manufacturing, Energy Harvesting, and Wireless Connectivity," in *IEEE Journal of Microwaves*, vol. 3, no. 1, pp. 237-255, Jan. 2023, doi: 10.1109/JMW.2022.3228683.
22. R. R. Chandan, S. Soni, A. Raj, V. Veeraiah, D. Dhabliya, S. Pramanik, and A. Gupta, Genetic Algorithm and Machine Learning, in *Advanced Bioinspiration Methods for Healthcare Standards, Policies, and Reform*, Hadj Ahmed Bouarara, IGI Global, 2023, doi: 10.4018/978-1-6684-5656-9
23. Y. -B. Lin, Y. -W. Lin and L. -H. Kao, "Anomaly Detection for Electric Energy Consumption in Smart Farms," in *IEEE Transactions on AgriFood Electronics*, vol. 1, no. 1, pp. 2-14, June 2023, doi: 10.1109/TAFE.2022.3232280.
24. S. Pramanik, R. P. Singh and R. Ghosh, "Application of Bi-orthogonal Wavelet Transform and Genetic Algorithm in Image Steganography", *Multimedia Tools and Applications*, https://doi.org/10.1007/s11042-020-08676-2020, 2020.

25. R. K. Karn and A. Suresh, "Prediction of Crops Based on a Machine Learning Algorithm," *2023 International Conference on Computer Communication and Informatics (ICCCI)*, Coimbatore, India, 2023, pp. 1-8, doi: 10.1109/ICCCI56745.2023.10128446.
26. K. Hu, X. Zhang, D. Lee, D. Xiong, Y. Zhang and X. Gao, "Boundary-Guided and Region-Aware Network with Global Scale-Adaptive for Accurate Segmentation of Breast Tumors in Ultrasound Images," in *IEEE Journal of Biomedical and Health Informatics*, vol. 27, no. 9, pp. 4421-4432, Sept. 2023, doi: 10.1109/JBHI.2023.3285789.

18

Modeling the Impact of Agricultural Investment Incentives on Rural Development: Legal and Economic Perspectives

Dal Chandra[1], Manoj Kumar Mishra[2], Ankit Pant[3], Ahmadi Begum[4], Sukhvinder Singh Dari[5] and Dharmesh Dhabliya[6*]

[1]*College of Law, Teerthanker Mahaveer University, Moradabad, Uttar Pradesh, India*
[2]*Department of Agriculture, Sanskriti University, Mathura, Uttar Pradesh, India*
[3]*School of Agriculture, Dev Bhoomi Uttarakhand University, Dehradun, Uttrakhand, India*
[4]*Department of Basic Science, Faculty of Engineering and Technology, JAIN (Deemed-to-be University), Karnataka, India*
[5]*Symbiosis Law School, Nagpur Campus, Symbiosis International (Deemed University), Pune, India*
[6]*Department of Information Technology, Vishwakarma Institute of Information Technology, Pune, Maharashtra, India*

Abstract

In order to address the growing difficulties associated with agricultural production, it is imperative to enhance our comprehension of the intricate agricultural ecosystems. This may occur via the use of contemporary digital technologies which continually track the physical world, generating substantial quantities of data at a remarkable rate. An examination of this extensive dataset will empower agriculturists and organizations to get worth from it, so enhancing their output. Despite its potential, big data analysis was not extensively used in the agricultural

**Corresponding author*: dharmesh.dhabliya@viit.ac.in
Dal Chandra: ORCID: https://orcid.org/0000-0003-4950-0711
Manoj Kumar Mishra: ORCID: https://orcid.org/0009-0007-7321-9838
Ankit Pant: ORCID: https://orcid.org/0009-0007-8622-4492
Ahmadi Begum: ORCID: https://orcid.org/0000-0001-7727-5940

Sabyasachi Pramanik, Niranjanamurthy M., Ankur Gupta and Ahmed J. Obaid (eds.) Mathematical Modeling in Agriculture, (373–396) © 2024 Scrivener Publishing LLC

sector. The goal is to conduct a comprehensive examination of existing research endeavors in farming which use the contemporary approach of big data analysis, with the purpose of addressing diverse pertinent issues. The paper includes a total of 34 distinct studies, which analyze the specific issue they focus on, offer a solution, describe the tools, algorithms, and data utilized, discuss the type and size of the big data working, evaluate the level of its application, and assess the overall effect. In conclusion, our review emphasizes the major capacity of big data analysis in farming for more intelligent agriculture. It demonstrates that the presence of suitable hardware and software, as well as the development of approaches and techniques for analyzing big data, along with the growth accessibility of big data origins, will likely stimulate further academic researches, public sector activities, and organization advances in the farming industry. This discipline is now at its nascent phase of growth and several obstacles must be overcome.

Keywords: Analysis of large-scale data, agriculture survey, intelligent agriculture

18.1 Introduction

The increase in population, together with socioeconomic issues, has traditionally been linked to a scarcity of food. Over the last five decades, the global population has increased from three billion to a higher figure. There are a greater number of individuals, namely six billion, which leads to an increased need for food. In accordance with estimates from the Food and Farming Agency of the UN, the worldwide census is projected to rise by over 32% by 2055. This necessitates a 72% hike in food yield. Land decay and water pollution, climatic alteration, socio-cultural growth, official regulations, and market changes introduce unreliability to food security. Food security is defined as the availability of enough, safe, and nutritious food for all individuals worldwide. These uncertainties provide a challenge for agriculture to enhance production while simultaneously reducing its environmental impact. Currently, farming is needed for 25% of anthropogenic greenhouse gas emissions [1].

In order to meet these growing requirements, several research projects were initiated from 1995. Recent developments in crop development modeling and production tracking, along with the use of worldwide navigation satellite systems such as GPS [2], have made it possible to accurately determine the exact location of measurements taken in the field. It is possible to produce maps that show the differences in space, which is referred to as "precision agriculture". This notion was introduced by [3]. The use of technology, including remote sensing, cloud computation and IoT, has led to the concept of "smart farming". The implementation of novel ICT used in crop control at the field level expand the notion of precision farming, enhancing the current processes of management and decision taking depending on context, situation, and position recognition. Intelligent agriculture has a crucial responsibility in addressing the difficulties associated with farming yield, including growth, environmental consequence, food security, and sustainability. Sustainable agriculture, as defined by [4] is closely connected to smart farming. It improves the ecological quality and resource foundation on which farming relies, while still meeting essential human food requirements. The concept may be defined as an agricultural strategy that incorporates several scientific disciplines, such as biology, chemistry, physics, ecology, economics, and social sciences, in a holistic manner. Its objective is to create intelligent and secure farming methods.

In order to tackle the difficulties associated with smart farming and sustainable agriculture, it is crucial to enhance the analysis and comprehension of the intricate, multivariate, and unpredictable agricultural ecosystems, as highlighted by [5]. The aforementioned growing digital technologies enhance our comprehension by continuously tracking and calculating many elements of the physical environment, resulting in the rapid generation of massive volumes of data. As noted by [6], this indicates the need for extensive gathering, storage, preprocessing, modeling, and evaluation of vast quantities of data originating from diverse and dissimilar origins. The presence of farming "big data" necessitates significant expenditures in infrastructure for data storage and processing, as discussed by [7], which must function in near real-time for certain implementations such as weather forecast and crop tracking. The aim is to explain a set of modern techniques that have been developed to help agriculturists and linked corporations make financial gains from vast amounts of diverse data. These techniques enable the quick collection, identification, and examination of data.

The usage of big data [8] evaluation is seen to be effective in many areas, including banks, insurance, online user conduct, evaluation and customization, and environmental research. According to [9], governmental institutions usage of big data evaluation to improve their capacity in solving national concerns such as economics, healthcare, job creation, natural catastrophes, and terrorism, in order to better serve their population. Big data analysis has gained success and popularity in other fields, but its application to agriculture is relatively new. The capable benefits of big data analysis in farming have been recognized by stakeholders only lately. According to leading agricultural firms, customizing guidance to farmers via the analysis of large-scale data might potentially improve annual global crop gains by around $22 billion.

The motive for conducting this research is the increasing popularity of big data evaluation in agriculture, and the major potential shown by recent developments and uses of big data in other fields. The existing surveys mostly focus on the theoretical components of this approach, such as the conceptual framework. The primary significance of this survey is in its ability to provide a more concentrated and specific examination of the specific challenges faced in agriculture, as opposed to current surveys that prioritize data analysis and solutions within the world of big data. The study provides an in-depth examination of the data utilized; the methodologies and techniques applied, and offer particular insights from a technical standpoint about the possibilities and benefits of big data evaluation. It also addresses open challenges, impediments, and viable solutions to overcome them.

18.2 Approach

The bibliographic analysis in the specific field of research consisted of three sequential steps: (a) gathering relevant literature, (b) refining the selection to include only pertinent material, and (c) thorough examination and evaluation of the most current and advanced research in the field. A keyword-based search was conducted to get papers and manuscripts from the scientific databases and from scientific indexing services online. The authors excluded studies that discussed big data but did not apply it to the farming field. The study also reviewed previous surveys conducted for relevant research. A total of 1,330 articles were originally identified from this endeavor. By limiting the search to articles written in English and requiring a minimum of two citations, the total amount of papers was decreased to 243. During the second round, we meticulously examined these publications to see whether they met the criteria of practical usage of big data evaluation in many farming applications. The usage of big data evaluation was measured in terms of fulfilling certain attributes known as the five "V" characteristics. Our main focus was on the first three "V"s, namely volume, velocity, and variety.

The quantification of V5 (i.e., truthfulness and valorization) posed further challenges. Out of the 232 articles, only 34 met our criteria. We had to reject a few intriguing attempts that did not meet the criteria in matters related to the used data analysis and the proposed remedies. During the last stage, the 40 articles chosen from the earlier phase were individually examined. This analysis took into account the specific issue each paper addressed, the suggested solution, any quantifiable effect made, as well as the techniques and systems used. The methods used, data sources leveraged, and specific "V" dimensions. They are satisfied with the abundance of big data.

Utilization of big data in the field of agriculture

Five dimensions of big data are as follows:

Volume (V1): The magnitude of data gathered for evaluation.

Velocity (V2): The temporal interval during which data is valuable and pertinent. For instance, a few data must be reviewed within a suitable timeframe in order to accomplish a certain objective, such as identifying pests and animal illnesses.

Variety (V3): The data includes multiple sources such as images, videos, remote sensing data, and farm-dependent sensing data. It is also gathered at various dates and times, and has various spatial resolutions. Additionally, the data comes in various formats, from different origins and areas, and is used in multiple implementation domains.

Veracity (V4): The quality, dependability, and capability of the data, including its correctness, dependability, and final level of confidence.

Valorization (V5): The capacity to disseminate information, valuing, and invention.

While the five "V"s may be used to characterize big data, it is not necessary for big data analysis to satisfy all five dimensions. Big data is often termed for its tendency to be less reliable and stable. Settling the authenticity of V4 another pertinent "V" may be visualization, which refers to the need of displaying intricate data structures and comprehensive data presented in a clear and accessible manner.

As stated above, [10] argue that the concept of big data is not primarily about the amount of data, but rather about the capacity to efficiently find, combine, display, and compare enormous datasets within an acceptable timeframe. Data extraction and analysis refers to the capacity to gather information and get valuable insights from sources that were previously not economically or technically viable to access. The subsequent sections delve into the prominent research endeavors, case studies, and methodologies employed to address agricultural challenges through the utilization of big data analysis. Additionally, the origins of big data are explored, along with the particular approaches involved for analyzing big data.

The authors conducted an assessment of the first three dimensions of big data, namely volume, velocity, and variety, utilizing straightforward indicators categorized as lower (L), medium (M), or higher (H). The estimates were derived via analyzing big data-related information extracted from 34 reviewed publications. The articles were contrasted to establish comparable status, assigning a label of L, M, or H to each "V" dimension of the papers. As stated in Section 2, a total of 34 distinct implementations of big data evaluation in farming were chosen for additional examination. This investigation included the farming region under consideration, the specific issue addressed the remedy and/or effect achieved by the conducted evaluation, the tools/algorithms [11] utilized to solve the issue, and the origins of data utilized. According to the writers, the text provides an assessment of the first three "V"s of big data (e.g., volume, pace, diversity), utilizing just basic indications like lower (L), medium (M), or higher (H). The estimates were derived from the subsequent process: firstly, we documented all the significant data-specific information found in the articles were examined. Subsequently, the authors contrasted these articles to establish relative rankings, assigning a "V" label to each dimension.

The majority of the publications deal with data sets that have moderate to high quantities, together with moderate to low rates of data flow and diversity. The only projects that demand immediate attention are those linked to animals and weeds, due to the urgency of addressing issues related to pests and illnesses. Exceptions include weather and climatic alteration-based initiatives, biodiversity-dependent applications, and agriculturists' decision-taking apps that are distinguished by a significant diverse range of data (taking into account the requirement for varied distinct facts). Remote sensing applications exhibit the greatest magnitude of data. The land-related initiatives, publications on food availableness and security, biodiversity, and soil evaluation have the least velocity. Ultimately, the least diverse range is seen in studies pertaining to weeds, soil, crops, animals, and remote sensing. This suggests that these regions lack the need or ability to access diverse data sources to tackle their own issues.

Mapping, weather estimation, and food security need substantial amounts of data. The prompt execution of tasks such as identifying animal illnesses and addressing plant malnutrition, and making judgments on agriculturists' output, weather forecasts, and food safety and quality, need swift action and the ability to respond in (almost) real time. For these situations, the choices being made have a shorter time frame, hence they need functional or analytical planning rather than long-term calculated planning (i.e., lower speed). Implementations like insurance indices and agriculturists' sustained yield enhancement need diverse data from disparate sources. Certain applications need a high level of data dependability, such as the study of plant and animal illnesses, the culling of herds, farmers' efforts to enhance productivity, and yield projections.

Overall, the progress of big data evaluation in smart agriculture is in its initial stages, as seen by the small amount of scientific research and commercial projects now available. This idea is corroborated by the discoveries enumerated in Figure 18.1, generated by WoS (Thomson Reuters, 2017) using the following search criteria:

"Big Data" AND ["Agriculture" OR "Farm"] retrieving 114 publications (with just a small portion being really applicable) and 120 citations during the past five years. While there is a substantial amount of literature available on big data, the research specifically focused on its use in agriculture seems to be relatively new and restricted. Among study topics using big data evaluation, "agriculture" is ranked 57th.

Origins of big-scale data

Given the conducted research, it is valuable to investigate the origins of the big data sources used by the many publications being studied. These sources come from several origins, including the agriculturists' fields' utilization of big data in many farming applications.

"V" Dimension refers to a particular measurement or aspect. "V" Description provides a detailed explanation or characterization. These terms are often used in the context of agricultural applications that are specific in nature.

V1 Weather forecasting, dairy herd culling identification are examples of applications that deal with high volume and vast volumes of data.

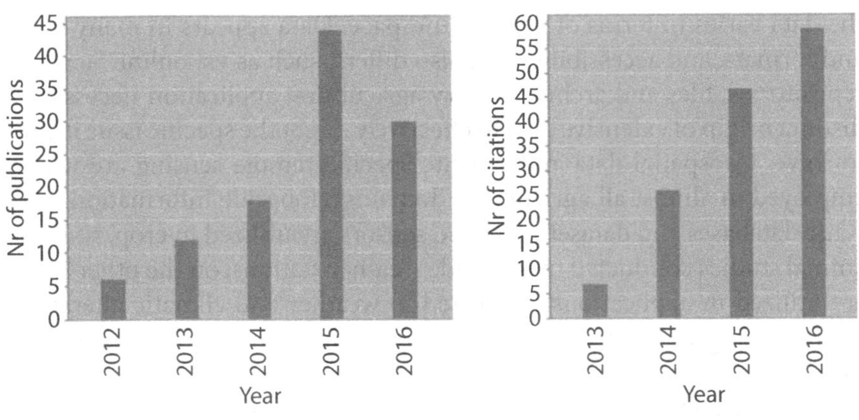

Figure 18.1 The research articles pertaining to "big data in agriculture" are on the left, and the corresponding amount of citations is on the right.

[12] conducted research on improving farmers' productivity. The GSMA (2014) focused on enhancing small farmers' insurance and protection. [13] worked on providing insurance and protection to small farmers. Global Envision (2006) explored financing options for farmers. [14] studied crop production estimations. The RIICE Partnership (2014) estimated food security utilizing remote sensing. [15] classified land usage and land cover changes. [16] focused on data sharing of earth observations.

V2 High velocity refers to situations when real time tends to be crucial and significant.

V3 Higher variety refers to the situation where data is received from diverse and dissimilar sources.

V4 is designed for situations when the precision and dependability of data are of utmost significance.

The applications of advanced technology in various fields include weather forecasting, wine fermentation, ensuring safety and quality of animal food, distinguishing between different types of weeds, identifying animal diseases, enhancing farmers' productivity, facilitating financial transactions for farmers in remote regions, and distributing data on earth observations. The topics covered in the text include estimating availability in poor countries, providing insurance and protection for small farmers, enhancing farmers' productivity, exchanging data on earth observations, and evaluating the viability of animal populations.

Mobile phones can provide valuable data on transportation, plant information, crop yields, weather conditions, and more. Social media can also contribute by reporting natural threats, pests, and illnesses observed in different farms. The sources described above are mainly diverse, and the data varies in terms of volume and pace. Data appears in many kinds and formats, and accessibility to it also differs, such as via online facilities, repositories, files and archives. Every agricultural application necessitates distinct origin of extensive data to effectively direct the specific issue it aims to solve. Geospatial data and satellite-specific remote sensing are widely employed in almost all agricultural domains, alongside information from static databases and datasets. Ground sensors are utilized in crop, soil, and animal studies conducted in the field. Weather stations, on the other hand, are utilized in applications connected to weather and climatic alteration, land mapping, agriculturists' decision taking, insurance and funding. With the exception of a limited number of studies according to [17], the majority of the publications that were reviewed integrate data from many sources to tackle their specific needs.

Methods and instruments for the analysis of big data

The 34 researches examine large datasets using specific techniques or a mix of methodologies, as outlined in [18]. Machine learning is used in 15 journals, cloud-based platforms in nine journals, image processing in eight journals, modeling and simulation in seven journals, statistical evaluation in six journals, and NDVI vegetation indices in six journals. Additionally, some approaches make use of online services such as messaging, online portals, and decision support in five papers, and GIS in four papers. With the exception of soil applications, the other methods use a variety of methodologies, mostly land applications with remote sensing.

The techniques mentioned by [19] are used for prediction, [20] for clustering, and [21] for classification issues. Additionally, Chi *et al.* apply these techniques in the field of image processing.

Sources and methods for analyzing large volumes of data specific to agricultural domains

1. Weather and climate data may be collected by weather stations, questionnaires, and static historical information. Analyzing this data can be done using machine learning techniques such as scalable vector machines and statistical analysis. The topics covered include climatic data, earth observation data, remote sensing using satellites, modeling, cloud platforms, MapReduce analytics, GIS geospatial analysis, and the works of [22].
2. Land remote sensing techniques such as satellites, synthetic aperture radar, and aircraft, together with geospatial analysis, are used in combination with machine learning algorithms such as scalable vector machines, K-means clustering, and random forest. These methods have been studied and documented by researchers such as [23]. The user's text contains references to several types of data, including historical records related to land characterization and crop phenology, rainfall and forests, as well as the use of severely randomized trees. It also mentions the use of NDVI vegetation indices and references to specific studies conducted. The methods used in this study include the use of temperature, elevation, and worldwide tree cover maps, camera sensor-based filtering, image processing, statistical analysis, and

spectral matching. Techniques such as multispectral photography and weather stations are used to perform estimates for reflectance and surface temperature.

3. Research on animals pertains to the study of historical knowledge on the physiological aspects of soils and animals, ground sensors (grazing actions, feed intake, weight, heat, ML (decision trees, neural networks, and scalable vector. The milk output of individual cows may be monitored via sound analysis, while camera sensors can capture multispectral data.
4. Crops may be monitored using ground sensors that measure metabolites, remote sensing techniques using satellites, and historical records. Machine learning algorithms such as scalable vector machines and K-means clustering can be used to analyze the collected data. The study conducted by [24] focuses on the use of wavelet-based filtering and Fourier transform techniques to analyze land usage, national land information, and statistical data on yields. The study also examines the use of NDVI vegetation indicators.
5. The data collected by ground sensors, including salinity, electrical conductivity, and moisture levels, together with camera images, are analyzed using machine learning techniques such as K-means clustering, Farthest First clustering, as described by [25].
6. The methods used for weed detection include remote sensing using airplanes and drones, as well as the utilization of historical data from digital libraries. Machine learning techniques such as neural networks and logistic regression are often used for picture analysis. The processing involves analyzing photos of plants and weeds, as well as plant-based data, using NDVI vegetation indices.
7. Food availability can be assessed using various methods such as surveys, past information, and databases like CIALCA, ENAR, and rice. Analytical techniques like ML with neural networks, statistical analysis, and modeling can also be employed. The datasets include security crop growth datasets, GIS geospatial data, statistical data, remote sensing simulation, network-dependent evaluation, GIS geospatial analysis, and synthetic aperture radar (SAR) processing.

8. The use of Biodiversity GIS geographical data involves the integration of past information and databases, such as the SER database of Statistics, together with the application of Bayesian belief networks as described by [26].
9. The farmers' decision involves utilizing static historical information and datasets, such as those provided by the US government survey. This is done through the use of cloud platforms, web services, mobile applications, and statistical analysis. Relevant sources for this decision-making process include the study by [27] and the Field to Market report from 2015. The methods used in data collection include remote sensing using satellites and drones, as well as the use of weather stations and human input. The data is then utilized for modeling, simulation and big data storage. Additionally, message-based systems are used for data processing. The user is referring to various types of data sources, including sensors, web-based data, GIS geospatial data, and feeds from social media. These data sources are connected through a middleware system.
10. Farmers' security utilizes web-based data, historical information, weather stations, and human expertise to assess risks and provide coverage. This is made possible through the use of cloud platforms, web services, and mobile applications. Notable organizations involved in this field include GSMA (2014), Syngenta (2010), and Global Envision. The use of sensors in agriculture allows for the monitoring of crops, yields, and financial transactions data. This technology was introduced in 2006 by the Syngenta Foundation for Sustainable Agriculture.
11. Remote sensing refers to the use of satellite, aircraft, and drone technology to gather information. This information may be combined with historical data and analyzed using cloud platforms, statistical evaluations, and GIS geospatial evaluation. The datasets mentioned include MODIS surface reflectance datasets, earth land surface processing, NDVI vegetation indices, DSS, and big data. Collection of photos, World Meteorological Organization weather datasets, and reservoir heights are obtained from storage, online and group portals, MapReduce analytics, and mobile devices.

The term "image-based data" [28] refers to data that is derived from pictures captured by cameras or obtained by remote sensing techniques. Image processing includes techniques for Fourier and harmonic analysis, wavelet decomposition, and curve fitting. These tools are used in conjunction with remote sensing and are often integrated, such as using the result of image processing as input for a machine learning model.

In addition, in relation to remote sensing applications, vegetation indices are often employed for crops/soils mapping defined as combinations of surface reflectance at two or more wavelengths chosen to emphasize a specific feature of plants. The most common one is the normalized difference vegetation index (NDVI), which is a graphical indicator used to interpret remote sensing measures, and to judge whether the object being seen has green vegetation or not. Finally, certain online services contain message-oriented middleware, which function in event-based systems when quick notifications and warnings are important, e.g., when there is some natural hazard.

18.3 Conversation

A recent trend is to approximate agricultural problems by utilizing image analysis, employing images obtained through remote sensing. These images can come from either airborne or satellite sources. Recent figures indicate a significant surge in the use of Landsat satellite data since 2008, coinciding with the establishment of the repository. This was made accessible free of charge. Remote sensing has several benefits in the context of agriculture. It is a well-recognized and non-invasive technique for systematically gathering information over vast geographical regions. Remote sensing is currently being used in agriculture to provide operational insurance products, such as crop damage insurance, flood and fire risk assessment, and protection against drought and excess rain.

Traditionally, when it comes to large areas like countries, the main sources of data have been satellites with low resolution but wide coverage, such as AVHRR, MODIS, MERIS, and SPOT-VEGETATION. Some of the relevant agricultural applications found in the research papers include weather forecasting and the characterization and mapping of crops. Conversely, data obtained from high spatial resolution satellites such as Landsat and SPOT have been used to assist with local and regional applications that need enhanced geographical accuracy, such as providing assistance for farmers' decision-making processes. Finally, several sheets use airplanes and drones to accomplish objectives, with a specific emphasis on identifying weeds. These studies have focused on cation analysis and grassland inventories. The integration of remote sensing with additional data sources such as GIS data, historical data, and field sensors greatly enhances the analysis conducted, particularly when it involves predictive tasks such as crop identification or the accurate differentiation of grasslands.

The limited recognition of recent research in the field of big data in smart farming suggests that this area is still in its early stages of development. This observation is consistent with the findings of [29]. However, there is a growing trend of adopting and applying big data in different agricultural domains. The growing quantity of scientifically peer-reviewed articles demonstrates the significant potential of using big data analysis in the field of agriculture. The field has a significant level of dynamism, characterized by a varied network of stakeholders and a continual influx of new participants, mostly in the form of high-tech businesses and start-ups.

Companies such as SenseFly (2012), aWhere Inc. (2015), PEAT UG (2016), and Blue River Technology (2011) are playing a significant role in the analysis and administration of big data in agriculture, both by using current roles and developing new ones. According to [30], high-tech organizations may develop new services and business analytics on a bigger scale and at a faster pace by investing in cloud platforms that provide more storage and computational capabilities. This can lead to the creation of innovative business models.

Unresolved challenges in the use of big data in agriculture

The use of big data analysis in agriculture has not universally yielded advantageous outcomes, as it has also engendered (or is anticipated to engender) some predicaments. Below, we enumerate the difficulties that have been identified and discussed.

Popular software programs used for agricultural big data analysis

Classification: Software tools

1. Image processing tools include the IM toolkit, VTK toolkit, and OpenCV library.
2. Some examples of machine learning (ML) tools are Google TensorFlow, R, Weka, Flavia, scikit-learn, SHOGUN, mlPy, Mlpack, Apache Mahout, Mllib, and Oryx.

3. Cloud-based systems designed for the storage, analysis, and computing of enormous amounts of information. The following are the companies and technologies in the field of big data: Cloudera, EMC Corporation, IBM InfoSphere BigInsights, IBM PureData system for analytics, Aster SQL MapReduce, Pivotal GemFire, Pivotal Greenplum, MapR convergent data platform, Hortonworks, and Apache Pig.
4. Geographic Information Systems (GIS) includes ArcGIS, Autodesk, MapInfo, and MiraMon.
5. The following are examples of large-scale databases: Hive, HadoopDB, MongoDB, ElasticSearch, Apache HAWQ, Google BigTable, Apache HBASE, Cassandra, Rasdaman, MonetDB/SciQL, PostGIS, Oracle GeoRaster, and SciDB.
6. Message-oriented middleware includes MQTT and RabbitMQ.
7. The following modeling and simulation tools are available: AgClimate, GLEAMS, LINTUL, MODAM, and OpenATK.
8. The statistical tools mentioned include Norsys Netica, R, and Weka.
9. Time-series analysis may be conducted using software such as Stata, RATS, MatLab, and BFAST.

From a sociopolitical standpoint, the establishment of major monopolies in the agri-food sector and the farmers' reliance on huge corporations for their agricultural activities may occur. The concentration of big data in the hands of large agri-businesses restricts the potential of this technology, further strengthening the capabilities and commercial benefits of just a few organizations.

The ownership and monetization of data give rise to concerns over privacy. Farmers express apprehension over the possible misapplication of data pertaining to their agricultural endeavors, by seed corporations or rival farms. [31] cautions that hedge funds may use up-to-the-minute data throughout the harvest season from many sources (such as weather data, yield projections, remote sensing, and equipment data) to engage in speculative activities inside commodities markets. The use of big data collecting and analytics has prompted concerns over its security, precision, and accessibility, as examined in the works of [32].

Furthermore, the use of big data varies across industrialized and developing nations, as stated by [33]. The RIICE Partnership (2014) assert that there is a digital divide between established and developing countries, mostly caused by unequal access to technology such as computer power, internet speed, and advanced software. Additionally, the lack of proficient analysts in the developing world exacerbates this divide. Big data in the developing world is characterized by lower volume and less diversity, as shown by the studies conducted by Rodriguez *et al.* (2017) and the articles assessed by [34]. The collecting of big data mostly favors large-scale, highly educated farmers who possess the resources and knowledge to gather it effectively and with precision. Technically, product makers have restricted access to accurate and reliable information, which is a problem.

Obstacles impeding the broader implementation of big data analysis

Prior research has identified several obstacles that impede the broader adoption of big data analysis. These include a shortage of skilled personnel and expertise as well as a scarcity of dependable infrastructures for gathering and analyzing large datasets. [35] emphasize that obtaining precise and useful data necessitates significant technical expertise in data mining and analysis techniques. Additionally, efficient data storage, management, and processing infrastructures are essential for handling complex datasets with multiple modes and high dimensions. These infrastructures should also support real-time processing in critical geospatial applications.

In addition, [36] have highlighted the absence of organization and control in relation to agricultural big data. This issue has also been acknowledged and tackled in certain revised papers, such as [37], which proposes that business models should be appealing to solution providers while also ensuring an equitable distribution of benefits among various stakeholders. [38] argue that the lack of clear semantics in big data makes it difficult for other academics and companies to comprehend and use it. According to the findings of this research, the majority of the emphasis on big data has mostly been on dealing with enormous amounts of data. This includes its applications in areas such as weather and climate change, land identification, farmers' decision-making, insurance and finance, and remote sensing. This has resulted in a distorted and limited viewpoint of the significance of big data for enterprises. In society, the factors of data velocity, diversity, validity, and valorization are all equally significant. This has been highlighted by [39].

In addition, there are still technical obstacles in remote sensing systems for farm management. These include challenges related to the timely collection and delivery of images, errors in sampling and the absence of high spatial resolution data, difficulties in interpreting images and extracting data, and the impact of weather conditions, among others. Common barriers to data usage include the lack of data or incomplete data, as well as data that is not reliable, diverse, or up-to-date. These barriers have been identified and discussed in several research papers

Solving unresolved issues and surmounting obstacles

From a sociopolitical perspective, farmers worldwide have begun to mobilize and organize themselves, for example, via cooperatives and online groups. This has allowed them to enhance their collective knowledge and experiences, as well as their comprehension of big data. According to [40], the data-driven economy has the capacity to provide relevant information for consumers of data ecosystems, including the agricultural sector. Policies pertaining to the management and security of data are necessary in order to promote the democratization of big data and enhance its potential effect and value.

Investments in cloud infrastructures are crucial for the storage, processing, and display of agricultural data on a wide scale. They also assist high-scale and high-speed commercial analytics. The infrastructures should possess a high level of accessibility for those without technological expertise and should not incur significant costs. Techniques such as data aggregation, data minimization, and thorough analysis may enhance user. [41] describe several methods for effectively handling large amounts of diverse data, specifically focusing on the many dimensions of big data. Furthermore, established and widely recognized technologies such as RDF, OWL, SPARQL, and Linked Data, as well as ontologies like AGROVOC, Agricultural Ontology Service, and AgOnt, can be employed as shared terminologies to achieve data interoperability. Additionally, the utilization of open-source software tools and libraries, such as crop type maps and calendars, biophysical measures and vegetation indices, yield models, crop area estimates, and seasonal weather forecasts, would be advantageous. These tools should have the capability to seamlessly integrate with other platforms in order to provide extensive and diverse data analysis, which is a key objective of the GEOSS platform and MERRA Services.

There is a need for more large datasets to be made accessible to the public. This is already a developing trend, as seen by the efforts of organizations like as OADA (2014), GODAN (2015), and AgGateway (2005). In addition, several online organizations have begun to provide diverse and extensive datasets across a broad range of agricultural domains.

Possible domains where big data analysis may be used

This article enumerates prospective domains where big data analysis might be used to tackle diverse agricultural issues in the future. The authors believe that the present research and publications being studied have not sufficiently addressed or completely ignored these topics. These areas include the following possibilities:

1. Platforms facilitate the access of supply chain participants to superior products and processes (RIICE Partnership, 2014), allowing crops to be incorporated into the global supply chain based on global.
2. Given the instances of farmers facing difficulties in selling their harvests due to either excessive supply or not achieving the expected yield (Syngenta Foundation for Sustainable Agriculture, 2016), it is imperative to develop tools that can provide more accurate predictions of crop yield and demand.
3. Offering counsel and direction to farmers on the fertilization requirements of their crops is expected to result in a more suitable approach to the use of fertilizers. This might also be relevant for optimizing the use of herbicides and insecticides.
4. The use of scanning technology in factories, shipping tracking, and retail monitoring of customers' purchases has the potential to improve the traceability of goods across the supply chain. This, in turn, may promote food safety. Ensuring the avoidance of food-borne infections is a crucial endeavor. This is a problem that needs cooperation and financial support from both local and global organizations and governments, in order to guarantee the safety of food. Furthermore, due to the susceptibility of agricultural products to deterioration after harvest, it is crucial to use optimization strategies to decrease losses and increase quality. Current advancements in food processing optimization include the use of (meta-) heuristics and evolutionary algorithms, along with neural networking.
5. Remote sensing is essential for monitoring the effects of different countries and regions on land and crop mapping at a large scale. This is crucial for measuring and attaining their productivity and environmental sustainability goals.

6. Developing more sophisticated and comprehensive scientific models and simulations for environmental phenomena might serve as a foundation for creating platforms that aid policy-makers in making decisions that promote the long-term viability of physical ecosystems.
7. High-throughput screening techniques that can provide precise and accurate quantitative measurement of the interaction between plants and their environment have been developed.
8. Autonomous agricultural robots have the potential to greatly transform agriculture and increase production. These robots can autonomously detect and eliminate weeds, combat pests, and harvest crops, among other tasks.
9. Fully automated and data-intensive enclosed production systems, including greenhouses and other indoor aeroponics systems lighted by LED lights. The emergence of [42] within the circular economy paradigm is characterized by reduced pesticide use, water and nutrient recycling, and increased proximity to the customer.
10. Genome editing, a kind of genetic engineering, enables the modification of a crop or animal's genome at the level of an individual genetic "letter". As [43] suggest that this approach may be more favorable to consumers since it closely replicates the natural mutation process that traditional crop breeding has historically relied on, without involving the creation of genetically modified plants or animals. This technology would complement the current studies in epigenetics.

The mentioned programs, namely #2, #7, #8, and #10, contribute to enhancing security. Additionally, applications #8, #9, and #10 also contribute to improving productivity. The prospective accessibility of this data to the general public has the capacity to generate significant prospects for study and advancement in the direction of more intelligent and environmentally friendly agriculture.

18.4 Conclusion

This research conducted a comprehensive examination of big data analysis in the field of agriculture, primarily focusing on the technical aspects. A total of 34 research publications were identified and assessed. These papers were examined in terms of the issue they addressed, the suggested solution, the tools or methods applied, and the data used. These projects provide information on the current agricultural applications that utilize big data analysis, the specific characteristics of big data used in these different scenarios, the common sources of big data, and the general methods and techniques employed for big data analysis. Identified are unresolved issues, together with obstacles that hinder the broader implementation of big data practices. Several strategies for tackling these issues and reducing obstacles have been considered. As demonstrated in this survey, the accessibility and transparency of hardware and software, methodologies, tools, and approaches for analyzing large amounts of data, along with the growing availability of big data sources and datasets, will likely stimulate more endeavors, projects, and new businesses in the agricultural industry. These initiatives may either tackle existing issues that were addressed in the 34 examples we provided, concentrate on emerging application areas we have identified, or even develop innovative services and products for new agricultural domains. The growing accessibility of large datasets and advanced data analysis methods, which are well-defined by shared meanings and ontologies, along with the adoption of universal protocols, have the capacity to further enhance research and development in the field of intelligent agriculture. This will help tackle the significant task of producing higher-quality food on a larger scale and in a more sustainable manner, while also safeguarding physical ecosystems and preserving natural resources.

References

1. A. Roy and S. Pramanik, A Review of the Hydrogen Fuel Path to Emission Reduction in the Surface Transport Industry, *International Journal of Hydrogen Energy*, https://doi.org/10.1016/j.ijhydene.2023.07.010, 2023
2. R. Jayasingh, J. Kumar R.J.S, D. B. Telagathoti, K. M. Sagayam and S. Pramanik, "Speckle noise removal by SORAMA segmentation in Digital Image Processing to facilitate precise robotic surgery", *International Journal of Reliable and Quality E-Healthcare*, vol. 11, issue 1, doi: 10.4018/IJRQEH.295083, 2022
3. P. T. Khanh, T. H. Ngọc, and S. Pramanik, Future of Smart Agriculture Techniques and Applications, in *Advanced Technologies and AI-Equipped IoT Applications in High Tech Agriculture*, Ed., A. Khang, IGI Global, 2023.
4. A. A. AlZubi and K. Galyna, "Artificial Intelligence and Internet of Things for Sustainable Farming and Smart Agriculture," in *IEEE Access*, vol. 11, pp. 78686-78692, 2023, doi: 10.1109/ACCESS.2023.3298215.

5. N. Mangra; F. Behmann; A. Thakur; A. Popescu; G. Suciu Jr; G. Giannattasio; R. Uppal; and W. Montlouis, "5G Enabled Agriculture Ecosystem: Food Supply Chain, Rural Development, and Climate Resiliency," in *5G Enabled Agriculture Ecosystem: Food Supply Chain, Rural Development, and Climate Resiliency*, vol., no., pp.1-40, 30 Jan. 2023.
6. T. H. Ngọc, P. T. Khanh and S. Pramanik, Smart Agriculture using a Soil Monitoring System, in *Advanced Technologies and AI-Equipped IoT Applications in High Tech Agriculture*, Ed., A. Khang, IGI Global, 2023.
7. Reepu, S. Kumar, M. G. Chaudhary, K. G. Gupta, S. Pramanik and A. Gupta, Information Security and Privacy in IoT, in *Handbook of Research in Advancements in AI and IoT Convergence Technologies*, Eds., J. Zhao, V. V. Kumar, R. Natarajan and T. R. Mahesh, IGI Global, 2023.
8. V. Veeraiah, V. Talukdar, Manikandan K., S. B. Talukdar, V. D. Solavande, S. Pramanik, and A. Gupta, Machine Learning Frameworks in Carpooling, in *Handbook of Research on AI and Machine Learning Applications in Customer Support and Analytics*, Eds., Md S. Hossain, R. C. Ho and G. Trajkovski, IGI Global, 2023.
9. N. Manikandan, S. N. Tadiboina, M. S. Khan, V. P. R., R. Singh and K. K. Gupta, "Automation of Smart Home for the Wellbeing of Elders Using Empirical Big Data Analysis," *2023 3rd International Conference on Advance Computing and Innovative Technologies in Engineering (ICACITE)*, Greater Noida, India, 2023, pp. 1164-1168, doi: 10.1109/ICACITE57410.2023.10183103.
10. S. Pramanik and S. Bandyopadhyay, Analysis of Big Data, in *Encyclopedia of Data Science and Machine Learning*, Ed. John Wang, IGI Global, 2023, doi: 10.4018/978-1-7998-9220-5.ch006
11. K. Dushyant, G. Muskan, A. Gupta and S. Pramanik, "Utilizing Machine Learning and Deep Learning in Cyber Security: An Innovative Approach", in *Cyber Security and Digital Forensics*, M. M. Ghonge, S. Pramanik, R. Mangrulkar, D. N. Le, Eds., Wiley, 2022, https://doi.org/10.1002/9781119795667.ch12
12. A. A. AlZubi and K. Galyna, "Artificial Intelligence and Internet of Things for Sustainable Farming and Smart Agriculture," in *IEEE Access*, vol. 11, pp. 78686-78692, 2023, doi: 10.1109/ACCESS.2023.3298215.
13. I. A. Omar, R. Jayaraman, K. Salah, H. R. Hasan, J. Antony and M. Omar, "Blockchain-Based Approach for Crop Index Insurance in Agricultural Supply Chain," in *IEEE Access*, vol. 11, pp. 118660-118675, 2023, doi: 10.1109/ACCESS.2023.3327286.
14. A. Mateo-Sanchis, J. E. Adsuara, M. Piles, J. Munoz-Marí, A. Perez-Suay and G. Camps-Valls, "Interpretable Long Short-Term Memory Networks for Crop Yield Estimation," in *IEEE Geoscience and Remote Sensing Letters*, vol. 20, pp. 1-5, 2023, Art no. 2501105, doi: 10.1109/LGRS.2023.3244064.
15. Q. Liu, X. Meng, X. Li and F. Shao, "Detail Injection-Based Spatio-Temporal Fusion for Remote Sensing Images with Land Cover Changes," in *IEEE*

Transactions on Geoscience and Remote Sensing, vol. 61, pp. 1-14, 2023, Art no. 5401514, doi: 10.1109/TGRS.2023.3252054.

16. B. K. Pandey, D. Pandey, V. K. Nassa, A. S. George, S. Pramanik and P. Dadheech, Applications for the Text Extraction Method of Complex Degraded Images, in *The Impact of Thrust Technologies on Image Processing*, Nova Publishers, 2023.
17. V. Talukdar, A. Z. Kurniullah, P. Keshwani, H. Khan, S. Pramanik, A. Gupta, D. Pandey, Load Balancing Techniques in Cloud Computing, *Emerging Trends in Cloud Computing Analytics, Scalability, and Service Models*, IGI Global, 2023.
18. D. Mondal, A. Ratnaparkhi, A. Deshpande, V. Deshpande, A. P. Kshirsagar and S. Pramanik, Applications, Modern Trends and Challenges of Multiscale Modelling in Smart Cities, in *Data-Driven Mathematical Modeling in Smart Cities*, IGI Global, 2023, doi: 10.4018/978-1-6684-6408-3.ch001
19. S. Pramanik, "Carpooling Solutions Using Machine Learning Tools", in *Handbook of Research on Evolving Designs and Innovation in ICT and Intelligent Systems for Real-World Applications*, K. K. Sarma, N. Saikia and M. Sharma, IGI Global, 2022, doi: 10.4018/978-1-7998-9795-8.ch002.
20. M. Sinha, E. Chacko, P. Makhija and S. Pramanik, Energy Efficient Smart Cities with Green IoT, in *Green Technological Innovation for Sustainable Smart Societies: Post Pandemic Era*, C. Chakrabarty, Ed., Springer, 2021, doi: 10.1007/978-3-030-73295-0_16
21. R. R. Chandan, S. Soni, A. Raj, V. Veeraiah, D. Dhabliya, S. Pramanik, A. Gupta, Genetic Algorithm and Machine Learning, in *Advanced Bioinspiration Methods for Healthcare Standards, Policies, and Reform*, Hadj Ahmed Bouarara, IGI Global, 2023, doi: 10.4018/978-1-6684-5656-9
22. D. Kaushik, M. Garg, Annu, A. Gupta and S. Pramanik, Application of Machine Learning and Deep Learning in Cyber Security: An Innovative Approach, in *Cybersecurity and Digital Forensics: Challenges and Future Trends*, M. Ghonge, S. Pramanik, R. Mangrulkar and D. N. Le, Eds., Wiley, 2022, doi: 10.1002/9781119795667.ch12
23. A. Gupta, A. Verma and S. Pramanik, Security Aspects in Advanced Image Processing Techniques for COVID-19, in *An Interdisciplinary Approach to Modern Network Security*, S. Pramanik, A. Sharma, S. Bhatia and D. N. Le, Eds., CRC Press, 2022.
24. S. Pramanik, R. P. Singh and R. Ghosh, "Application of Bi-orthogonal Wavelet Transform and Genetic Algorithm in Image Steganography", *Multimedia Tools and Applications*, https://doi.org/10.1007/s11042-020-08676-2020, 2020.
25. S. Pramanik, K. M. Sagayam and O. P. Jena, Machine Learning Frameworks in Cancer Detection, *ICCSRE*, 2021, Morocco, 2021.
26. S. Pramanik, M. G. Galety, D. Samanta, N. P Joseph, Data Mining Approaches for Decision Support Systems, *3rd International Conference on Emerging Technologies in Data Mining and Information Security*, 2022.

27. V. VidyaChellam, V. Veeraiah, A. Khanna, T. H. Sheikh, S. Pramanik and D. Dhabliya, A Machine Vision-based Approach for Tuberculosis Identification in Chest X-Rays Images of Patients, *ICICC*, 2023, Springer, doi.org/10.1007/978-981-99-3315-0_3
28. S. Praveenkumar, V. Veeraiah, S. Pramanik, S. M. Basha, A. V. Lira Neto, V. H. C. De Albuquerque and A. Gupta, Prediction of Patients' Incurable Diseases Utilizing Deep Learning Approaches, *ICICC*, 2023, Springer, doi.org/10.1007/978-981-99-3315-0_4
29. S. Pramanik, "An Effective Secured Privacy-Protecting Data Aggregation Method in IoT", in *Achieving Full Realization and Mitigating the Challenges of the Internet of Things*, Eds., M. O. Odhiambo and W. Mwashita, IGI Global, 2022, doi: 10.4018/978-1-7998-9312-7.ch008
30. R. Bansal, A. J. Obaid, A. Gupta, R. Singh, S. Pramanik "Impact of Big Data on Digital Transformation in 5G Era", *2nd International Conference on Physics and Applied Sciences (ICPAS 2021)*, doi:10.1088/1742-6596/1963/1/012170, 2021.
31. N. Lumoring, D. Chandra and A. A. S. Gunawan, "A Systematic Literature Review: Forecasting Stock Price Using Machine Learning Approach," *2023 International Conference on Data Science and Its Applications (ICoDSA)*, Bandung, Indonesia, 2023, pp. 129-133, doi: 10.1109/ICoDSA58501.2023.10277318.
32. V. Veeraiah, K. O. Thejaswini, R. Dilip, S. K. Jain, A. Sahu, S. Pramanik, and A. Gupta, The Suggested Use of Big Data in Medical Analytics by Fortis Healthcare Hospital, in *Adoption and Use of Technology Tools and Services by Economically Disadvantaged Communities: Implications for Growth and Sustainability*, IGI Global, 2023.
33. S. Ahamad, S. Janani, V. Talukdar, T. Sharma, A. Sahu, S. Pramanik, and A. Gupta, Apollo Hospital's Proposed Use of Big Data Healthcare Analytics, in *Big Data Analytics Techniques for Market Intelligence*, IGI Global, 2023.
34. R. Bansal, B. Jenipher, V. Nisha, Jain, Makhan R., Dilip, Kumbhkar, S. Pramanik, S. Roy and A. Gupta, "Big Data Architecture for Network Security", in *Cyber Security and Network Security*, Wiley, 2022, https://doi.org/10.1002/9781119812555.ch11
35. A. Bhattacharya, A. Ghosal, A. J. Obaid, S. Krit, V. K. Shukla, K. Mandal and S. Pramanik, "Unsupervised Summarization Approach with Computational Statistics of Microblog Data", in *Methodologies and Applications of Computational Statistics for Machine Learning*, D. Samanta, R. R. Althar, S. Pramanik and S. Dutta, Eds, IGI Global, 2021, pp. 23-37, doi: 10.4018/978-1-7998-7701-1.ch002
36. N. P. Nagendra, G. Narayanamurthy, R. Moser, E. Hartmann and T. Sengupta, "Technology Assessment Using Satellite Big Data Analytics for India's Agri-Insurance Sector," in *IEEE Transactions on Engineering Management*, vol. 70, no. 3, pp. 1099-1113, March 2023, doi: 10.1109/TEM.2022.3159451.

37. A. Mandal, S. Dutta, S. Pramanik, "Machine Intelligence of Pi from Geometrical Figures with Variable Parameters using SCILab", in *Methodologies and Applications of Computational Statistics for Machine Learning*, D. Samanta, R. R. Althar, S. Pramanik and S. Dutta, Eds., IGI Global, 2021, pp. 38-63, doi: 10.4018/978-1-7998-7701-1.ch003
38. P. K. Mall, S. Pramanik, S. Srivastava, M. Faiz, S. Sriramulu and M. N. Kumar, FuzzyNet-Based Modelling Smart Traffic System in Smart Cities Using Deep Learning Models, in *Data-Driven Mathematical Modeling in Smart Cities*, IGI Global, 2023, doi: 10.4018/978-1-6684-6408-3.ch005
39. A. Namdev, V. Veeraiah, S. Dhamodaran, S. Islam, S.Pramanik, and A. Gupta, Cloud Computing Applications in Biomedicine, in *Emerging Trends in Cloud Computing Analytics, Scalability, and Service Models*, IGI Global, 2023.
40. Dhamodaran S, S. Ahamad, J.V.N. Ramesh, S. Sathappan, A. Namdev, R. R. Kanse, S. Pramanik, Fire Detection System Utilizing an Aggregate Technique in UAV and Cloud Computing, in *Handbook of Research on Thrust Technologies' Effect on Image Processing*, IGI Global, 2023.
41. V. Veeraiah, D. J. Shiju, J. V. N. Ramesh, Ganesh K. R., S. Pramanik, D. Pandey, and A, Gupta, Healthcare Cloud Services in Image Processing, in *Handbook of Research on Thrust Technologies' Effect on Image Processing*, IGI Global, 2023.
42. S. Ghosh, R. D. Raut, M. C. Mandal and A. Ray, "Are the Small- and Medium-Sized Enterprises in Developing Countries Ready to Implement Circular Economy Practices? An Empirical Quest for Key Readiness Factors," in *IEEE Transactions on Engineering Management*, doi: 10.1109/TEM.2023.3257237.
43. S. Balasubramaniam, S. Somathilaka, S. Sun, A. Ratwatte and M. Pierobon, "Realizing Molecular Machine Learning through Communications for Biological AI," in *IEEE Nanotechnology Magazine*, vol. 17, no. 3, pp. 10-20, June 2023, doi: 10.1109/MNANO.2023.3262099.

19

Optimizing Harvest Scheduling in Agriculture: A Mathematical Modeling Approach Considering Legal Restrictions

Heejeebu Shanmukha Viswanath[1], Umesh Kumar Tripathi[2], Minnu Sasi[3], Kishore Kumar Pedapenki[4], Prashant Dhage[5] and Ankur Gupta[6*]

[1]*College of Agriculture Sciences, Teerthanker Mahaveer University, Moradabad, Uttar Pradesh, India*
[2]*Department of Agriculture, Sanskriti University, Mathura, Uttar Pradesh, India*
[3]*School of Agriculture, Dev Bhoomi Uttarakhand University, Dehradun, Uttrakhand, India*
[4]*Department of Electrical and Electronics Engineering, Faculty of Engineering and Technology, JAIN (Deemed-to-be University), Karnataka, India*
[5]*Symbiosis Law School, Nagpur Campus, Symbiosis International (Deemed University), Pune, India*
[6]*Department of Computer Science and Engineering, Vaish College of Engineering, Rohtak, Haryana, India*

Abstract
The inclusion and adaption of the most recent technology are justified by the agriculture sector's ongoing progress. These days, technologies dependent on the Internet facilitate crop management feasible. Their usage facilitates the making use of data and the design of stand-alone applications that although successful, made it difficult to have scaled projections throughout the time of a field's functional life. In order to achieve better management of agricultural plots in irrigation communities and concurrently monitor crop demands, a data model was developed to solve this issue. As a result, the goal of this research is to develop

**Corresponding author*: ankurdujana@gmail.com
Heejeebu Shanmukha Viswanath: ORCID: https://orcid.org/0009-0006-0347-0712
Umesh Kumar Tripathi: ORCID: https://orcid.org/0009-0005-9467-5547
Minnu Sasi: ORCID: https://orcid.org/0000-0002-0176-7890
Kishore Kumar Pedapenki: ORCID: https://orcid.org/0000-0002-4192-5101

Sabyasachi Pramanik, Niranjanamurthy M., Ankur Gupta and Ahmed J. Obaid (eds.) Mathematical Modeling in Agriculture, (397–420) © 2024 Scrivener Publishing LLC

an open and interoperable platform depending on industry-quality interfaces and protocols which would permit for the aggregation of diverse information origins while ensuring compatibility with different third-party remedies for exchanging and utilizing such information. The platform is built on open, standardized interfaces and protocols that facilitate the unification of all data into a single data model for improved information usage and distribution. The program provided farmers linked to the platform's platform with better water irrigation management thanks to the system's complete instantiation in a genuine prototype irrigation community.

Keywords: IoT, data model, precision irrigation, smart agriculture, irrigation

19.1 Initialization

Uncertainty and ongoing technological change define society. Productivity tends to increase more quickly in more computerized industries than in less digitized industries. For many years, irrigation agricultural communities have taken advantage from the usage of information and communication technology (ICT) [1] that enables them to advance the state of agriculture [2], make it easier to complete daily activities, and transform fields into productive and sustainable frameworks. By providing irrigation groups and different agricultural organizations with tools for analytics and estimation, these newer technologies not only allow the digital transformation technique required by the industry but also improve optimization.

Aerial images [3], big data [4], cloud computation [5], AI [6], and the Internet of Things (IoT) [7], to name a few, are examples of the current technological advancements and the digital transformation that are giving the agricultural industry newer tools to aid assess the true requirements of fields and boost their productivity. In particular, the introduction of IoT has altered the conventional model for controlling access to and administration of sensors and actuators, facilitating everything online-accessible and integrating and managing information and real-world data into the digital environment. Improvements in this alteration over previously used methods have made it possible for new farming management tools to be developed. These benefits mostly center on the following:

- Problem recognition: The application of newer technologies like satellite images [8], application algorithms [9], drones [10], high-tech sensors [11], mobile applications [12], and GPS [13] that enable the evaluation of crop status and the estimation of issues (inappropriate fertilizer usage, water stress, altering weather situations, and pest controlling) before they begin to affect crop production.

- Increased yield: The sensorization of crops enables the design of planting and fertilization schedules based on elements like seed kind and soil conditions, which increase agricultural output levels.
- Better decision-making: It is feasible to determine which parts of land need more water by evaluating and monitoring agronomic factors, crop water requirements, and precipitation projections. Then, by scheduling irrigation, it is able to modify the amount of water delivered without putting the crop under stress.
- Behavior analysis: The most advantageous activities and those that need to be deleted or adjusted may be determined as a consequence of continual crop tracking and the examination of the links between various components, like production, energy efficiency, and the farming [14] techniques utilized.

The farming industry is supplying its fields with newer technology (sensors, actuators, weather data, drones, and satellite images) as a result of the digital transformation in order to optimize [15] resources, increase productivity, and simultaneously lessen environmental impact. A standard integration architecture that unifies the complete dataset is required for the digital transformation of the farming industry in order to provide new services that are tailored to the sector's requirements. A special semantic model which arranges and assesses every data gathered must include the data produced by the farms since they are so important.

The Next Generation Service Interface (NGSI), promoted by the Open Mobile Alliance (OMA), for Context Information Management (ETSI ISG CIM), provides a framework for defining a global set of standards depending on the context at the European level. This framework is provided by the open initiative FIWARE. As a result, following the standards and giving up proprietary technology and code may help the agriculture industry become more competitive. Isolated fixes are inappropriate in today's interconnected society. Contrarily, interoperability must be the core characteristic all solutions embrace in order to promote a more productive setting where information is shared for everyone's benefit. The platform that executes standards, open interfaces, and data models depending on NGSI-LD is the suggested solution. It enables the integration of devices from many providers. The platform also makes use of the data that has been saved to assess the variables that affect crop growth, production processes, and irrigation water consumption in real time. The platform's main characteristics are as follows:

1. Scalability and flexibility: The platform is current with protocols, technologies, and characteristics which change quickly rather than being tied to a single supplier. It can be incorporated to function with all key technological systems and is network independent.
2. Interoperable: The platform provides a variety of IoT agents which ease communication with devices which employ standardized IoT protocols, like LWM2M [16] over Constrained Application Protocol (CoAP), JavaScript Object Notation (JSON), with the option of utilizing parameterizing agents to aggregate any different category of future protocols. The platform is available to application-level interaction with various third-party platforms utilizing both its API and its NGSI-LD-based data architecture.
3. Semantic enrichment: The use of the data model to standardize the administration of the many components of a farm.
4. Competitively and effectiveness: Related to management and farming operations, the model enables accurate and fast judgments. The capacity to spontaneously record the health condition of the crop or other natural resources gives managers a fast and effortless means to diagnose problems.

The rest of this chapter is as follows: Section 19.2 introduces the suggested remedy and lists its most distinctive components. The implementation in an irrigation-related neighborhood is described in Section 19.3. The system's assessment at various stages of implementation is shown in Section 19.4, along with the outcomes. The conclusion and fresh ideas based on these findings are discussed in Section 19.5.

Application of Information Models in Agriculture

Scalability, heterogeneity, and dynamism are the characteristics of the data gathered by all linked devices;

Scalability: There are millions of sensors that are constantly producing large amounts of data;

Heterogeneity: There are many different types of sensors;

Dynamism: The higher rate of formation necessitated the creation of data models which permit for superior usage and distribution of the data from these devices.

Shown the aforementioned elements, interoperability techniques that enable the integration of various tools, programs, or platforms are necessary to filter the massive volumes of data which several systems are creating. The shared data frameworks are helpful because they are crucial tools for enhancing information systems' communication, knowledge recovery, and interoperability in order to create more effective implementations for the agro-food industry.

There are several data frameworks accessible today, most of which were created by the groups responsible for creating their standards. The Open Geospatial Consortium (OGC) data frameworks are utilized mainly in the geosciences and ecological areas. The automotive, healthcare, industrial, and smart home sectors are just a few of the vertical industries for which the Open Connectivity Foundation defines data formats. Although it does not place much emphasis on the data, the WWW [17] Consortium Thing Description offers various languages to report actual objects. The standards governing the interoperability of geographical data sets and their related facilities are defined by different vocabularies, like those provided by the INSPIRE directive; these are required.

Currently, a few tools can expedite the data modeling of a particular area or industry. [18] provided a variety of agricultural instruments to aid in the development of this job:

- Controlled vocabulary (a group of keywords or phrases predetermined for a certain domain. One such is the multilingual AGROVOC vocabulary that is supported by the Foodstuff and Farming Agency of the UN.
- Other ontologies include CIARD Ring, Crop Ontology, AgroPortal, Dairy Farming Ontology, AgOnt.
- Specified ontologies: SSN (in sensor discovery) and Cotton Ontology (diseases and pests which influence cotton); data interchange standards: AgriOpenLink, AgroXML, and AgroRDF.

The uniformity of architectures serves as the foundation for many choices. ETSI ISG CIM offers the appropriate interfaces and context information management, and it enables the consistent description of the information via next-generation service interfaces with linked data.

Public governments support certain models, like the systematic water management framework applicable to irrigation framework SEGA, a project of the Spanish Ministry of Farming that is incorporated into the ISO 21624 standard and is run by TRAGSA Agent. The goal of SEGA is to provide a standardized framework for irrigation which, when used, enables interoperability between various frameworks which coexist in the similar ability and effectively controls irrigation water usage. The use of the systematized approach enhances the specification of the control system requirements and irrigation schedules by clearly separating decision-making from execution.

Solutions for IoT in the Agriculture Sector

Agriculture is a field with several issues, including irrigation, fertilizer and pesticide use, and crop, land, and livestock monitoring. To provide the best answers possible, several researchers are engaged in the following work:

- Productivity gains: The agriculturist may relate irrigation, herbicides, and fertilizers only while necessary thanks to the sensorization of crops to offer gain in the real world. For instance, [19] suggested using machine learning with the Internet of Things (IoT) to anticipate illnesses in crops prior to manifesting, enabling the farmer to implement the required defensive mechanisms, enhancing production and lowering the usage of pesticides. A lower-price sensor-based platform called SEenviro, introduced by [20] uses a disease framework to handle alerts in orchards.
- Finding previously untreated damaging agronomic elements in crops via remote sensing methods is possible with the use of satellite pictures or drones.
- Monitoring plant behavior: The PLANTAE platform is a framework that uses AI to evaluate the full dataset received from crops in order to make future estimations; as an illustration, [21] tracked crops utilizing the images supplied by a drone to check the availability of parasites and to determine whether future assistance was required. Other studies, like that of [22], kept an eye on plant behavior to spot pests and diseases early. Farmers submit incidents that enhance the models of illnesses utilized through mobile apps.
- Effective water management: Water stress should not be caused by crop water modifications. The usage of cloud computing using the FIWARE platform, as shown by [23], provides for an improvement in the control of water utilized for irrigation in places with water shortages. The SWAMP platform, which is similarly based on FIWARE, should be

taken into consideration. With the help of a context engine built on the SPARQL Protocol and RDF Query Language (SPARQL) Event Processing Farming, its main goal is to create novel IoT-based techniques for the smart control of irrigation water;

- Improving greenhouse management: [24] suggested a flexible platform competent to satisfy the demands of hydroponic crops in a greenhouse with full recirculation by monitoring its many components. NGSIs were used to represent the data in these FIWARE-dependent arrangement MQTT connections. In order to estimate the growth rate of potatoes in various circumstances, [25] built a system to track both greenhouse conditions and plant behavior.

IoT platforms may be thought of as the foundation of every industrial sector that seeks to integrate smart technology into its operations. For the filtering, evaluation, computation, decision-taking, administration and visualization of data in newer facilities, these platforms gather and store data in a distributed database. A results contrast is challenging due to the absence of a standardized platform; there exist several evaluations of the various kinds of platforms for the aggregation of IoT facilities and their features that may be obtained, like that offered by [26].

A proposed system for irrigation community administration, this paper suggests an open and programmable platform. The platform supports the aggregation of diverse information origins and compatibility with various third-party remedies for sharing and using this data. It is built on open, standard interfaces and protocols. Additionally, this platform makes use of the data that has been saved to assess in real time the elements that affect crop development, the production process, and irrigation water use.

Draft Data Model

Farming officers should be given standard data frameworks which enable agriculturists to make the best decisions by enabling them to have the benefit of the obtainable data and knowledge because of the need of similarity required to handle the massive quantity of data generated by the farming section. As a result, a data framework is suggested here which satisfies the consumers' useful needs.

Discussions with agriculturists, accomplished operators, and field officers regarding the important data that has to be maintained on a farm served as the foundation for the suggested data model. Although they

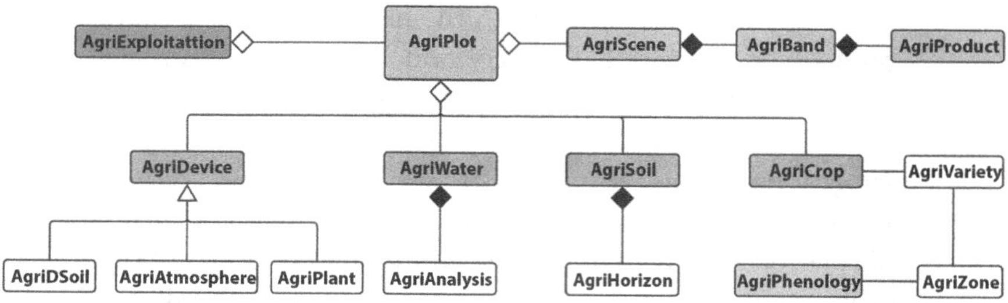

Figure 19.1 A specific data model for managing agricultural plots.

encompass the information that needs to be maintained, the current models are very specialized, making their implementation time-consuming. We thus made the decision to create a newer framework which will satisfy the requirements of our consumers.

Basically, the framework that is being given is dependent on the harmonized data frameworks that FIWARE has suggested, notably the AgriFood framework which takes irrigation and crop management needs into account. The model was developed as a flexible tool to meet the demands of farmers, enabling them to describe the plots, lower the crops' water requirements, and manage the quality of the soil and water as well as the accompanying meteorological conditions. As shown in the class diagram in Figure 19.1, the initial iteration of the framework is based on the effective control of the crop plots and various associated components.

The administration of crop plots, constituted by the AgriPlot system and a portion of a farming property, shown by the AgriExploitation element, is the main emphasis of the suggested data model. Four characteristics of the plot enable it to be specifically described:

- Devices, constituted by AgriDevice, are those that provide the plot with data and are categorized according to the parameters examined: AgriDSoil, which measures soil properties (temperature, soil dissolution, and soil moisture probes); AgriPlant, which gauges crop development (dendrometers, leaf/trunk diameter) and AgriAtmosphere, instruments that take note of the atmospheric parameters (warmth, humidity, radiation, and wind velocity). The best possible agricultural choices are made with these facts.
- Water, constituted by AgriWater, provides information on the kind of water utilized and the variables which influence irrigation. AgriAnalysis oversees the water testing done at

various locations across the farm's water delivery system. As they represent the kind and amount of nourishments transported by the water, the evaluations are often required for germination and irrigation.

- Soil, constituted by AgriSoil, shows the type of soil present in the plot. AgriHorizon, which provides the many properties of the levels (horizons) that define the ground, may be used to provide further information about this entity. Cultivation, represented by AgriCrop, describes the crop and the variety, AgriVariety, with which the plot is related. The several phenological phases that control the crop's development dictate it; AgriPhenology offers the crop coefficients and their length throughout time. The crop variety and the AgriZone climate zone, which affects the ideal climatic conditions for the growth of the crops, dictate the phenology.

Another component to take into account is the management of aerial photographs, which is handled by the AgriScene entity and includes any satellite or drone photos or sceneries made up of bands called AgriBands with various wavelengths. After processing, these bands may provide AgriProduct derivatives like mosaics (which combine two or further neighboring landscapes) or indices of vegetation or water, which are helpful for spotting issues in the farms.

The capacity to exchange information uniformly across various systems or applications is the key component of IoT-based interoperability protocols. The employment of data models is one of the components that make this continual interchange of information possible. The collection of traits that make up the entities and their interactions, which constitute the specialized lexicon of the sector, must be unified in order for it to be of greater significance. We enhanced the model's interoperability by linking data from various vocabularies or ontologies using NGSI-LD. For these reasons, it is deemed advantageous to employ information models to enable the integration of any device or distinguishing feature of the research industry.

NGSI-LD is used to create an implementation of this paradigm in Section 19.2. The following are some benefits of using NGSI-LD over NGSI:

- The information model focuses on information and is built on graphs. Relationship as a notion arises. Properties and relations are possible for entities. Every data type in NGSI-LD may be linked with unique Uniform Resource Identifiers (URIs) that correspond to well-accepted semantic identifiers. It permits one to refer to vocabularies because every term is unambiguously explained. Events of every entity may be the motive of resources or connections. Users may now mention their data declarations, the framework and query language are now better restricted, we can now work with linked data to synchronize vocabularies thanks to the usage of JSON-LD.

- Syntactic changes include the removal of the metadata dictionary, the use of GeoProperty in place of geo:json, the inclusion of JSON-LD @context, the substitution of TemporalProperty for DateTime, and the usage of a "object" field to represent the relation target.

19.2 Structure of the System

We suggest a cheap platform that does not depend on the provider and interoperable as the foundation for every link between applications and the irrigation components because newer IoT devices and newer communications technologies enable the user to communicate more quickly and effectively. By providing a common framework in this way, we enable applications to operate on any device utilizing the similar group of commands, enabling a major interoperability between various services. Applications can then interact with any component of a farm in a similar manner.

Figure 19.2 shows the platform's architecture, which is suggested in this study. It has a tiered modular structure that extends from the installation of sensors and the observation of data extraction methods to the smart processing of data. Every one of these layers is dependent on open-source projects that adhere to standards, such as those offered by the FIWARE fraternity group.

The system's many information sources, including sensors, actuators, publicly accessible information on the Internet, and databases with up-to-date relevant information, are represented by the first layer, Device and Data Acquisition. Various communication protocols are used by these devices to convey data to an IoT backend framework built by IoT Agents. This framework serves as a conduit between the IoT devices and the platform, transforming data in a unidirectional or bidirectional (in the case of actuators) fashion.

Due to the limitations placed by the frameworks on the amount of processing power, memory, and also electricity accessible to the devices, a particular module is necessary. Since these limitations often preclude them from employing heavy protocols at the application level, they instead

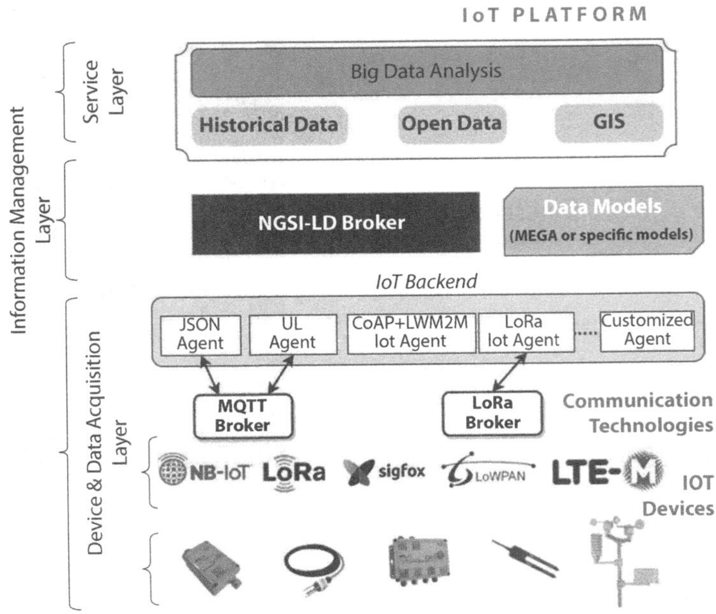

Figure 19.2 The planned design in broad strokes.

incorporate other technologies like LoRa and Sigfox or utilize lightweight protocols like CoAP and MQTT [27].

Data storage, processing, and dissemination are handled by the software components that make up the second layer, information management. The cloud-based computers that make up the distributed architecture cooperate to handle enormous volumes of data and make them accessible to the top layer. Data updates and alerts of data changes are sent and received using the NGSI-LD interface. Figure 19.3 shows an example of how data from an IoT gateway that uses JSON via MQTT is incorporated into the system.

The equipment that analyzes water as it leaves a wastewater treatment facility to ascertain its turbidity and pH levels was selected as the element. Customers may obtain this data using one of two methods when the element is formed in the broker utilizing the NGSI-LD data framework query or subscription.

The NGSI-LD broker serves as the platform's main component. It is an information broker that offers a subscription/notification strategy, as well as an HTTP REST API dependent on NGSI-LD for both registration and discussion. Interoperability is important at this second step as well. The usage of NGSI-LD offers a variety of potential bilateral connections

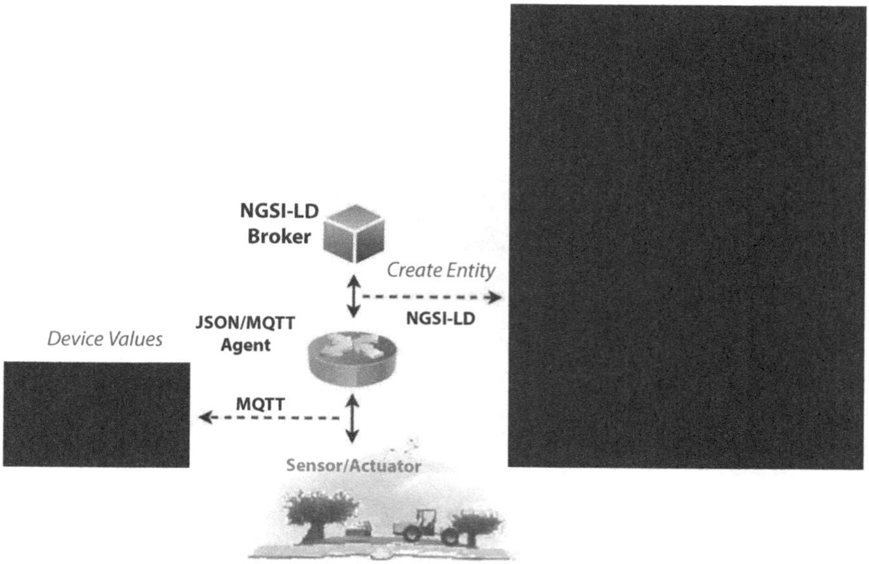

Figure 19.3 The data framework for applications at the IoT level.

with third-party systems. MEGA is a solution which was included into the platform by the creation of a connection. The MEGA standard, which enhances the design of irrigation schedules and the management of system needs, offers instructions for the use of a standardized approach to irrigation. The data is held in a non-relational database called MongoDB [28] and is rewritten for each object inside this management layer, leaving just the current state. This database includes data about users, organizations, and users' interactions with the infrastructure and information flow. All of this data is represented using data models, which enable the standardization of data structures and guarantee the generality required for any further extension.

In order to provide various remedies to the issues which arise in a farming execution, like water control, irrigation design, data evaluation, and tracking of ecological indicators, the final layer, referred to as "Service," acts as an interface between users and the core layer being incorporated as a result of the ability to connect additional components via connectors with NGSI-LD, we are able to provide our solution extra capabilities through the use of other modules. Data updates and alerts of data changes are sent and received using the NGSI-LD interface. Control actions are triggered by changes in configuration settings and are handled by user-level subsystems.

The data is consulted or passed to upper modules responsible for distributing it for various purposes, like historical data, larger-scale data control, or aggregation with GIS [29], as a consequence of the employment of specialized connectors. This is accomplished by using a representational state transfer interface (REST) that employs NGSI-LD for links between the analytic modules and the end applications.

19.3 Irrigation Community Event

This part verifies the platform's architecture and our data model. It was used in an irrigation community where the platform was used to combine many systems, including SCADA, GIS, and sensors from various manufacturers. Irrigation fraternities are entities affiliated with basin establishments that are in charge of overseeing the joint use of public water resources. For years, they depended on automatons and certain irrigation systems to do this; these systems were closely connected, permitting for complete interactivities between the hardware and software, still foregoing a chance of contact with hardware or software developed by other parties.

Nearly 1,000 people live in the Miraflores irrigation fraternity, which is situated in the Segura river valley in the region of Jumilla. The region has roughly 1,330 acres of agricultural land, most of which is used for woody crops and is irrigated locally. Fruit trees, notably pear, peach, and apricot, are the principal crops grown. The town consumes 1.5 hm^3 of water from the Jumilla sewage plant per year and 3.8 hm^3 of surface water. The community then uses the available flows to push the recovered water to six linked regulatory rafts with a combined volume of 1 hm^3 at the treatment plant's outlet. Integration of the many technical components (SCADA, drawing software, independent sensors) which offer data to the group in a lone control point is one of the major issues which have to be handled. Figure 19.4 depicts the overall plan that was used to handle the platform rollout. Channeling water from the headwater reservoir with its related water analyses was the initial phase.

The system's most important component is the plot. Dataloggers and Internet of Things connections are used to combine all the devices required for its specification. This procedure involves integrating

SCADA data as well as the sensors which communicate in the plot (hydrants, farming weather stations, and soil moisture sensors) into the system. The data model receives the information for further viewing and analysis by the irrigation community's users utilizing mobile or online apps.

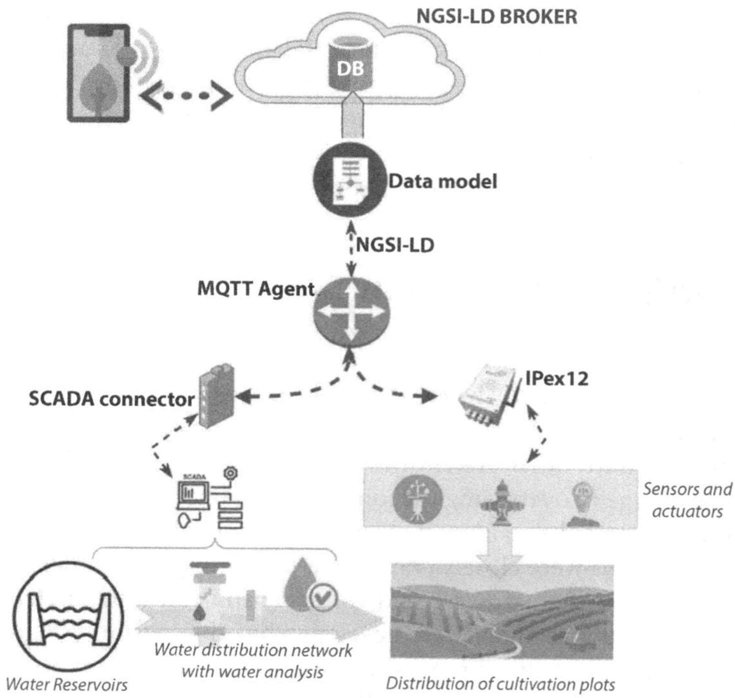

Figure 19.4 The case study's layout.

Odin Solutions supplied the controllers for the implementation, which included the installation of controllers for tracking and remote control in outside areas. Because the platform is open source, the cost of implementation was decreased. The cost of data loggers handling several platform devices or sensors was reduced by this functionality. Here, the price of the selected datalogger varies depending on the quantity of sensors incorporated into the framework, with an average cost of about 650 euros (consisting of setup and installation). The controllers' web server makes it simple to setup them, and this lets us set up tasks independently of the MQTT instructions that come from the cloud computation layer. The IPex12 is specifically framed for outside environments like that are seen in farming settings, and comes with a water- and dust-proof casing. A battery and a solar panel are included.

Web services are used to host management procedures. 400 GB of hard drive space and 28 GB of RAM are used by the virtual machine. Currently, a decent performance is feasible, but because of the adaptability of our design, more resources may be provided by altering or shifting virtual pictures.

The entities were created in the broker once the platform's various modules had been setup so that subsequent subscriptions can be obtained on their behalf. The broker oversees subscriptions for entity establishment or modification. All of these modifications are saved in MongoDB, a non-relational JSON-LD object-based database, to allow for quick and adaptable information accessibility. Processing speed is essential since, in the absence of it, a bottleneck would be established, slowing down the remaining implementations.

The data framework suggested in Section 19.2 was utilized to enhance the irrigation plot data in accordance with the requests made by the irrigation community's users. Through the use of NGSI-LD, a model instantiation of an apricot tree plot with a soil moisture sensor was created. Listing 1 lists the soil moisture sensor, whereas Listing 2 defines the crop.

Listing 1 shows an example of an NGSI-LD (next generation service interfaces with linked data) interface for an agricultural device

"id": "urn:ngsi-ld:AgriDevice:AgriDSoil:01LaPuebla", "type": "AgriDevice", "category": "type: "Property", "value": "Soil", "s4ee:deviceName": "type": "Property", "value": "EnviroSCAN", "s4ee:

Listing 2 is an example of an AgriCrop using NGSI-LD

"id": "urn:ngsi-ld:AgriCrop:Apricot", "type": "AgriCrop", "taxon:commonName": "Property", "value": "Apricot", "taxon:scientificName": "taxon:Plant", "type": "Property", "value": "Prunus armeniaca" "type": "Property", "value": "Fruit tree", and "demandIrrigation": "@context": ["http://uri.etsi.org/ngsi-ld/v1/ngsi-ld-core-context.jsonld", "http://aims.fao.org/aos/agrovoc/c_7951", "type": "Property", "value": "Yes"], "agroVocConcept"

["demandIrrigation": "http://aims.fao.org/aos/agrovoc/c_25352.html" and "taxon": "http://purl.org/biodiversity/taxon"

}

We outlined the traits that characterize the plot after specifying the crop grown in the plot and the tools which aid in improving irrigation control to increase productivity. The characteristics needed to determine the crop's water requirements, such as plantation density, emitter flowRate, number of emitters, and tree crown diameter, are shown in Listing 3.

Listing 3 shows an example of an NGSI-LD showing an AgriPlot and associated elements

"id": "urn:ngsi-ld:AgriPlot:CRM0001", "type": "AgriPlot", "name": "type": "Property", "value": "La Puebla 1"; "location": "type": "GeoProperty", "value": "type": "Polygon"; "coordinates": [[38.461229, -1.320572], Name, density, flow rate, diameter, and emitters are all listed at the following

19.4 Assessment and Authentication

The platform's authentication in a group of irrigators is discussed from three distinct angles in this section. The equipment mentioned in Section 3.2 was used to perform the experiments in a controlled setting.

Analysis of Information Brokers

The core component of the module is an NGSI-LD broker that is in charge of storing every data and handling various requests, responses, subscriptions, and provider management. In light of this, an evaluation was carried out by choosing the metrics of CPU utilization, memory consumption, and reaction times in accordance with the activities carried out. First, several execution tests were carried out to see whether the platform could make the modifications without affecting the level of service. The tests were run using the identical queries, and their response times related to CPU and RAM were examined. Every test was dependent on a 1-minute implementation with a different quantity of concurrent messages or links. The results of tests conducted on the processor's and memory's behavior are shown in Figure 19.5.

Figure 19.5a demonstrates that the first intervals (2, 4, and 8 threads) required a 20% increase in CPU use. The CPU remained constant beyond this point and up to 1,024 threads, having average values closer to 92%. Figure 19.5b demonstrates that, without changing the amount of resources utilized, the behavior in the usage of RAM is almost constant from start to finish. The range of the RAM use increase throughout the testing fluctuated between 5% and 9%, which are ideal levels.

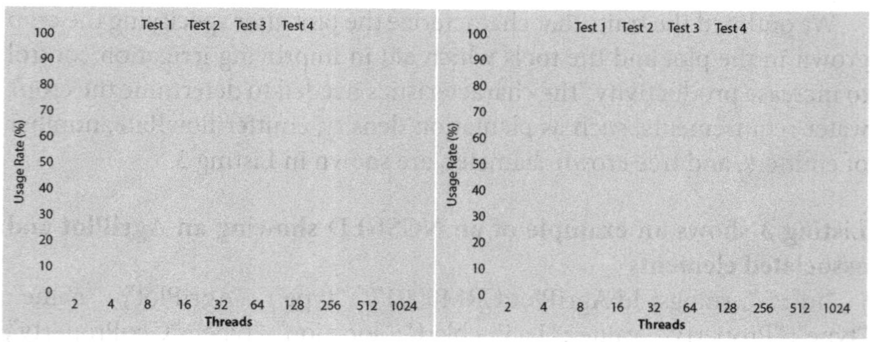

(a) (b)

Figure 19.5 Analysis of module scalability. CPU utilization (a) RAM usage (b).

Based on the feature of information dissemination that the broker had gained, the platform's behavior was assessed. In order to do this, the reaction time acquired after carrying out different actions on the various broker operating modes is studied. The graphs in Figure 19.6 show the timings for the various modes, including context provider control, entity management, and subscription management.

- Because changes to the data kept on the platform do not need routine inquiries, the publication/subscription approach offers a decrease in the number of queries.
- The control of external context providers is a procedure which facilitates access to the data and makes the technique clearer for the end client. As a result, the creation time is large as it must be modified to the suggested data framework. The broker's role is to stand in for the customer while dealing with the context provider. Because of this, search and dialogue times are quicker than they are for other procedures.

Based on the three layers that make up the platform's design, this section evaluates the platform's reaction times. Figure 19.7 displays the system's timings after 100 requests were processed. Figure 19.7a displays the notification timings at its two levels: at the NGSI-LD broker level and the MQTT level (where the MQTT broker alerts the IoT Agent). Since the database (MongoDB) handles all of the semantic enrichment offered by the usage

Figure 19.6 Reaction time for activities using various techniques of operation.

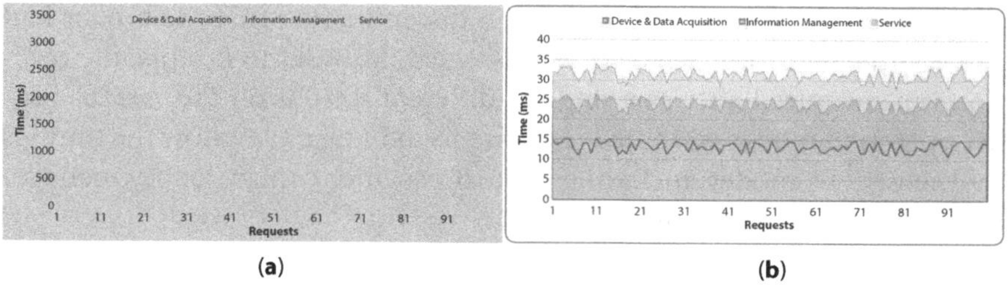

Figure 19.7 Reaction time in accordance with the platform's tiers at the level of notification (a) and actuation (b).

of the data framework, the Information Management Layer takes longer, on average 21.69 s. However, it is feasible to consist of all the pertinent data of the irrigation community homogenized and merged into a lone location in exchange for this extension of time. The system's performance management features are shown in Figure 19.7b. According to the graph, a connected device responds more quickly than one that isn't; a low reaction time is defined as one that is less than 31.18 milliseconds on average.

Data interoperability is crucial to the agricultural level use of the platform. Thus, the suggested remedy (data framework and platform) combines the usage of a data framework with regulated solutions, like offered by NGSI-LD, to provide the essential data translation processes. According to the shifting needs of farms, this approach facilitates the aggregation of numerous systems/devices or data sources, producing an open and interoperable data integration architecture. Data exchange among other farms and the gathering of performance indicators are all made possible via the usage of data models.

Figure 19.8 depicts the development of irrigation prior to the installation of the platform. The abnormality of the irrigation system was discovered via data analysis. We considered it important to specify irrigation thresholds given by the agricultural professionals in order to have a continuous and uniform irrigation system. We came to the conclusion that the quantity of water between these two values is the functional water or humidity that is accessible to the plant since the farm capacity and the wilt point dictate the greatest and lowest limits of the soil humidity which may be utilized by the crops. On an apricot tree plot, tests were conducted to gauge irrigation effectiveness once the community engineers had established the crop limits. The first test was carried out between March and April after the platform had been deployed, and the second test was carried out between June and July after the irrigation programming had been improved in response to the first test's findings.

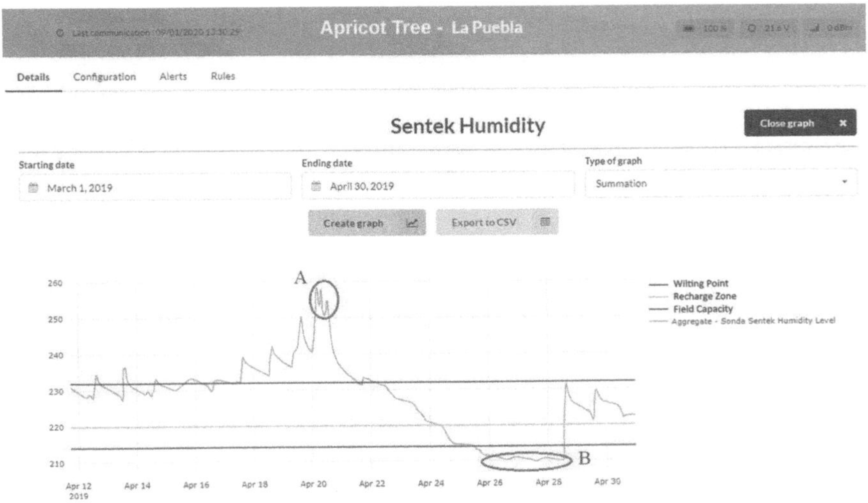

Figure 19.8 Growth of apricot crop humidity depending on predetermined water levels.

After analyzing the first test, we discovered that during irrigation, two crucial zones, A and B, were created with severe peaks which were outside of the advised limits, suggesting that since water was limited, it was not being used effectively. Zone A denotes root asphyxiation, or the restriction of a plant's ability to breathe through its roots, in the crop; Zone B, on the other hand, denotes a plant's inability to provide itself with water, which lowers the crop's quality of production. The maximum and lowest criteria may be established after which regulations for effective irrigation can be determined. The platform produces alerts for users through the mobile application whether the thresholds are met or not. These guidelines may also be used to start any equipment that increases irrigation efficiency by activating digital inputs.

The appropriate irrigation restrictions are set or modified as a consequence of using the platform, analyzing the data acquired, and working with an agricultural technician. Other helpful characteristics may also be determined, such as the soil type, the soil type's composition, or the placement of the probes. Figure 19.9 demonstrates how well the irrigation was used when the irrigation programming was altered. The vertical column and water contributions match the humidity curves, indicating consistent and uniform sharing. Irrigation was regulated to keep the humidity within the appropriate ranges by managing the hydrants and authorizing filters depending on the limitations specified, or to advice of potential warnings to operate scheduled irrigation to prevent a decrease in productivity.

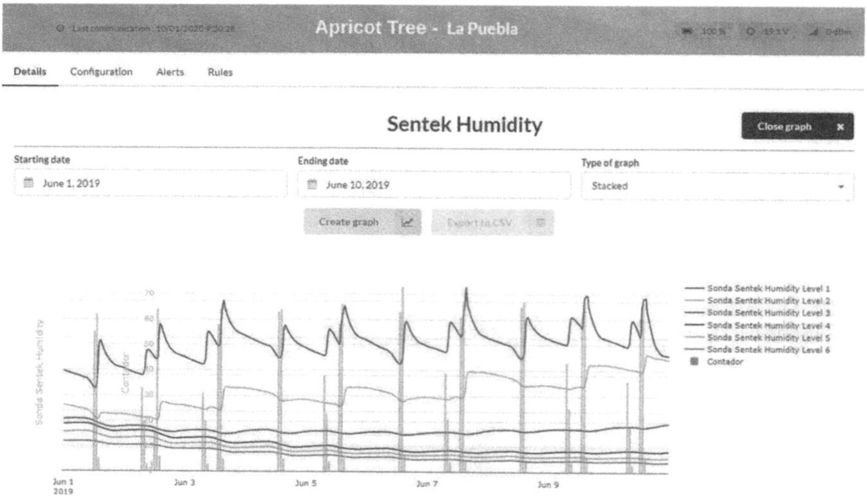

Figure 19.9 The development of effective irrigation while contrasting soil moisture and hydrant water input.

The platform's flexibility is shown by the capability to export specific data for enhanced processing or analysis, as seen in Figures 19.8 and 19.9 by selecting the export to CSV choice.

19.5 Conclusions

This project served as a demonstration of the value of interoperability in irrigation, water management, and agriculture. An open, interoperable platform that can integrate diverse data origins at the IoT level is described here to support and enhance the decision-taking in a group of irrigators. The platform serves as a single access point for all of this data since it facilitates the aggregation of controllers and sensors from many vendors. As a result, it is feasible to enhance agricultural processes in order to sustainably provide greater economic, environmental, and social yields. The platform's ability to integrate with other systems allows for the expansion of its service offering. The compatibility feature was examined in a real-world setting where this trait enabled effective integration with other systems previously installed in the irrigation community's facilities. The platform incorporates a variety of IoT protocols, including MQTT, CoAP, LoRa and HTTP. They transmit the data in a variety of data formats. However, some of them, like Lightweight Machine to Machine (LWM2M), JSON, and UltraLight, provide compatibility. The provided platform adopts these representation formats to make it easier for it to be integrated into IoT devices that employ this technology. The adaption of this data to the NGSI-LD interface and data framework is carried out by the agents of the IoT backend of our platform.

The following findings were reached after studying the data in the irrigation community: 1) It was suggested to use a homogeneous data model to address the unique requirements of farming, like effective water control. Utilizing an NGSI-LD broker, this framework was verified on the platform. 2) The platform was verified utilizing metrics to examine its nature. The deployment of this framework in an irrigation group gives its managers the ability to manage the agronomic data and its interaction with the devices which supply information to optimize irrigation water. First, the NGSI-LD broker's scalability was evaluated using average CPU and memory use of 83.21% and 56.39%, respectively. Measurements on the most important broker actions were taken at the latency level, with the construction of a context provider taking an average of 510.28 milliseconds on average. The platform's overall validity was shown by the time delays between the moment the information is received by the device and when it is accepted by the platform (2.97 s) and between the moments the action is done (31.17 ms). 3. The platform was evaluated at the user level in a number of apricot tree plots, which improved the control of the water utilized. This enhancement was made possible by specifying the ideal irrigation thresholds for every crop and by producing filter alerts which in specific circumstances; let the hydrants to be modified, facilitating effective water usage. In order to determine which platform offers the best performance and reaction times, the speed of computing and latency must be compared in the future by reproducing the edge computing platform with the present cloud computing. Processing of satellite photos is one of them. With applications dependent on remote sensors, this feature would keep track of the crop's primary characteristics at the plot level. This newer feature would concentrate on irrigation management, diagnostics, and management. In the short term, the PAZ satellite (the first Spanish radar earth observation satellite that is present in the National Earth Observation Program) would be utilized in place of the Sentinel 2 satellite that is more concentrated on farming concerns. The creation of a data analysis module which will enable us to cross-reference and examine the past of the data collected is another area of future study. By using less water, fertilizers, and phytosanitary chemicals, for example, this will improve agricultural management outcomes, boost output, and lower costs.

References

1. S. Pramanik, M. G. Galety, D. Samanta, and N. P. Joseph, Data Mining Approaches for Decision Support Systems, *3rd International Conference on Emerging Technologies in Data Mining and Information Security*, 2022.
2. S. Pramanik, Intelligent Farming Utilizing a Soil Tracking Device, in *Convergence of Cloud Computing, AI and Agricultural Science*, Eds., A. K. Sharma, N. Chanderwal and R. Khan, IGI Global, 2023.
3. S. Dhamodaran, S. Ahamad, J. V. N. Ramesh, G. Muthugurunathan, K. Manikandan, S. Pramanik, and D. Pandey, Food Quality Assessment Using Image Processing Technique, in *Handbook of Research on Thrust Technologies' Effect on Image Processing*, IGI Global, 2023.
4. G. Taviti Naidu, KVB Ganesh, V. VidyaChellam, S Praveenkumar, D. Dhabliya, S. Pramanik, and A. Gupta, Technological Innovation Driven by Big Data, in *Advanced Bioinspiration Methods for Healthcare Standards, Policies, and Reform*, Hadj Ahmed Bouarara, IGI Global, 2023, doi: 10.4018/978-1-6684-5656-9
5. S. Ahamad, V. Veeraiah, J. V. N. Ramesh, R. Rajadevi, S. R. Reeja, S. Pramanik and A. Gupta, Deep Learning-based Cancer Detection Technique, in *Handbook of Research on Thrust Technologies' Effect on Image Processing*, IGI Global, 2023.
6. S. Dhamodaran, S. Ahamad, J. V. N. Ramesh, S. Sathappan, A. Namdev, R. R. Kanse, and S. Pramanik, Fire Detection System Utilizing an Aggregate Technique in UAV and Cloud Computing, in *Handbook of Research on Thrust Technologies' Effect on Image Processing*, IGI Global, 2023.
7. A. M. Ray, S. Pramanik, B. Das, A. Khanna, Hybrid Cryptography and Steganography Method to Provide Safe Data Transmission in IoT, *ICDAM*, 2023.
8. V. Veeraiah, D. J. Shiju, J. V. N. Ramesh, K. R. Ganesh, S. Pramanik, D. Pandey, and A. Gupta, Healthcare Cloud Services in Image Processing, in *Handbook of Research on Thrust Technologies' Effect on Image Processing*, IGI Global, 2023.
9. A. Namdev, V. Veeraiah, S. Dhamodaran, S. Islam, S. Pramanik, and A. Gupta, Cloud Computing Applications in Biomedicine, in *Emerging Trends in Cloud Computing Analytics, Scalability, and Service Models*, IGI Global, 2023.
10. V. Talukdar, A. Z. Kurniullah, P. Keshwani, H. Khan, S. Pramanik, A. Gupta, and D. Pandey, Load Balancing Techniques in Cloud Computing, in *Emerging Trends in Cloud Computing Analytics, Scalability, and Service Models*, IGI Global, 2023.
11. S. Pramanik, Intelligent Farming, Utilizing a Soil Tracking Device, in *Convergence of Cloud Computing, AI and Agricultural Science*, IGI Global, 2023, doi: 10.4018/979-8-3693-0200-2.ch009

12. P. K. Mall, S. Pramanik, S. Srivastava, M. Faiz, S. Sriramulu and M. N. Kumar, FuzzyNet-Based Modeling Smart Traffic System in Smart Cities Using Deep Learning Models, in *Data-Driven Mathematical Modeling in Smart Cities*, IGI Global, 2023, doi: 10.4018/978-1-6684-6408-3.ch005
13. D. Mondal, A. Ratnaparkhi, A. Deshpande, V. Deshpande, A. P. Kshirsagar and S. Pramanik, Applications, Modern Trends and Challenges of Multiscale Modelling in Smart Cities, in *Data-Driven Mathematical Modeling in Smart Cities*, IGI Global, 2023, doi: 10.4018/978-1-6684-6408-3.ch001
14. P. T. Khanh, T. H. Ngọc, and S. Pramanik, Future of Smart Agriculture Techniques and Applications, in *Advanced Technologies and AI-Equipped IoT Applications in High Tech Agriculture*, Ed., A. Khang, IGI Global, 2023.
15. Reepu, S. Kumar, M. G. Chaudhary, K. G. Gupta, S. Pramanik and A. Gupta, Information Security and Privacy in IoT, in *Handbook of Research in Advancements in AI and IoT Convergence Technologies*, Eds., J. Zhao, V. V. Kumar, R. Natarajan and T. R. Mahesh, IGI Global, 2023.
16. V. Veeraiah, V. Talukdar, Manikandan K., S. B. Talukdar, V. D. Solavande, S. Pramanik, and A. Gupta, Machine Learning Frameworks in Carpooling, in *Handbook of Research on AI and Machine Learning Applications in Customer Support and Analytics*, Eds., Md S. Hossain, R. C. Ho and G. Trajkovski, IGI Global, 2023.
17. V. Jain, M. Rastogi, J. V. N. Ramesh, A. Chauhan, P. Agarwal, S. Pramanik, and A. Gupta, FinTech and Artificial Intelligence in Relationship Banking and Computer Technology, in *AI, IoT, and Blockchain Breakthroughs in E-Governance*, Eds, K. Saini, A. Mummoorthy, R. Chandrika and N. S. Gowri Ganesh, IGI Global, 2023.
18. T. H. Ngọc, P. T. Khanh and S. Pramanik, Smart Agriculture Using a Soil Monitoring System, in *Advanced Technologies and AI-Equipped IoT Applications in High Tech Agriculture*, Ed., A. Khang, IGI Global, 2023.
19. R. Bansal, B. Jenipher, V. Nisha, Jain, Makhan R., Dilip, Kumbhkar, S. Pramanik, S. Roy and A. Gupta, "Big Data Architecture for Network Security", in *Cyber Security and Network Security*, Eds., Wiley, 2022, https://doi.org/10.1002/9781119812555.ch11
20. R. Anand, J. Singh, D. Pandey, B. K. Pandey, V. K. Nassa and S. Pramanik, "Modern Technique for Interactive Communication in LEACH-Based Ad Hoc Wireless Sensor Network", in *Software Defined Networking for Ad Hoc Networks*, M. M. Ghonge, S. Pramanik and A. D. Potgantwar, Eds., Springer, 2022, https://doi.org/10.1007/978-3-030-91149-2_3
21. X. Li, W. Pei, Y. Wang, Z. He, H. Lu and M.-H. Yang, "Self-Supervised Tracking via Target-Aware Data Synthesis," in *IEEE Transactions on Neural Networks and Learning Systems*, doi: 10.1109/TNNLS.2022.3231537.
22. A. Bhattacharya, A. Ghosal, A. J. Obaid, S. Krit, V. K. Shukla, K. Mandal and S. Pramanik, "Unsupervised Summarization Approach with Computational Statistics of Microblog Data", in *Methodologies and Applications of Computational Statistics for Machine Learning*, D. Samanta, R. R. Althar,

S. Pramanik and S. Dutta, Eds., IGI Global, 2021, pp. 23-37, doi: 10.4018/978-1-7998-7701-1.ch002

23. V. Veeraiah, D. J. Shiju, J. V. N. Ramesh, Ganesh Kumar R, S. Pramanik, A. Gupta, and D. Pandey, Healthcare Cloud Services in Image Processing, in *Handbook of Research on Thrust Technologies' Effect on Image Processing*, IGI Global, 2023.
24. G. L. Priya, C. Baskar, S. S. Deshmane, C. Adithya and S. Das, "Revolutionizing Holy-Basil Cultivation with AI-Enabled Hydroponics System," in *IEEE Access*, vol. 11, pp. 82624-82639, 2023, doi: 10.1109/ACCESS.2023.3300912.
25. A. Roy and S. Pramanik, A Review of the Hydrogen Fuel Path to Emission Reduction in the Surface Transport Industry, *International Journal of Hydrogen Energy*, https://doi.org/10.1016/j.ijhydene.2023.07.010, 2023
26. A. D. Andrushia, T. M. Neebha, A. T. Patricia, K. M. Sagayam, S. Pramanik, Capsule Network-based Disease Classification for VitisVinifera Leaves, *Neural Computing and Applications*, doi: 10.1007/s00521-023-09058-y, 2023.
27. C. Zunino, G. Cena, S. Scanzio and A. Valenzano, "Adaptive Seamless Redundancy to Achieve Highly Dependable MQTT Communication," in *IEEE Transactions on Industrial Informatics*, vol. 20, no. 1, pp. 984-994, Jan. 2024, doi: 10.1109/TII.2023.3271708.
28. R. Jayasingh, J. Kumar R.J.S, D. B. Telagathoti, K. M. Sagayam and S. Pramanik, "Speckle noise removal by SORAMA segmentation in Digital Image Processing to facilitate precise robotic surgery", *International Journal of Reliable and Quality E-Healthcare*, vol. 11, issue 1, DOI: 10.4018/IJRQEH.295083, 2022.
29. R. Bansal, A. J. Obaid, A. Gupta, R. Singh, S. Pramanik "Impact of Big Data on Digital Transformation in 5G Era", *2nd International Conference on Physics and Applied Sciences (ICPAS 2021)*, doi:10.1088/1742-6596/1963/1/012170, 2021.

20

Quantifying the Economic Benefits of Agricultural Data Sharing: A Mathematical Modeling Perspective

Aruno Raj Singh[1], Vinaya Kumar Yadav[2], Laishram Zurika[3], Dasarathy A. K.[4], Abhishekh Benedict[5] and Dharmesh Dhabliya[6]*

[1]College of Law, Teerthanker Mahaveer University, Moradabad, Uttar Pradesh, India
[2]Department of Agriculture, Sanskriti University, Mathura, Uttar Pradesh, India
[3]School of Agriculture, Dev Bhoomi Uttarakhand University, Dehradun, Uttrakhand, India
[4]Department of Civil Engineering, Faculty of Engineering and Technology, JAIN (Deemed-to-be University), Karnataka, India
[5]Symbiosis Law School, Nagpur Campus, Symbiosis International (Deemed University), Pune, India
[6]Department of Information Technology, Vishwakarma Institute of Information Technology, Pune, Maharashtra, India

Abstract

A full collection of tools is available via the WEKA (Waikato Environment for Knowledge Analysis) framework for using data mining methods on huge data sets. This chapter outlines a process framework for data analysis and the assistance which WEKA offers to the framework. The domain framework that the data mining approach has "learned" may then be easily included into a software program. An application in the farming sector—mushroom grading—is used to demonstrate the WEKA-based analysis and application building process.

Keywords: Application development, data mining, and machine learning

**Corresponding author*: dharmesh.dhabliya@viit.ac.in
Aruno Raj Singh: ORCID: https://orcid.org/0009-0008-4742-2397
Vinaya Kumar Yadav: ORCID: https://orcid.org/0009-0005-6050-7756
Dasarathy A. K.: ORCID: https://orcid.org/0000-0002-4222-8607

Sabyasachi Pramanik, Niranjanamurthy M., Ankur Gupta and Ahmed J. Obaid (eds.) Mathematical Modeling in Agriculture, (421–436)

20.1 Introduction

The practice of finding previously undiscovered and possibly intriguing patterns in huge datasets [1] is known as data mining [2]. The mined data is often shown as a framework of the semantic structure of the dataset, and the framework can be applied to fresh data to make predictions or categorize it. As an alternative, human domain specialists may decide to personally inspect the framework in search of elements which clarify formerly unclear or unrecognized aspects of the domain under investigation. In this research, the authors focus on ML methods for developing domain frameworks or evaluating datasets (more information is provided in Section 20.3). Although less understandable to humans than a neural model, ML [3] techniques may produce frameworks with classification/prediction accuracy equivalent to, for instance, ANN [4].

The WEKA research team has two goals: first, to conduct fundamental data mining research by creating new machine learning algorithms, and second, to mine data from current farming datasets supplied by Australian academics and research institutions. We have created the WEKA system, a data mining workbench, to help achieve these objectives. It includes the following tools: classifiers and different data mining approaches that can handle categorical and numerical learning jobs, meta classifiers for improving the performance of classification data mining approaches (like, boosting and bagging rules), feature selection tools that are necessary for recognizing unrelated attributes to keep out from the dataset.

The WEKA source is offered with extensive, automatically produced documentation to help the user navigate their interactions with the program. More information on the past, Unix-based versions of the WEKA system may be found in [5]. WEKA's present implementation takes the form of a collection of Java class libraries. The program is accompanied with a new data mining manual that completely describes and records all of the data mining methods used in WEKA. Any machine with a WWW browser may execute application applications created using the WEKA class libraries. Section 20.2 provides a broad description of data mining. Weka's current version of its machine learning algorithms are described in Section 20.3; the WEKA tools that enable the data mining process model are described in Section 20.4. Section 20.5 includes a case study that demonstrates data mining utilizing WEKA, and Section 20.6 summarizes our findings.

20.2 Model for Data Mining Process

More than 60 real-world data sets, mostly farming data sets given by research organizations and companies in Australia, have been studied for this project. In order to incorporate the induced domain knowledge into a software module, the authors have created a process framework for using data mining methods on data based on our experience (Figure 20.1). The main components of this model are as follows:

- Communication in both directions between the data source and the data mining specialist. The domain expert provides data regarding data semantics and "legal" changes which may be registered to the data, and the data mining specialist guides the technique to enhance the understandability and accuracy of the outcomes. Together, they convert the raw data into the final data set(s) input to the ML techniques.
- A method that is iterative. Finding a satisfactory "fit" between a portrayal of the data and a data mining method requires numerous cycles through the process model in machine learning, which is an exploratory process. Additionally, even when unique attribute combinations are processed through multiple techniques, the resulting data models might be vastly dissimilar even though the findings have the same level of predictive accuracy. These diverse viewpoints might provide insightful information about trends across various data subsets.

The model shown in Figure 20.1 shows a clockwise flow of activity. As needed by the data mining algorithms included into WEKA, the raw data is first regarded as a lone table in the pre-processing step. The table is converted into the attribute/value table description known as the ARFF format, which contains header information on the data types of the attributes. To manage missing numbers, find incorrect values, and eliminate outliers, and other issues, the data may also need extensive "cleansing."

Currently, the data provider and the data mining specialist work together to convert the cleaned data into a format which will enable a data mining algorithm to build a comprehensible, precise data model. For instance, these two analysts can assume that one or more qualities are unimportant

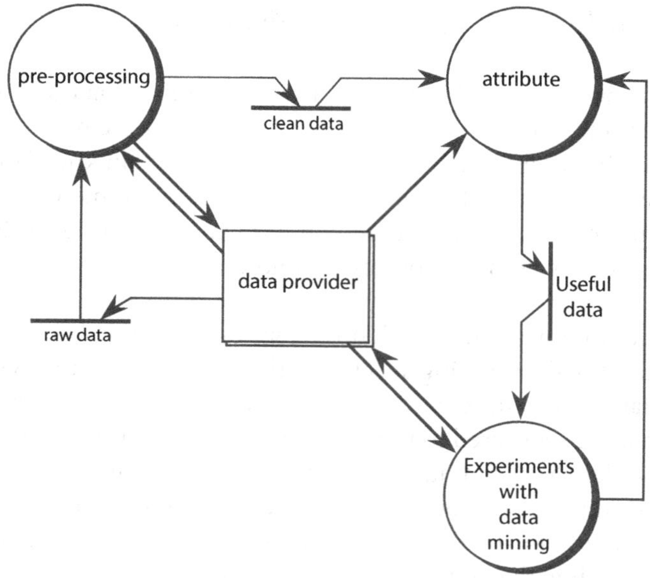

Figure 20.1 Data flow diagram for the process framework for an ML implementation.

and disregard the irrelevant columns. Mathematical manipulation of attributes is possible, for instance, to normalize values in a particular column or to merge two or more columns into a lone derived attribute. Other examples include converting all columns holding temperature readings to a common scale.

The data mining algorithms then process one or more iterations of the cleaned data. The subject matter expert chooses whether parts of the result are sufficiently unique or intriguing to merit further investigation and which parts reflect accepted wisdom in that discipline. The algorithm's output is interpreted by a data mining specialist, who also offers suggestions for other experiments that may be conducted using the data.

20.3 Techniques for Machine Learning

Twelve learning schemes, including 10 classifiers, a clustering [6] algorithm, and an association rule learner, are implemented in the current version of WEKA. The software architecture is adaptable enough to accommodate the integration of different learning schemes and learning scheme types into WEKA. The kinds of learning that WEKA presently supports are discussed in this section.

Classifiers

A literal classifier—typically in the formation of a decision tree or collection of rules which may be utilized to estimate the categorization of incoming data instances—is the outcome of this kind of learning technique. The remaining characteristics may be found in the "if" sections of the rules (or the non-leaf nodes of the decision tree), as one attribute in the input table is specified as the category or class for estimation.

ZeroR is WEKA's most basic learning system. This method uses a single rule to represent the dataset. ZeroR estimates the average class value for numeric prediction problems or the most common class value in the training data for issues with a nominal class value when supplied a newer data item for classification. When establishing a benchmark performance against which other learning strategies may be measured, ZeroR is helpful. It is feasible for different learning approaches to produce models in certain datasets.

OneR, the following technique, generates extremely basic rules based on a single characteristic. In fact, OneR was shown to outperform more complex algorithms on most of the common ML test datasets, making it valuable for creating a baseline for classification performance. The fact that most of the common test databases have extremely basic fundamental connections in the data seems to be at least partially to blame for this outcome. Simple connections may be parsimoniously recognized and represented using OneR. Real-world databases can moreover include extremely basic organized data regarding a topic.

When given a newer data item, the Naïve Bayes [7] framework estimates the likelihood that it falls under each of the potential class categories. Naïve Bayes applies a Naïve Bayesian classifier that generates probabilistic rules. In that each attribute is handled as if it were totally independent and as if it contributed equitably to the framework, the Bayesian classifier may be described as "naive" in this sense. Extraneous characteristics will distort the model if they are present in the dataset. Naïve Bayes, like OneR, may provide unexpectedly strong results on a variety of real-world datasets despite how simple it is.

The dataset is compiled into a "decision table" using Decision Table [8]. A decision table, in its most basic form, has the similar quantity of characteristics as the earliest dataset. A newer data item is given a categorization by locating the line in the decision table which corresponds to the non-class estimates of the data item. The wrapper approach is used in this version to choose a suitable subset of characteristics for the table. The approach decreases the possibility of over-fitting and generates a smaller, more condensed decision table by removing features that make little to no contribution to a dataset model.

Instance-based learning techniques just store the dataset and then build a model. Using a distance measure, a new data item is categorized by contrasting it to these "memorized" data pieces. The closest original data item is designated as the new item's category, or its "nearest neighbor." For numeric characteristics, the distance-weighted average of the k closest items may be allocated, or the majority class of the k nearest data elements can be chosen. IBk is a k-nearest-neighbors classifier implementation. There are two ways to set the number of closest neighbors (k): manually or automatically through cross-validation.

The standard method C4.5 version 8, which is often used in real-world machine learning applications, is implemented in j48. Decision tree models are produced by this implementation. Part of it is a more contemporary method for creating ordered collections of rules called "decision lists" of rules. A newer data element is contrasted to every rule in the list one at a time and it is given the class of the first rule that matches it (a default category is used if no rule matches strongly). Using C4.5's heuristics, this program creates trimmed partial decision trees and promptly converts them into appropriate rules.

Support vector machines (SVMs) [9], a crucial new paradigm in machine learning, are implemented by SMO using the "sequential minimal optimization" approach. SVMs have been widely used in text categorization learning models. Despite being a quickest method for learning SVMs, SMO is sometimes sluggish to converge, especially when dealing with noisy data.

Three techniques for numerical prediction are included in WEKA. Linear regression is the most basic. Locally weighted regression (LWR) is an application of a much complex learning strategy for numerical estimation. According to [10], M5Prime is a logical reconstruction of Quinlan's M5 model tree inducer. The leaf nodes of the decision tree may be modified to hold a numeric value that is the mean of all the values in the dataset which the leaf employs to, even though decision trees were created to assign nominal classes. This depiction can then be expanded to numerical estimation.

Finally, for both numerical and nominal classification issues, DecisionStump constructs straightforward binary decision "stumps" (1-level decision trees). By growing a third branch from the stump, or by considering "missing" as a different attribute value, it deals with missing values. The LogitBoost boosting technique, covered in the next section, is mostly used in combination with DecisionStump.

Meta-Classifiers

These fundamental learning schemes may now be improved upon or have their capabilities expanded thanks to recent advancements in computational learning theory. These performance boosters are referred to as "meta-learning schemes" or "meta-classifiers" [11] since they employ the output from different students as their input. Why not set up a committee of classifiers to vote on the categorization of an example rather than utilizing a single classifier to make predictions? This is the fundamental notion underlying creating an ensemble or meta classifier by mixing several models.

Boosting [12] and bagging [13] are two of the most popular techniques for building ensemble classifiers [14]. These classifiers may often improve predicted accuracy compared to a single classifier. The cost of this improvement in performance, however, is that it is sometimes impossible to comprehend what motivates the better decision-taking. Every framework in the ensemble estimates a class and gives the forecast a confidence rating. This voting method is used for both bagging and boosting votes on classifications. The class with the highest value (or greatest degree of confidence) is picked. These values are added together.

The sets of models representing the data used by the two approaches are derived in quite different ways. Bagging creates distinct models from the training data by using a sampling strategy that duplicates certain occurrences while deleting others. The number of iterations defines how many individual models are produced for the ensemble; this method involves using a new training set for each iteration.

Algorithmically:

- Assuming n to be the quantity of occurrences in the training data, perform
- Using deletion and replication randomly choose n instances
- Using a learning approach, create a framework from the specimen.
- End

Boosting is an iterative process similar to bagging, except sampling new training data, every newer framework is impacted by the performance of earlier models. Instances which were misclassified in earlier versions are elevated, while those that were properly classified are demoted. The trick is to weight the instances and apply a learning approach which may build models while accounting for these weights. The weights are initially equal, and a model is built. Less weight is placed on examples that this model properly categorizes in order to give erroneously categorized cases greater "importance" in the next iteration.

Algorithmically:

- For each of the t iterations, give every instance the same amount of weight
- Utilize a learning method to create a framework from the weighted examples and then store it
- Reduce the weight of every case that the model properly identified End

The AdaBoost.M1 boosting method allows the user choice over the number of boosting rounds carried out. LogitBoost, which is appropriate for issues involving two-class scenarios, such as the SMO class mentioned above, implements another boosting method. It is important to split the multi-class issue into many two-class problems before combining the outcomes in order to use these approaches on datasets with multiple classes. That is precisely what the MultiClassClassifier boosting method does.

Clustering

Instead of producing predicting rules for a specific class, clustering algorithms look for logical groups (or "clusters") inside the dataset. Most often, this method is utilized in an exploratory manner to produce hypotheses on the connections between different data points. A second learning step that uses a classifier to generate rules or decision trees [15] to assign every event in the dataset to the cluster that the clustering algorithm allocated it to is often done after clustering. Then, to comprehend the clusters semantically, one may look at these 'cluster descriptions' produced by the classifier.

An EM clustering algorithm implementation is part of WEKA. The underlying premise of this technique is the same as that of previous clustering techniques: the qualities in the dataset are independent random variables. Few clustering methods require every record to follow precisely a single cluster; however, EM allows each event to consist of many clusters, which is a beneficial addition that, in reality, may offer better flexible and "fuzzy" representations of the underlying form of the dataset.

Rule of association

An application of the Apriori algorithm, a learning strategy often used in "market basket analysis" (MBA), is available in WEKA. MBA algorithms are now commonly used in consumer buying pattern analysis, especially in identifying goods that are often bought in tandem. These approaches were created in reaction to the massive amount of transaction data that barcode-dependent instruction and purchase frameworks were producing. The corporate community immediately realized the enormous potential worth of this data for marketing, but conventional data analysis approaches were unable to handle the breadth of the hypothesis space that these datasets create.

Data is logically sorted into "baskets" (often records where the things bought by a certain customer at a specific period are grouped together) for this kind of analysis. MBA algorithms like Apriori analyze patterns of purchases by identifying "association rules" that indicate that the existence of one element in a basket implies the existence of one or more other goods. An illustration of this rule may be that consumers who buy toothpaste are keener to purchase bananas during the similar grocery store visit. The outcome may thereby be utilized to recommend product pairings for various promotions or discounts, make a methodical retail plan, and facilitate data on co-branding and brand commitment.

20.4 Website Tools

WEKA offers tools for pre-processing data and assessing the execution of various learning approaches in addition to the learning methods outlined above.

Dataset preparation

A broad collection of procedures known as filters, which allow data to be handled at the instance and attribute value levels, house WEKA's preprocessing functionality. These filters share a common set of command-line arguments and a consistent command-line interface. Numerous filter methods include tools for general attribute editing, such as the ability to add and remove characteristics from the dataset. Building models with various subsets of the whole dataset is one of the most frequent tasks while testing learning techniques in the creation of a data mining implementation. Three feature selection techniques are provided by WEKA to assist in selecting features for including in an investigation: Relief, a locally generated correlation-based strategy, and the wrapper technique.

In certain circumstances, such as when normalizing every value in an attribute, it might be favorable to implement a transformation function to a complete column in the dataset. To simplify the categorization problem to binary classification, it can be easy to constitute multi-class nominal attributes as two-class nominal attributes. Moreover, the quantity of classes can be decreased by combining two values of a nominal attribute into a lone value. These changes, which may be used on non-class characteristics as well, are supported by filters. A filter is offered that converts a multi-valued nominal property into a binary valued attribute as certain learning methods (like SMO) can only accept binary attributes. A lot of algorithms [16] can't deal with real-valued properties, therefore there is a discretization filter offered. It can carry out supervised discretization or unsupervised discretization (with equal width binning).

In real-world datasets, missing values are a common occurrence that many data mining algorithms struggle to deal with. In fact, most algorithms simply ignore data lines with missing values, which, in certain situations, may drastically limit the quantity of useful data in a training set and prevent the development of a viable framework. Globally replacing missing values with approximated values prior to a learning process is one method for handling missing variables. WEKA offers a filter that restores every missing value with the mean (for numeric characteristics) or the mode (for nominal attributes). A model may be significantly skewed by the existence of outliers in a dataset. A filter may eliminate outliers by eliminating all events which display a certain set of nominal attribute values or a numeric value lower than a specified threshold.

Comparison of algorithmic performance

The WKA suite's ability to reliably assess learning schemes is one of its primary features. A researcher may, for instance, produce a "league table" that compares the effectiveness of various strategies across various datasets. The results of using 10 classifiers on 37 datasets from the UCI repository [17], a huge collection of benchmark datasets for the ML research group.

The quantity of datasets for which the plan substantially outperformed another scheme (at the 95% confidence level) is shown in Column 2 under Wins. Loss is the total quantity of datasets for that one method severely underperformed another. The win-loss differential (W-L) determines the final score. Logit boosting basic stumps for 15 or 105 iterations seems to be the finest overall for these 37 test sets among the WEKA systems that are available.

20.5 Case Study: Grading of Mushrooms

The WEKA evaluation tools, the WEKA learning techniques, and the data mining approach framework have proven beneficial in focusing the examination of real-world datasets. Here, the authors discuss a lone implementation—a framework for classifying mushrooms according to grade. The purpose of the study was to create a classification framework that could grade mushrooms according to quality with accuracy comparable to that of human inspectors. Members of the Postharvest Group, Ruakura Research Center, who served as the data supplier and subject-matter expert in the process model for this study, were involved in its execution.

In the instance, the data pre-processing step required creating a test dataset in partnership with the agricultural experts in addition to just cleaning the raw data. There are 282 mushrooms included in this dataset's descriptions. Weight, hardness, and the percentage of the cap opening were among the features. Subjective measurements included the Likert scale estimations of the degree of dirt, stalk damage. Utilizing the three general commercial ratings (first, second, and third), three inspectors independently rated the mushrooms. Digital pictures were also taken for each of the 282 mushrooms. Frequency bin values (0-4) from the analysis of the R,G,B and H,S,V histograms for the top (t) and bottom (b) pictures of the specimen mushrooms were extracted from these photos to offer an additional 60 image-based characteristics.

Many of these 68 features were successfully removed using the wrapper technique in combination with framework construction utilizing the J4.8 classifier (matching to the cycle of "attribute analysis" and "experiments with machine learning schemes" in the process framework, Figure 20.1). For each of the three inspectors, a unique model was created using J4.8 and wrapper search. The models created indicated that each inspector graded mushrooms based on unique combinations of criteria. The top and bottom photos' properties were employed in every prediction model. For the three classes of mushroom categorization, only Inspector 2 utilized weight. The wrapper approach omitted the subjective measures (dirt, stalk damage, bruising, shrivel, bacterial blotch, and P. gingerii) since they didn't improve the precision of any estimation frameworks. Finally, compared to the initial 68 qualities, each model only included four to seven of them. The degree of consent with the human specialists was, on average, adequate and the models' average accuracy was contrasted agreeably to that of the human examiners.

These findings suggest that visual characteristics, which may be automatically retrieved from digital photos, are adequate for a decent detachment of mushrooms into three wide grade bands (wherein "good" is judged in contrast to human grading criteria). The subjective qualities, which are often thought to be very important in grading, seem to have little impact on the job. This unexpected nugget of mined data supports the findings of a basic ML paper that generated a group of rules for determining soybean diseases which were remarkably different from specialist beliefs on the best course of action—but that were so precise that a single specialist approved the disclosed regulations in place of his own.

Using boosting, a much more precise but incomprehensible framework was produced. Boosting is a "black box" method for creating an ensemble of frameworks which broadly attain improved accuracy, as was explained in Section 20.3 of this article. Fifty models were built and AdaBoost scored in one experiment. The different frameworks receive a vote proportionate to their accuracy on the training data, which is how this model was able to grade mushrooms with a better accuracy but at the sacrifice of readability.

20.6 Conclusion

Information mined from data might facilitate patterns about the area that is researched which are at odds with conventional thinking in a field, as shown by the case study offered in Section 20.5 as an example. A data mining study might concentrate on finding these unexpected or surprising parts of the model so that the outcomes can be used in the original domain from whence the data was derived. The findings in this example suggest that the subjective characteristics used to grade mushrooms may not be practical, and as such, they may not need to be evaluated or documented. The growth of better objective standards for mushroom quality grading and market price may benefit from criteria dependent on the characteristics discovered in the J4.8 frameworks. Using a model productively to enable automatic categorization of fresh instances may be the aim of various data mining applications. The learning component in these apps will probably make up a very tiny portion of a much bigger software system. A learning module may be inserted into a bigger framework with the minimal amount of new code since WEKA learning schemes are available from other applications. For instance, Figure 20.2 displays a WEKA applet based on the Section 20.5-described J4.8 mushroom grading model. The model

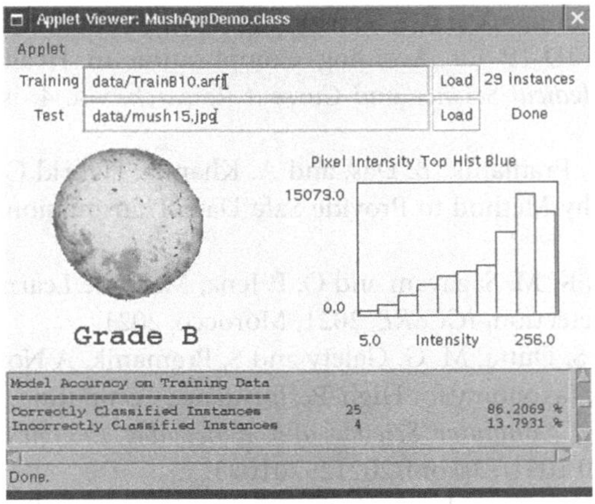

Figure 20.2 Applet for grading mushrooms.

can distinguish between mushrooms of grades A, B, and C using data from image processing of a photograph of a mushroom cap (Figure 20.2). The applet may simply be modified to use other models produced by WEKA as needed.

Learning approaches require making available on the desktops of individuals which deal with data and are familiar with the application domain from which it comes as machine learning technology develops and matures. The algorithms must leave the lab and enter the working environments of persons who may utilize them. The introduction of machine learning technologies into the workplace has advanced significantly with WEKA.

References

1. Ashisha G. R., Anitha Mary X., T. George, Martin Sagayam K., Unai Fernandez-Gamiz, Hatıra Günerhan, M. N. Uddin and S. Pramanik, Analysis of Diabetes Disease Using Machine Learning Techniques: A Review, *Journal of Information Technology Management*, 2023.
2. A. D. Andrushia, T. M. Neebha, A. T. Patricia, K. M. Sagayam, S. Pramanik, Capsule Network-based Disease Classification for VitisVinifera Leaves, *Neural Computing and Applications*, doi: 10.1007/s00521-023-09058-y, 2023
3. R. Jayasingh, J. Kumar R. J. S., D. B. Telagathoti, K. M. Sagayam and S. Pramanik, "Speckle noise removal by SORAMA segmentation in Digital Image Processing to facilitate precise robotic surgery", *International Journal of Reliable and Quality E-Healthcare*, vol. 11, issue 1, doi: 10.4018/IJRQEH.295083, 2022

4. S. Dutta, S. Pramanik and S. K. Bandyopadhyay, "Prediction of Weight Gain during COVID-19 for Avoiding Complication in Health", *International Journal of Medical Science and Current Research*, vol. 4, issue 3, pp. 1042-1052, 2021.
5. A. M. Ray, S. Pramanik, B. Das, and A. Khanna, Hybrid Cryptography and Steganography Method to Provide Safe Data Transmission in IoT, *ICDAM*, 2023.
6. S. Pramanik, K. M. Sagayam and O. P. Jena, Machine Learning Frameworks in Cancer Detection, *ICCSRE*, 2021, Morocco, 2021.
7. D. Samanta, S. Dutta, M. G. Galety and S. Pramanik, A Novel Approach for Web Mining Taxonomy for High-Performance Computing, *4th International Conference of Computer Science and Renewable Energies (ICCSRE'2021)*, 2021, DOI:10.1051/e3sconf/202129701073.
8. R. Bansal, A. J. Obaid, A. Gupta, R. Singh, and S. Pramanik, "Impact of Big Data on Digital Transformation in 5G Era", *2nd International Conference on Physics and Applied Sciences (ICPAS 2021)*, doi:10.1088/1742-6596/1963/1/012170, 2021.
9. V. VidyaChellam, V. Veeraiah, A. Khanna, T. H. Sheikh, S. Pramanik and D. Dhabliya, A Machine Vision-based Approach for Tuberculosis Identification in Chest X-Rays Images of Patients, *ICICC*, 2023, Springer, doi.org/10.1007/978-981-99-3315-0_3
10. S. Praveenkumar, V. Veeraiah, S. Pramanik, S. M. Basha, A. V. Lira Neto, V. H. C. De Albuquerque and A. Gupta, Prediction of Patients' Incurable Diseases Utilizing Deep Learning Approaches, *ICICC*, 2023, Springer, doi.org/10.1007/978-981-99-3315-0_4
11. V. Veeraiah, K. O. Thejaswini, R. Dilip, S. K. Jain, A. Sahu, S. Pramanik, and A. Gupta, The Suggested Use of Big Data in Medical Analytics by Fortis Healthcare Hospital, in *Adoption and Use of Technology Tools and Services by Economically Disadvantaged Communities: Implications for Growth and Sustainability*, IGI Global, 2023.
12. S. Ahamad, S. Janani, V. Talukdar, T. Sharma, A. Sahu, S. Pramanik, and A. Gupta, Apollo Hospital's Proposed Use of Big Data Healthcare Analytics, in *Big Data Analytics Techniques for Market Intelligence*, IGI Global, 2023.
13. P. T. Khanh, T. H. Ngọc, and S. Pramanik, Future of Smart Agriculture Techniques and Applications, in *Advanced Technologies and AI-Equipped IoT Applications in High Tech Agriculture*, Ed., A. Khang, IGI Global, 2023.
14. T. H. Ngọc, P. T. Khanh and S. Pramanik, Smart Agriculture Using a Soil Monitoring System, in *Advanced Technologies and AI-Equipped IoT Applications in High Tech Agriculture*, Ed., A. Khang, IGI Global, 2023.
15. Reepu, S. Kumar, M. G. Chaudhary, K. G. Gupta, S. Pramanik and A. Gupta, Information Security and Privacy in IoT, in *Handbook of Research in Advancements in AI and IoT Convergence Technologies*, Eds., J. Zhao, V. V. Kumar, R. Natarajan and T. R. Mahesh, IGI Global, 2023.

16. V. Veeraiah, V. Talukdar, Manikandan K., S. B. Talukdar, V. D. Solavande, S. Pramanik, and A. Gupta, Machine Learning Frameworks in Carpooling, in *Handbook of Research on AI and Machine Learning Applications in Customer Support and Analytics*, Eds., Md S. Hossain, R. C. Ho and G. Trajkovski, IGI Global, 2023.
17. V. Jain, M. Rastogi, J. V. N. Ramesh, A. Chauhan, P. Agarwal, S. Pramanik, and A. Gupta, FinTech and Artificial Intelligence in Relationship Banking and Computer Technology, in AI, IoT, and Blockchain Breakthroughs in E-Governance, Eds., K. Saini, A. Mummoorthy, R. Chandrika and N. S. Gowri Ganesh, IGI Global, 2023.

Index

Also of Interest

Check out these other related titles from Scrivener Publishing

Agroforestry, Edited by Abhishek Raj, Manoj Kumar Jhariya, Arnab Banerjee, Ramesh Kumar Jha, and Krishan Pal Singh, ISBN: 9781394231133. This book offers comprehensive insights into the management of agroforestry for livelihood security and sustainable development in the tropics, addressing ecological interactions, productivity, and the monetization of carbon credits, while also outlining a future roadmap and policy challenges.

Smart Agritech: Robotics, AI, and Internet of Things (IoT) in Agriculture, Edited by Santosh Kumar Srivastava, Durgesh Srivastava, Korhan Cengiz, and Pramod Gaur, ISBN:9781394302959. The main goal of *Smart Agritech: Robotics, AI, and the Internet of Things (IoT) in Agriculture* is to explore how emerging technologies such as robotics, artificial intelligence (AI), and IoT can be leveraged to improve efficiency, sustainability, and productivity in agriculture.

DIGITAL AGRICULTURAL ECOSYSTEM: Revolutionary Advancements in Agriculture, Edited by Kuldeep Singh and Prasanna Kolar, ISBN: 9781394242931. The book comprehensively explores the dynamic synergy between modern technology and agriculture, showcasing how advancements such as artificial intelligence, data analytics, and smart farming practices are reshaping the landscape to ensure food security in the era of climate change, as well as bridging the gap between cutting-edge research and practical implementation.

THE DIGITAL AGRICULTURAL REVOLUTION: Innovations and Challenges in Agriculture through Technology Disruptions, Edited by Roheet Bhatnagar, Nitin Kumar Tripathi, Nitu Bhatnagar and Chandan Kumar Panda, ISBN: 9781119823339. The book integrates computational intelligence, applied artificial intelligence, and modern agricultural practices and will appeal to scientists, agriculturists, and those in plant and crop science and management.

HANDBOOK OF AGRICULTURAL BIOTECHNOLOGY: Volume I Nanopesticides, Edited by Charles Oluwaseun Adetunji and Julius Kola Oloke, ISBN: 9781119836148. The book provides detailed information on the application of nanopesticides for the management of numerous agricultural pests and pathogens, as well as to foster an innovative pathway toward future sustainable agriculture and food systems.

HANDBOOK OF AGRICULTURAL BIOTECHNOLOGY: Volume II Nanobioherbicides, Edited by Charles Oluwaseun Adetunji, Julius Kola Oloke, ISBN: 9781119836155. The book provides a detailed examination of the application of nanobioherbicides that come from plants including information on the different metabolites derived from numerous plants that could become bioherbicides.

HANDBOOK OF AGRICULTURAL BIOTECHNOLOGY: Volume III Nanofungicides, Edited by Charles Oluwaseun Adetunji and Julius Kola Oloke, ISBN: 9781119836162. This book provides state-of-the-art information about recent advancements in the application of nanobiofungicides for effective management of post-harvest diseases and agricultural fungal diseases, including blights, mildews, molds, and rusts.

HANDBOOK OF AGRICULTURAL BIOTECHNOLOGY: Volume IV Nanoinsecticides, Edited by Charles Oluwaseun Adetunji and Julius Kola Oloke, ISBN: 9781119836179. The book provides detailed information about the application of repellent products that contain plant-based ingredients known as nanobioinsecticides, including the pesticide evaluation scheme guidelines for repellent testing, relevant information about the procedures to evaluate several repellent compounds, the development of new products that offer high repellency, and guidelines for consumer safety.

ECORESTORATION FOR SUSTAINABILITY, Edited by Arnab Banerjee, Manoj Kumar Jhariya, Surendra Singh Bargali, and Debnath Palit, ISBN: 9781119879718. With climate change and land degradation advancing at an alarming rate, the state of the global ecosystem and how it can be restored in our march toward sustainability is incredibly important, and this new, groundbreaking volume tackles these important issues head on.

Books in the series, "Bioprocessing in Food Science"

NUTRACEUTICALS FROM FRUIT AND VEGETABLE WASTE, Edited by Vidisha Tomer, Navnidhi Chhikara, Ashwani Kumar, and Anil Panghal, ISBN: 9781119803508. Nutraceuticals from Fruit and Vegetable Waste, the latest volume in the series, "Bioprocessing in Food Science," is an essential reference on the valorization of fruit and vegetable waste and the state of the art for fruit and vegetable processing.

NUTRITIONAL SCIENCE AND TECHNOLOGY: Concept to Application, Edited by Tejpal Dhewa, Anil Kumar Puniya, and Anil Panghal, ISBN: 97811198089. This new volume in the groundbreaking series, "Bioprocessing in Food Science," Nutritional Science and Technology, is a comprehensive resource covering topics such as human nutrition, food safety concerns, and emerging technologies like nanotechnology, and it also addresses the need for innovative food products, contamination, and regulatory challenges in the global food market.

HARVESTING FOOD FROM WEEDS, Edited by Prerna Gupta, Navnidhi Chhikara, and Anil Panghal, ISBN: 9781119791973. An exciting new volume in the groundbreaking new series, "Bioprocessing in Food Science," Harvesting Food from Weeds focuses on the latest processes, industrial applications, and leading research on this important, evolving new area, for engineers, scientists, students, and other industry professionals.

MICROBES IN THE FOOD INDUSTRY, Edited by Navnidhi Chhikara, Anil Panghal, and Gaurav Chaudhary, ISBN: 9781119775584. This newest volume in the groundbreaking new series, "Bioprocessing in Food Science," focuses on the latest processes, industrial applications, and leading research on microbes in the food industry, for engineers, scientists, students, and other industry professionals.

NOVEL TECHNOLOGIES IN FOOD SCIENCE, Edited by Navnidhi Chhikara, Anil Panghal, and Gaurav Chaudhary, ISBN: 9781119775584. The third book in the series, "Bioprocessing in Food Science," covers the latest and greatest new technologies in food science.

FUNCTIONAL FOODS, Edited by Navnidhi Chhikara, Anil Panghal, and Gaurav Chaudhary, ISBN: 9781119775560. Presenting cutting-edge information on new and emerging food engineering processes, *Functional Foods*, the second volume in the groundbreaking new series, "Bioprocessing in Food Science," is an essential reference on the modeling, quality, safety, and technologies associated with food processing operations today.

THERMAL FOOD ENGINEERING OPERATIONS, Edited by Nitin Kumar, Anil Panghal, and M. K. Garg, ISBN: 9781119775591. Presenting cutting-edge information on new and emerging food engineering processes, Thermal Food Engineering Operations, the first volume in the new series, "Bioprocessing in Food Science," is an essential reference on the modeling, quality, safety, and technologies associated with food processing operations today.